Heavy-Fermion Systems

HANDBOOK OF METAL PHYSICS

SERIES EDITOR

Prasanta Misra
Department of Physics, University of Houston,
Houston, TX, 77204-5005, USA

Heavy-Fermion Systems

PRASANTA MISRA
Department of Physics, University of Houston,
Houston, TX, 77204-5005, USA

ELSEVIER

AMSTERDAM • BOSTON • HEIDELBERG • LONDON • NEW YORK • OXFORD
PARIS • SAN DIEGO • SAN FRANCISCO • SINGAPORE • SYDNEY • TOKYO

Elsevier
Radarweg 29, PO Box 211, 1000 AE Amsterdam, The Netherlands
Linacre House, Jordan Hill, Oxford OX2 8DP, UK

First edition 2008

Library of Congress Cataloging-in-Publication Data
A catalog record for this book is available from the Library of Congress

British Library Cataloguing in Publication Data
A catalogue record for this book is available from the British Library

ISBN: 978-0-444-51503-2
ISSN: 1570-002X

Transferred to Digital Printing in 2008

08 09 10 11 12 10 9 8 7 6 5 4 3 2 1

The Book Series 'Handbook of Metal Physics' is dedicated to my wife

Swayamprava

and to our children

Debasis, Mimi and Sandeep

Preface

Metal Physics is an interdisciplinary area covering Physics, Chemistry, Materials Science, and Engineering. Due to the variety of exciting topics and the wide range of technological applications, this field is growing very rapidly. It encompasses a variety of fundamental properties of metals such as electronic structure, magnetism, superconductivity, as well as the properties of semimetals, defects and alloys, and surface physics of metals. Metal physics also includes the properties of exotic materials such as high-T_c superconductors, heavy fermion systems, quasicrystals, metallic nanoparticles, metallic multilayers, metallic wires/chains of metals, novel-doped semimetals, photonic crystals, low-dimensional metals, and mesoscopic systems. This is by no means an exhaustive list and more books in other areas will be published. I have taken a broader view and other topics, which are widely used to study the various properties of metals, will be included in the Book Series. During the past 25 years, there has been extensive theoretical and experimental research in each of the areas mentioned above. Each volume of this Book Series, which is self-contained and independent of the other volumes, is an attempt to highlight the significant work in that field. Therefore, the order in which the different volumes will be published has no significance and depends only on the timeline in which the manuscripts are received.

The Book Series *Handbook of Metal Physics* is designed to facilitate the research of Ph.D. students, faculty, and other researchers in a specific area in Metal Physics. The books will be published by Elsevier in hard cover copy. Each book will be either written by one or two authors who are experts and active researchers in that specific area covered by the book or by multiple authors with a volume editor who will coordinate the progress of the book and edit it before submission for final editing. This choice has been made according to the complexity of the topic covered in a volume as well as the time that the experts in the respective fields were willing to commit. Each volume is essentially a summary as well as a critical review of the theoretical and experimental work in the topics covered by the book. There are extensive references at the end of each chapter to facilitate researchers in this rapidly growing interdisciplinary field. Since research in various sub-fields in Metal Physics is a rapidly growing area, it is planned that each book will be updated periodically to include the results of the latest research. Even though these books are primarily designed as reference books, some of these books can be used as advance graduate-level textbooks.

The outstanding features of this Book Series are the extensive research references at the end of each chapter, comprehensive review of the significant theoretical work, a summary of all important experiments, illustrations wherever necessary, and discussion of possible technological applications. This would spare the active researcher in a field to do extensive search of the literature before she or he would start planning to work on a new research topic or in writing a research paper on a piece of work already completed.

Since each volume will have an introductory chapter written either by the author(s) or the volume editor, it is not my intention to write an introduction for each topic (except for the book being written by me). In fact, they are much better experts than me to write such introductory remarks.

Finally, I invite all students, faculty, and other researchers, who would be reading the book(s) to communicate to me their comments. I would particularly welcome suggestions for improvement as well as any errors in references and printing.

Acknowledgments

I am grateful to all the eminent scientists who have agreed to contribute to the Book Series. All of them are active researchers and obviously extremely busy in teaching, supervising graduate students, publishing research papers, writing grant proposals, and serving on committees. It is indeed gratifying that they have accepted my request to be either an author or a volume editor of a book in the Series. The success of this Series lies in their hands and I am confident that each one of them will do a great job.

The idea of editing a Book Series on Metal Physics was conceived during a meeting with Dr. Charon Duermeijer, publisher of Elsevier (she was Physics Editor at that time). After several rounds of discussions (via e-mail), the Book Series took shape in another meeting where she met some of the prospective authors/volume editors. She has been a constant source of encouragement, inspiration, and great support while I was identifying and contacting various experts in the different areas covered by this extensive field of Metal Physics. It is indeed not easy to persuade active researchers (scattered around the globe) to write or even edit an advance research-level book. She had enough patience to wait for me to finalize a list of authors and volume editors. I am indeed grateful to her for her confidence in me.

I am also grateful to Dr. Anita Koch, Manager, Editorial Services, Books of Elsevier, who has helped me whenever I have requested her, i.e., in arranging to write new contracts, postponing submission deadlines, as well as making many helpful suggestions. She has been very gracious and prompt in her replies to my numerous questions. I have profited from conversations with my friends who have helped me in identifying potential authors as well as suitable topics in my endeavor to edit such an ambitious Book Series. I am particularly grateful to Professor Larry Pinsky (Chair) and Professor Gemunu Gunaratne (Associate Chair) of the Department of Physics of University of Houston for their hospitality, encouragement, and continuing help.

Finally, I express my gratitude to my wife and children who have loved me all these years even though I have spent most of my time in the physics department(s) learning physics, doing research, supervising graduate students, publishing research papers, and writing grant proposals. There is no way I can compensate for the lost time except to dedicate this Book Series to them. I am thankful to my daughter-in-law Roopa who has tried her best to make me computer literate and in the process has helped me a lot in my present endeavor. My fondest dream is that when my grandchildren Annika and

Millan attend college in 2021 and Kishen and Nirvaan in 2024, this Book Series would have grown both in quantity and quality (obviously with a new Series Editor in place) and at least one of them would be attracted to study the subject after reading a few of these books.

Prasanta K. Misra
Department of Physics, University of Houston,
Houston, Texas, USA

Contents

Preface vii

Chapter 1. Overview of Heavy Fermion Systems 1

References 6

Chapter 2. Kondo Lattice, Mixed Valence, and Heavy Fermions 9

2.1. Periodic Anderson and Kondo-lattice models 9

2.2. Theoretical approaches (for solution of renormalized Hamiltonian) 11

2.3. Cluster calculations 14

References 19

Chapter 3. Dynamical, Extended Dynamical, and Cluster Dynamical Mean-Field Theories: (DMFT, EDMFT, and Cluster DMFT) 23

3.1. The local impurity self-consistent approximation (LISA) 23

3.2. Brief discussions of the dynamical mean-field equations 26

3.3. Methods of solution 28

3.4. Application of LISA to periodic Anderson model 30

3.5. Kondo insulators 32

3.6. The multichannel Kondo lattice 32

3.7. RKKY interaction 34

3.8. Extended dynamical mean-field theory (EDMFT) 35

3.9. Quantum cluster theories 46

References 48

Chapter 4. Fermi-Liquid, Heavy Fermi-Liquid, and Non-Fermi-Liquid Models 51

4.1. Fermi-liquid theory of Landau 51

4.2. Fermi-liquid model for Kondo-lattice systems 53

4.3. Heavy Fermi liquids 57

4.4. Non-Fermi-liquid behavior in f electron metals 63

4.5. The quadrupolar Kondo model 66

4.6. Quantum-critical point theories 67

4.7. Weak-coupling theories 68
4.8. Strong-coupling theories 71
References 84

Chapter 5. Metamagnetism in Heavy Fermions (Experimental Review) 87
5.1. Introduction 87
5.2. $CeRu_2Si_2$ 88
5.3. $Sr_3Ru_2O_7$ 92
5.4. $CeCu_{6-x}Au_x$ 98
5.5. UPt_3 99
5.6. UPd_2Al_3 101
5.7. URu_2Si_2 102
5.8. $CePd_2Si_2$ 106
5.9. $YbRh_2Si_2$ 108
5.10. $CeIr_3Si_2$ 110
References 111

Chapter 6. Theory of Metamagnetism in Heavy Fermions 115
6.1. Review of theoretical models 115
6.2. Strong-coupling spin-fluctuation theory in the high-field state 115
6.3. Metamagnetic transition in a small cluster t–J model 117
6.4. Competition between local quantum spin fluctuations and magnetic-exchange interaction 120
6.5. Itinerant electrons and local moments in high and low magnetic fields 124
References 137

Chapter 7. Heavy Fermion Superconductors (Ce-Based Compounds) 141
7.1. Overview 141
7.2. $CeCu_2Si_2$ 142
7.3. $CeCu_2Ge_2$ 150
7.4. $CePd_2Si_2$ 150
7.5. $CePd_2Ge_2$ 152
7.6. $CeRh_2Si_2$ 154
7.7. $CeNi_2Ge_2$ 155
7.8. $CeIn_3$ 156
7.9. $CePt_3Si$ 159
7.10. $CeCoIn_5$ 163

7.11. $CeRhIn_5$ 167
7.12. $CeIrIn_5$ 168
7.13. $CeNiGe_3$ 171
7.14. $Ce_2Ni_3Ge_5$ 175
7.15. Summary and conclusion 175
References 178

Chapter 8. U-Based Superconducting Compounds **183**
8.1. Overview 183
8.2. UBe_{13} 184
8.3. UPt_3 188
8.4. URu_2Si_2 201
8.5. UPd_2Al_3 208
8.6. UNi_2Al_3 215
8.7. UGe_2 219
8.8. URhGe 221
8.9. UIr 225
References 228

Chapter 9. Filled Skutterdites and Transuranium Heavy Fermion Superconductors **235**
9.1. Filled skutterdites 235
9.2. $PrOs_4Sb_{12}$ 235
9.3. $PuCoGa_5$ 244
9.4. $PuRhGa_5$ 247
9.5. Similarities between Cu and Pu containing "high T_c" superconductors 251
References 252

Chapter 10. Brief Review of Theories of Heavy Fermion Superconductivity **255**
10.1. Introduction 255
10.2. BCS theory of anisotropic superconductivity 256
10.3. Symmetry classifications and generalized Ginzburg–Landau theory 260
10.4. Density of states of quasiparticles 264
10.5. Collective modes 265
10.6. Coexistence of antiferromagnetism and superconductivity 266
10.7. Influence of antiferromagnetic fluctuations in superconductivity 269
10.8. Fulde–Ferrell–Larkin–Ovchinnikov (FFLO) superconducting state 271
10.9. Magnetically mediated superconductivity 273

10.10. Superconductivity due to valence fluctuations 276
10.11. Magnetic-exciton-mediated superconductivity 281
10.12. Quadrupolar exciton exchange 284
References 287

Chapter 11. Kondo Insulators **291**
11.1. Introduction 291
11.2. $Ce_3Bi_4Pt_3$ 293
11.3. CeRhAs 296
11.4. CeRhSb 300
11.5. CeNiSn 300
11.6. $CeRu_4Sn_6$ 306
11.7. U_2Ru_2Sn 308
11.8. $CeFe_4P_{12}$ and $CeRu_4P_{12}$ 311
11.9. $CeOs_4Sb_{12}$ 313
11.10. UFe_4P_{12} 316
11.11. TmSe 317
11.12. U_2Ru_2Sn 320
11.13. YbB_{12} 322
11.14. SmB_6 324
11.15. SmS 325
11.16. Theory of Kondo insulators 326
References 330

Subject Index **335**

Chapter 1

Overview of Heavy Fermion Systems

In the 1970s and early 1980s, there was an intense excitement about the mixed-valence (MV) systems [1–4]. It was well known that many rare earth and actinide compounds exhibited MV phenomena in the sense that the valence of the quasilocalized f electron was approximately 3.67 instead of 4. In fact, different experimental results would yield different fractional numbers. The valence fluctuations in such systems are highly correlated and have many remarkable properties. Some of them exhibit Fermi-liquid behavior with a large effective mass at low temperatures but have free moments at high temperatures. The interest in these materials became more widespread after quite a few cerium and uranium compounds were classified as heavy fermion systems [5–7].

The heavy fermion systems, which are also sometimes referred to as heavy electron systems, were originally assumed to be a loosely defined collection of intermetallic compounds containing lanthanide (mostly Ce, Yb) or actinide (mostly U, Np) elements. Recently, other compounds such as quasi two-dimensional $CeCoIn_5$ and "Skutterdites" such as $PrOs_4Sb_{12}$ have been shown to exhibit such behavior. I will present an overview of the "exotic" properties of these systems, the only common feature of which is that they have large effective mass (50–1000 times greater than the mass of a free electron) below a coherence temperature T^*. The effective mass is estimated through the electronic specific heat. In general, for temperatures much smaller than the Debye temperature and Fermi energy, the specific heat C of a metal can be expressed as

$$\frac{C}{T} = \gamma + \beta T^2, \tag{1.1}$$

where $\gamma = V_m k_F k_B{}^2 m^*/3\hbar^2$, V_m is the molar volume, m^* the effective mass of the electron, k_F the Fermi vector, and T is the absolute temperature. Here, γ is the electronic contribution and β the contribution of the phonons to the specific heat. It may be noted that there is an additional spin-fluctuation term $\delta T^3 \ln T$ in the specific heat of UPt_3 and UAl_2.

For normal metals like copper or aluminum, γ is of the order 1 mJ/mol K^2 at low temperatures. For heavy fermion systems, γ abruptly increases to a very large value below the coherence temperature T^*. In fact, since γ increases abruptly at low temperatures for many materials, an arbitrary but generally accepted definition of heavy fermions [5] is those systems which has $\gamma > 400$ mJ/f atom mol K^2. In addition, γ is generally normalized to a mole of f atoms so that there can be a comparison between systems with different structure. Some of the other properties of heavy fermions include: (i) an enhanced

Pauli spin susceptibility indicating a large effective mass; (ii) the Wilson ratio is approximately one; (iii) a huge T^2 term in the electrical resistivity; and (iv) highly temperature-dependent de Haas–van Alphen oscillations amplitudes at very low temperatures.

$CeAl_3$ was the first heavy fermion system discovered by Andres et al. [8]. They measured the specific heat and electrical resistivity of $CeAl_3$ and found that below 0.2 K, there was enormous magnitudes of the linear specific heat term $C = \gamma T$ ($\gamma = 1620$ mJ/mol K^2) and the T^2 term in $\rho = AT^2$ ($A = 35$ μΩ cm/K^2). Earlier, $CeAl_3$ had been cited as an example of MV compound. The seminal results of Andres et al. [8] for variation of C and ρ with temperature are presented in Figures 1.1 and 1.2.

The intense interest in heavy fermion systems started with the discovery of superconductivity in $CeCu_2Si_2$ by Steglich et al. [9]. The big surprise was that, up to that time, magnetism and superconductivity were considered to be contradictory phenomena. However, in $CeCu_2Si_2$, the 4f electrons which are responsible for local magnetic moments at higher temperatures are also responsible for superconductivity below the critical temperature T_c. After the discovery of high-T_c superconductivity in 1986 and the realization that both high-T_c superconductors and heavy fermions are not only highly correlated electron systems but also exhibit "nearness" to magnetism at the superconducting phase, the interest in these systems grew rapidly. In general, the traditional BCS theory of superconductivity does not apply to either of these two "exotic" systems.

The experimental results of Steglich et al. [9] are presented in Figures 1.3 and 1.4. The main part of Figure 1.3 shows the resistivity and the inset shows the low-field ac susceptibility of $CeCu_2Si_2$ as a function of temperature. Figure 1.4 shows in a C/T versus T plot, the specific heat jumps of two other $CeCu_2Si_2$ samples. One can see that these jumps do not look very profound.

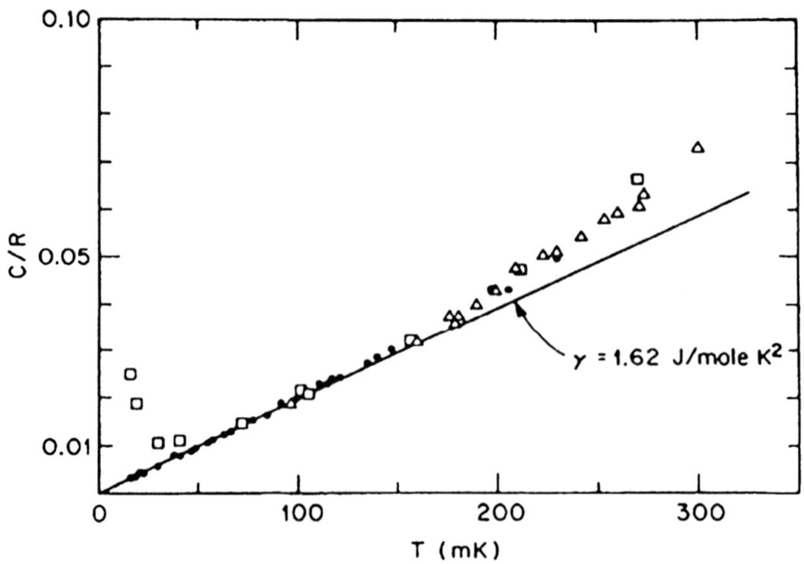

Figure 1.1. Specific heat of $CeAl_3$ at very low temperatures in zero field (●, Δ) and in 10 kOe (□). Reproduced with the permission of the American Physical Society from Ref. [8].

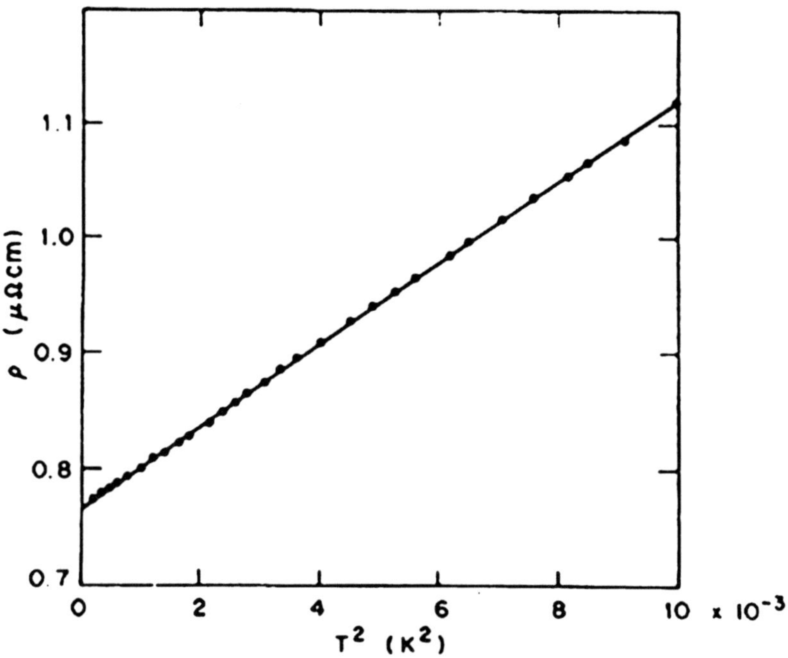

Figure 1.2. Electrical resistivity of $CeAl_3$ below 100 mK, plotted against T^2. Reproduced with the permission of the American Physical Society from Ref. [8].

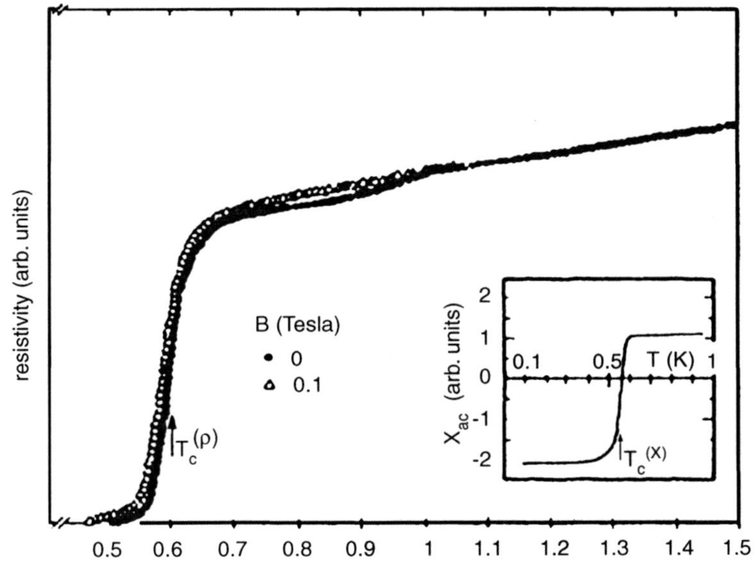

Figure 1.3. Resistivity (main part) and low-field ac susceptibility (inset) of $CeCu_2Si_2$ as a function of temperature. Arrows give transition temperatures $T_c^{(\rho)} = 0.60 \pm 0.03$ K and $T_c^{(\chi)} = 0.54 \pm 0.03$ K. Reproduced with the permission of the American Physical Society from Ref. [9].

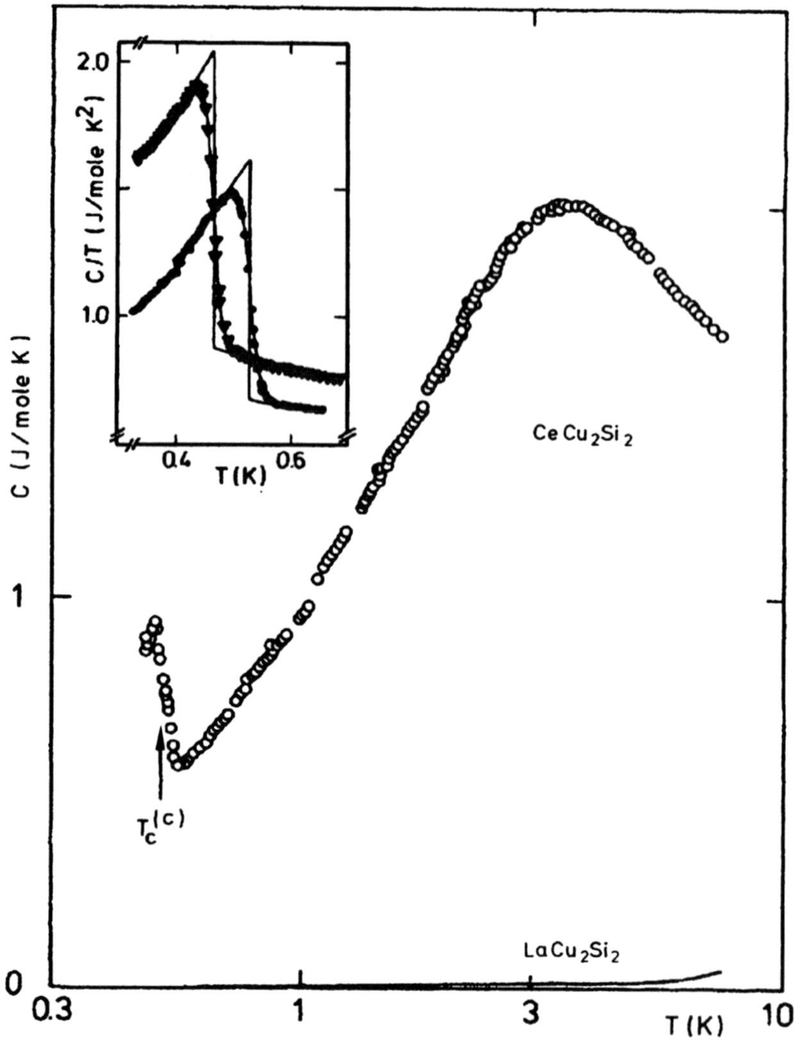

Figure 1.4. Molar specific heat of $CeCu_2Si_2$ at $B = 0$ as a function of temperature on logarithmic scale. Arrow marks transition temperature $T_c^{\copyright} = 0.51 \pm 0.04\,K$. Inset shows in a C/T versus T plot, the specific heat jumps of two other $CeCu_2Si_2$ samples. Reproduced with the permission of the American Physical Society from Ref. [9].

The main part of Figure 1.4 shows, on a logarithmic scale, the molar specific heat of $CeCu_2Si_2$ at $B = 0$ as a function of temperature. The inset in Figure 1.4 shows in a C/T plot the specific heat jumps of two other $CeCu_2Si_2$ samples, which do not look very profound.

However, the specific heat jumps below the coherence temperature T^*, which is characteristic of heavy fermion systems, is elegantly displayed when one plots C/T versus T^2. Stewart [5] has plotted C/T versus T^2 of non-superconducting single crystals of $CeCu_2Si_2$ and a piece of a superconducting single crystal of UBe_{13}. These results are reproduced in Figure 1.5 in which the line through UBe_{13} serves as a guide to the eye.

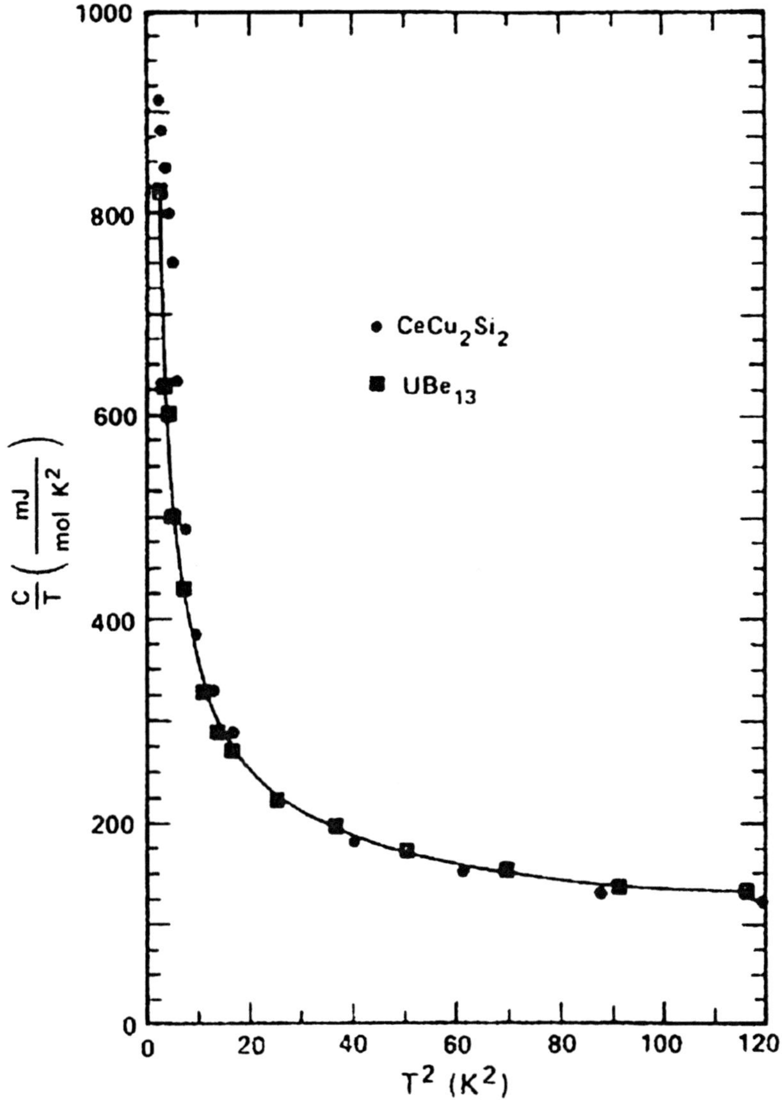

Figure 1.5. Specific heat of non-superconducting single crystals of $CeCu_2Si_2$ (●) and a piece of superconducting crystal (■) of UBe_{13}. The line through the UBe_{13} serves as a guide to the eye. Reproduced with the permission of the American Physical Society from Ref. [5].

The heavy fermions have a wide variety of ground states such as superconductors UBe_{13} [10] with non-Fermi-liquid properties in their normal state, and UPt_3 [11] which orders antiferromagnetically below the Neel temperature (T_N), exhibits a heavy Fermi-liquid state well below T_N, and has unconventional superconductivity with a multicomponent superconducting parameter [12]. At very low temperatures, some heavy fermions are antiferromagnets with weak moments ($CeAl_2$ [13], U_2Zn_{17} [14]), narrow-gap semiconductors (CeNiSn [15], $Ce_3Bi_4Pt_3$ [16]) with quasiparticles having large effective masses, while a few are Fermi liquids with no long-range ordering ($CeAl_3$ [8], $CeCu_6$ [17]). Some other heavy fermions superconductors like $CeCoIn_5$ [18] are quasi two-dimensional. At present, approximately 50 heavy fermion systems have been discovered and there is no uniformity in their properties. As we have noted, some heavy fermions are Fermi liquids with no ordering while some others are non-Fermi liquids. Both magnetic and superconducting quantum critical points [19] have been observed in some of these systems. The only common feature is the large effective mass and the fact that they are highly correlated electron systems. It has also been found that $CeRu_2Si_2$ exhibits metamagnetism [20–22] which has a wide variety of technological applications. The intense interest is due to the many unsolved theoretical problems and a variety of possible technological applications.

It is interesting to note that the magnitude of the nuclear relaxation rate [23] in UBe_{13} and of the ultrasonic attenuation [24] in UPt_3 in the normal state are of the same order as ordinary metals. The thermal conductivity measurements in $CeCu_2Si_2$ [9], UBe_{13} [24], and UPt_3 [24] yield results similar to ordinary metals. The local-moment relaxation rate of rare earth impurities in UBe_{13} [25] is approximately of the same order as in materials with normal effective mass. Hence, it is easy to conclude that the large effective mass of heavy fermions is not due to band-structure renormalization.

The theory of these systems lags behind the experiment although several powerful techniques have been applied. These include Bethe ansatz method, $1/N$ expansion method, renormalization group technique, and exotic theories such as Dynamical Mean Field theory and Quantum Critical Point. There are several theories for heavy fermion superconductivity in which there is widespread interest because of some similarities with theory of high-T_c superconductivity. There are a large number of review articles and proceedings of international conferences on heavy fermions.

It is impossible to discuss in detail the properties of these various heavy fermion systems in a monograph. In fact, it requires a series of volumes (may be titled as *Hand Book of Heavy Fermion Physics*) to discuss all the properties of these systems, one of the most intense areas of research in Condensed Matter Physics. I will summarize the relevant experimental results and provide extensive references for each system. Similarly, I will briefly summarize the various theories including the recently proposed exotic ones. It may be mentioned that extensive research is making the field grow very rapidly. This monograph is intended to introduce researchers to heavy fermion systems as well as to serve as a reference book for researchers already working in this area.

References

[1] R.D. Parks (ed.), *Valence Instabilities and Related Narrow-Band Phenomena* (Plenum, New York, 1977).

[2] J.M. Lawrence, P.S. Riseborough, and R.D. Parks, Rep. Prog. Phys. **44**, 1 (1981).

[3] L.M. Falicov, W. Hanke, and M.B. Maple (eds.), *Valence Fluctuation in Solids* (North-Holland, Amsterdam, 1981).

[4] P. Wachter and H. Boppart (eds.), *Valence Instabilities* (North-Holland, Amsterdam, 1982).

[5] G.R. Stewart, Rev. Mod. Phys. **56**, 755 (1984).

[6] C.M. Varma, Rev. Mod. Phys. **48**, 219 (1976), Comm. Solid State Phys. **11**, 211 (1985).
[7] C.M. Varma, J. Magn. Magn. Mater. **47–48**, 606 (1985).
[8] K. Andres, J.E. Graebner, and H.R. Ott, Phys. Rev. Lett. **35**, 1779 (1975).
[9] F. Steglich, J. Aarts, C.D. Bredl, W. Lieke. D. Meschede, W. Franz, and J. Schafer, Phys. Rev. Lett. **43**, 1892 (1979).
[10] H.R. Ott, H. Rudigier, Z. Fisk, and J.L. Smith, Phys. Rev. Lett. **50**, 1595 (1983).
[11] G.R. Stewart, Z. Fisk, J.O. Willis, and J.L. Smith, Phys. Rev. Lett. **52**, 679 (1984).
[12] R.A. Fisher, S. Kim, B.F. Woodfield, N.E. Phillips, L. Taillefer, K. Hasselbach, J. Flouquet, A.L. Georgi, and J.L. Smith, Phys. Rev. Lett. **62**, 1411 (1989).
[13] B. Barbara, J. Boucherle, J. Buevoz, M. Rossignol, and J. Schweizer, Solid State Commun. **24**, 481 (1977).
[14] H.R. Ott, H. Rudigier, P. Delsing, and J. Fisk, Phys. Rev. Lett. **52**, 1551 (1984).
[15] T. Takabatake, F. Teshima, H. Fuji, S. Nishigori, T. Suzuki, T. Fujita, Y. Yamaguchi, and J. Sakuri, Phys. Rev. B **41**, 9607 (1990).
[16] M.F. Hundley, P.C. Canfield, K.D. Thompson, and Z. Fisk, Phys. Rev. B **42**, 6842 (1990).
[17] G.R. Stewart, Z. Fisk, and M.S. Wise, Phys. Rev. B **30**, 482 (1984).
[18] C. Martin, C.C. Agosta, S.W. Tozer, H.A. Radovan, E.C. Palm, T.P. Murphy, and J.L. Sarrao, Phys. Rev. B **71**, 020503 (2005).
[19] P. Haen, J. Floquet, F. Lapierre, P. Lejay, and G. Remenyi, J. Low Temp. Phys. **67**, 391 (1987); L. Taileffer, J. Flouquet, and G.G. Lonzarich, Phys. B **169**, 257 (1991).
[20] L. Puech, J.M. Mignot, P. Lejay, P. Haen, and J. Flouquet, J. Low Temp. Phys. **70**, 237 (1988).
[21] J.M. Mignot, J. Flouquet, P. Haen, F. Lapierre, L. Puech, and A. Voiron, J. Magn. Magn. Mater. **76–77**, 97 (1988).
[22] D.E. MacLauglin, C. Tien, W.G. Clark, K. Glover, M.D. Lan, Z. Fisk, J.L. Smith, and H.R. Ott, Phys. Rev. Lett. **53**, 1833 (1984).
[23] D.J. Bishop, C.M. Varma, B. Batlogg, E. Bucher, Z. Fisk, and J.L. Smith, Phys. Rev. Lett. **53**, 1009 (1984).
[24] D. Jaccard, J. Flouquet, P. Lejay, and J.L. Tholence, J. Appl. Phys. **57**, 2719 (1985).
[25] F. Gandra, S. Schultz, S.B. Oseroff, Z. Fisk, and J.L. Smith, Phys. Rev. Lett. **55**, 2719 (1985).

Chapter 2

Kondo Lattice, Mixed Valence, and Heavy Fermions

2.1. Periodic Anderson and Kondo-lattice models

It has been noted that the majority of the rare earth and actinide compounds have local moments and can be classified as systems in the magnetic regime. The f orbitals have no charge fluctuation in this region and have integral valence. Therefore, they can be considered to be in a Mott-insulating stage. Weak residual spin polarization of the conduction electrons, Rudderman–Kittel–Kasuya–Yosida (RKKY) interactions between the local moments, magnetic transition at low temperatures, and spin-wave excitations occur. The spin waves scatter the conduction electrons at low temperatures. Those rare earth compounds which have either localized 4f electrons (Ce, Yb) or 5f electrons (U, Np) can be considered as a Kondo lattice [1], where at each lattice site, a local moment interacts via an exchange coupling J_K with the spin of any conduction electron sitting at the site. It may be noted that the Kondo temperature T_K is defined as the temperature below which an isolated local moment would be screened by the spins of the conduction electrons while the RKKY interaction characterizes the induced coupling between two local moments. It is a long-range exchange mediated by the conduction electrons near the Fermi surface and oscillates with $k = 2k_F$ asymptotically. As mentioned, earlier, most other rare earth compounds with 4f or 5f electrons are mixed-valence (MV) systems. However, a few acquire an incredibly large effective mass below the coherence temperature and these are known as heavy fermion systems.

The MV and Kondo systems have been theoretically analyzed extensively. In fact, the single magnetic impurity in the MV [2–4] and the Kondo [5, 6] regime are well understood. The magnetic susceptibility, specific heat, transport, and excitation properties of the Anderson model [7] have been extensively studied by using renormalization group technique [8] and the Bethe ansatz method [9–19]. The lattice problem in the Kondo and MV regimes is much more complicated. One can consider the properties in the Kondo and MV regimes to be very similar to a collection of non-interacting f impurities in an sd band.

In the 1980s, it became obvious from experimental results [20–22] that the three phenomena of MV, Kondo, and heavy fermion behavior, at least in Ce systems, correspond to different regimes of one fundamental phenomenon. In order to correlate these regimes and to study their dependence on the various relevant parameters, the simplest Hamiltonian is the orbitally non-degenerate periodic Anderson model. The periodic

Anderson model for a system consisting of a set of N sites denoted by sites i, j is described by the Hamiltonian

$$H = t\sum_{i\neq j,\sigma} C^{\dagger}_{i\sigma} C_{j\sigma} + E_{\mathrm{f}} \sum_{i,\sigma} f^{\dagger}_{i,\sigma} f_{i,\sigma} + U \sum_{i} f^{\dagger}_{i\uparrow} f_{i\uparrow} f^{\dagger}_{i\downarrow} f_{i\downarrow} \\ + V\sum_{i\neq j,\sigma} (C^{\dagger}_{i\sigma} f_{j\sigma} + f^{\dagger}_{j\sigma} C_{i\sigma}). \tag{2.1}$$

Here t (which can be positive or negative) is the transfer (hopping) integral of the extended orthogonal orbitals between sites i and j (restricted to nearest neighbors). $C^{\dagger}_{i\sigma}$ and $C_{i\sigma}$ are the creation and annihilation operators for these extended orbitals at sites i and j with spin σ. There is one extended orbital per site per spin with a mean energy which is the origin of the energy scale. $f^{\dagger}_{i\sigma}$ and $f_{i\sigma}$ are the creation and annihilation operators for the localized f orbitals with energy E_{f}. U is the Coulomb repulsion between two electrons of opposite spin in the f orbital and describes a short-range interaction between them. V is a positive hybridization parameter between the localized and the band orbitals in neighboring sites. U is positive while t and E_{f} can have either sign. When we consider the f orbital only on a single site, the model (2.1) is reduced to a single-impurity Anderson model [7].

The Hamiltonian (2.1) can be augmented by additional terms such as second-neighbor hopping, Coulomb repulsion between extended orbitals, and f electrons or between electrons on different sites. Since the orbital degeneracy is neglected, there is no Hund's rule coupling between the f orbitals in this model. However important such terms are in applications to real systems, they are ignored here in the belief that they would contribute nothing really essential to the qualitative physics.

When each f orbital is occupied by a single electron (either up spin or down spin), the system is described as the Kondo regime. The empty sites and doubly occupied sites become virtual states. The low-energy physics of the periodic Anderson model (2.1) can be described by an effective model where the f electron degrees of freedom are represented by localized spins. Schrieffer and Wolff [23] used a second-order perturbation with respect to V to obtain an effective Hamiltonian

$$H = t\sum_{j\neq i,\sigma} (C^{\dagger}_{i\sigma} C_{j\sigma} + \mathrm{H.C.}) + J\sum_{i} S_i S^{\mathrm{c}}_i, \tag{2.2}$$

where

$$S_i = \tfrac{1}{2}\sum_{\sigma,\sigma'} \tau_{\sigma,\sigma'} f^{\dagger}_{i\sigma} f_{i\sigma'}, \tag{2.3}$$

$$S^{\mathrm{c}}_i = \tfrac{1}{2}\sum_{\sigma,\sigma'} \tau_{\sigma,\sigma'} C^{\dagger}_{i\sigma} C_{i\sigma'}, \tag{2.4}$$

and τ are the Pauli spin matrices. Thus, S^{c}_i are the spin-density operators of the conduction electrons and S_i are the localized spins.

The Hamiltonian in equation (2.2) also known as the Kondo-lattice model and the exchange interaction is antiferromagnetic ($J > 0$) and inversely proportional to U. In fact,

$J = 8V^2/U$ under symmetric conditions. The exchange coupling is the source of interesting many-body effects in the Kondo-lattice model.

The complexity of solving the Kondo-lattice model arises due to complex correlation effect involving both the localized spin and the itinerant electron degrees of freedom. In fact, a conduction electron undergoes a spin-flip process with a localized spin if its spin is antiparallel. The conduction electron leaves a trace of its spin-exchange processes with the localized spins while moving around the lattice. The direction of the localized spins is determined by the history of the electrons that had passed through this site. Thus, the conduction electrons are no more independent. There are similar correlation effects in the periodic Anderson model due to the dynamic aspects of the localized electrons. Since these systems are highly correlated, most of the theoretical models developed during the past 30 years are approximate treatments of the complex problem.

2.2. Theoretical approaches (for solution of renormalized Hamiltonian)

2.2.1. *1/N expansion method*

Considerable insight into these problems has been obtained by using the $1/N$ expansion method [24–33], where N is the orbital degeneracy of the f ion. A generic model which incorporates the N degrees of freedom of the localized moment was derived by Coqblin and Schrieffer [24] as a generalization of the Kondo model. The Hamiltonian H is given by

$$H = \sum_{m=1}^{N} \sum_k E_k C_{k,m}^{\dagger} C_{k,m} + \frac{J}{N} \sum_{m,m'=1}^{N} \sum_i f_{im}^{\dagger} C_{im} C_{im'}^{\dagger} f_{im'}, \tag{2.5}$$

where the localized moment is described by the fermion operators at site i with flavor m.

However, the $1/N$ expansion method is inadequate to explain the periodic features associated with asymptotically high orders of multisite expansion and does not explain the collective properties of heavy fermions. For example, transport measurements in magnetic fields [34, 35] show the difference between heavy fermion systems and dilute Kondo alloys.

2.2.2. *Gutzwiller approach*

There have been attempts to explain the mass renormalization by using the Hubbard–Gutzwiller model [36–38]. A variational wave function, similar to that of Gutzwiller [39] for the Hubbard model, was used to describe the ground state of the Anderson Hamiltonian in these models. This approach can be summarized as follows. An upper bound of the ground-state energy E_g can be written as

$$E_g = \langle \Psi_0 \mid H_{\text{eff}} \mid \Psi_0 \rangle + UDL, \tag{2.6}$$

where D is the density of doubly occupied f orbitals, L is the number of sites, and

$$H_{\text{eff}} = \sum_{k,\sigma} E_k C_{k\sigma}^{\dagger} C_{k\sigma} + E_f \sum_{k,\sigma} f_{k\sigma}^{\dagger} f_{k\sigma} + V \sum_{k,\sigma} q(n_{f\sigma})(C_{k\sigma}^{\dagger} f_{k\sigma} + f_{k\sigma}^{\dagger} C_{k\sigma}). \tag{2.7}$$

In equation (2.7), the renormalization factor $q(n_{f\sigma})$ is obtained by using the Gutzwiller approximation [36–39]. In the large-U limit, the double occupancy of the f states is almost completely suppressed and one obtains

$$q^2(n_{f\sigma}) = \frac{1 - n_f}{1 - n_{f\sigma}} \, (\sigma = \uparrow, \downarrow). \tag{2.8}$$

The scattering rate of the conduction electrons at a certain site, $\Gamma_{\text{eff}} = \pi V_{\text{eff}}^2 \times \rho(\varepsilon_f) \approx (1 - n_f)^2$. The transfer of a conduction electron into an f state is renormalized by the probability, $1 - n_f$, of finding the f level unoccupied. $q(n_f)$ is more complicated for intermediate U but consideration of the whole range of U is necessary to discuss the MV and the Kondo regime.

One can introduce a chemical potential μ for the f electrons:

$$H = H_{\text{eff}} - \mu \Sigma_{i,\sigma} n_{i\sigma}^{f}. \tag{2.9}$$

In Figure 2.1, the schematic shape of the hybridization band obtained by diagonalizing equation (2.9) is shown. While the bare value ε_f is deeply below the Fermi level, the effective f level, $\varepsilon_f - \mu$, is close to the Fermi level.

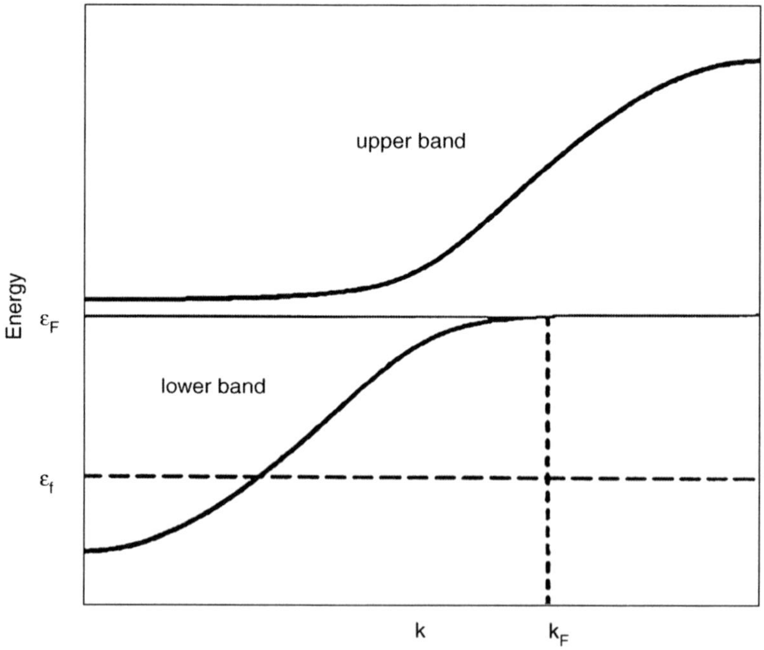

Figure 2.1. Schematic diagram of the upper and lower hybridization band, which are separated by a small gap obtained by Gutzwiller approximation or slave boson treatment. The Fermi level lies close to $\varepsilon_f - \mu$, while the position of the f level, ε_f, is far down the Fermi sea filling the lower band. Reproduced with the permission of the American Physical Society from Ref. [39].

2.2.3. Slave boson method

The slave boson method was developed for the $U = \infty$ periodic Anderson model which had the constraint $n_f \leq 1$ at each site [40, 41]. Kotliar and Ruckenstein [42] extended this method to the general case. The essential features of this method is that the localized electron operators are written as a composition of a fermion $\hat{f}$ and a boson $\hat{b}$ where we may consider the boson as an f vacancy. Every site is occupied either by an $\hat{f}$ fermion or a $\hat{b}$ boson. The localized electron operators are written as a composition of a boson $\hat{b}$ and a fermion $\hat{f}$. One defines

$$f_{i\sigma}^{\dagger} = \hat{f}_{i\sigma}^{\dagger}\hat{b}_i \text{ and } f_{i\sigma} = \hat{b}_i^{\dagger}\hat{f}_{i\sigma}. \tag{2.10}$$

The operator equality

$$\sum_{\sigma} \hat{f}_{i\sigma}^{\dagger}\hat{f}_{i\sigma} + \hat{b}_i^{\dagger}\hat{b}_i = 1, \tag{2.11}$$

satisfies the above condition. The Anderson lattice Hamiltonian can be written as

$$H = \sum_{\mathbf{k},\sigma} \epsilon_{\mathbf{k}} c_{\mathbf{k}\sigma}^{\dagger} c_{\mathbf{k}\sigma} + \epsilon_f \sum_{i,\sigma} \hat{f}_{i\sigma}^{\dagger}\hat{f}_{i\sigma} + V \sum_{i,\sigma} (c_{i\sigma}^{\dagger}\hat{b}_i^{\dagger}\hat{f}_{i\sigma} + \hat{f}_{i\sigma}^{\dagger}\hat{b}_i c_{i\sigma})$$
$$+ \sum_i \lambda_i \left(\sum_{\sigma} \hat{f}_{i\sigma}^{\dagger} f_{i\sigma} + \hat{b}_i^{\dagger} b_i - 1 \right). \tag{2.12}$$

Here, λ_i is a Lagrangian multiplier for the site i needed to impose the local constraints.

The properties of the Hamiltonian (2.12) are usually discussed in a mean-field approximation. It is assumed that the bosons have Bose condensations, $\langle \hat{b}_i \rangle = b_0$, and the Lagrange multiplier $\lambda_i = \lambda_0$ for all sites. Thus, the constraint is obeyed only on the average over the whole system. The resultant Hamiltonian is equivalent to the Hamiltonian H_{eff} obtained in equation (2.7) by using Gutzwiller approach.

2.2.4. Other theoretical methods

The Anderson lattice is essentially an extension of the single-impurity Anderson model [7] in the sense that there is one impurity at each site and therefore the solution becomes extremely complicated. A wide variety of techniques have been used to solve the Anderson-lattice problem. Considerable insight of the problem has been obtained by using the functional integration [43–47], perturbation expansion [48] in U, and real-space renormalization group methods [49, 50]. However, none of these techniques has been adequate to explain the wide range of physical properties.

It has also been argued [51–53] that the physics of heavy fermions is similar to that of liquid ^{3}He. Some authors [54, 55] have developed a Fermi-liquid theory which discards quasiparticle interactions. However, the experimental results on transport properties of heavy fermions contradict these theories by furnishing strong evidence [56] that the large effective mass of heavy fermions does not arise primarily due to band-structure renormalization.

Brandow [57] has proposed a variational theory of valence fluctuations but has ignored the fact that within the cluster-expansion terms, the interacting sites must all be distinct. In fact, his theory is valid when $U \to \infty$ and only at very low temperatures.

2.3. Cluster calculations

In order to correlate the MV, Kondo, and heavy fermion behavior and to study as to how they correspond to different regimes of one fundamental phenomenon, (at least in Ce systems), Misra et al. [58] have considered the application of the periodic Anderson model to finite clusters with periodic boundary conditions. Although the phrase "periodic Anderson model" is somewhat inappropriate when applied to a small system, we have used it in reference to a cluster in which each site has a localized orbital and an extended orbital with appropriate Coulomb repulsion, hybridization, and transfer-matrix elements. The value of small cluster calculations is that exact solutions of the Hamiltonian are obtained. One has confidence that the results do show the properties of the Hamiltonian and not of some particular approximate scheme. Such information is valuable in many-body systems where there is controversy about even the qualitative features of the solutions. However, it has to be recognized that in some respects, small clusters are not representative of bulk materials. For example, at sufficiently low temperatures the specific heat of a cluster model will vanish exponentially and the magnetic susceptibility will either be infinity or be zero. However, the large number of states obtained even for a small cluster suggests that statistical mechanics may give results which fairly represent a large system over a reasonable range of temperatures. The first such calculation for MV systems was made by Arnold and Stevens [59] who considered an isolated three-atom cluster. They calculated the thermodynamic properties which appear similar to MV systems but their calculation implicitly assumes an infinite value of U. Parlebas et al. [60] have applied the periodic Anderson model to a four-site tetrahedral cluster with periodic boundary conditions. They have calculated exactly the many-body eigenstates and eigenvalues of the Hamiltonian and hence the variation of f occupation number for different parameters. Hirsh and Fye [61] have developed an alternate numerical approach by using a Monte Carlo method to treat a small number of magnetic impurities interacting with the conduction electrons in a metal.

We have applied [58] the periodic Anderson model (2.1) to three different four-site clusters of equal length (square, rhombus, and tetrahedron) with periodic boundary conditions. The systems differ principally in the number of distinct neighbor pairs (4, 5, and 6, respectively) thereby including the importance of band-structure effects. For example, while there are no three-dimensional crystal structures equivalent to the square and the rhombus, our model Hamiltonian for a tetrahedron is identical to that of an fcc lattice if the Brillouin-zone sampling is restricted to four reciprocal-lattice points, the zone center Γ, and the three square-face center points X [62]. We studied the region of crossover between the magnetic, Kondo, and MV regimes by varying the different parameters $U/|t|$, $V/|t|$, and $E_f/|t|$ for the three geometries and our results for the tetrahedron should reflect the properties of cerium alloys.

One can distinguish the three regimes by considering n electrons per site. Consider the non-f electrons to constitute an electron reservoir. In Figure 2.2, E_d is the Fermi level

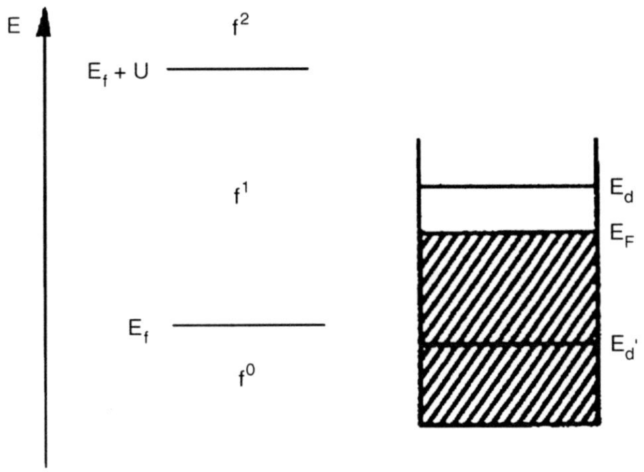

Figure 2.2. A schematic diagram illustrating the dependence of the three different regimes on the position of the f level.

when there are n non-f electrons per site and E_d' is the Fermi level when there are n–1 electrons per site (if there is no interaction with the f electrons). E_F is the chemical potential when the f electrons are in contact with the electron reservoir. Let E_f be the energy boundary such that at $T = 0$, the ion will be in the f^0 state if $E_F < E_f$ and in the f^1 state if $E_F > E_f$. $E_f + U$ is another ionization boundary separating the f^1 state from the f^2 state. The crossover from one regime to another depends sensitively on the various parameters U, V, and E_f as well as on the geometry (band structure).

The Hamiltonian (2.1) is conveniently considered on a basis of states diagonal in occupation numbers and we have calculated the many-body eigenstates and eigenvalues. We note that spin is a good quantum number and the states can be classified as spin singlets, triplets, and quintets. The method of calculation is explained in detail in Ref. [58]. We note that for $n = 4$, there are 784 singlet, 896 triplet, and 140 quintet states. These rather large number of states should tend to make the results somewhat representative of large systems except at extremely low temperatures (lower than the separation between the ground state and the first excited state). In a later paper [63], we considered an eight-electron system in a tetrahedron and there are 12 870 states available for eight particles. However, the results were similar to that obtained in Ref. [58] in which the method of calculation is described in detail. We shall project the significant results through some diagrams and then summarize the results. The details of our studies of model Hamiltonians on finite clusters have been described in Ref. [64].

We constructed a computer program to diagonalize the Hamiltonian (2.1) within subspaces of fixed values of S_z. We have calculated the f-state occupation (n_f), temperature dependence of specific heat (C_v), and the magnetic susceptibility (χ_f) of f electrons (by using a canonical ensemble) for a large number of parameters. In Figure 2.3, we have presented n_f in terms of E_f and various hybridization energies. In Figure 2.4, we have presented a typical example by plotting C_v/T against T for E_f ranging from -5.0 to -4.0 (n_f varies from 0.9943 to 0.9788).

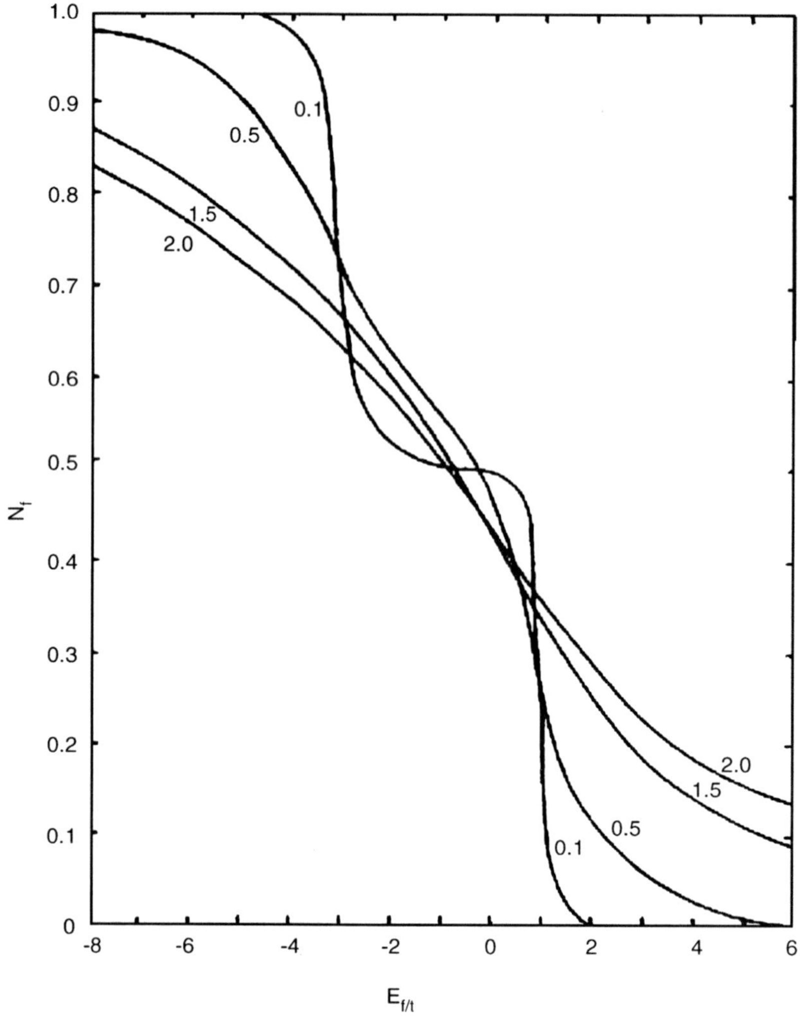

Figure 2.3. The f occupation number per site, n_f, in the four-electron ground state in terms of E_f and various hybridization energies V for $t = -1$ and $U = 50$ for a tetrahedron. Reproduced with the permission of the American Physical Society from Ref. [58].

We notice that for $E_f = -5.0$, C_v/T increases very rapidly at very low temperatures (which mimics the onset of heavy fermion behavior) but gradually decreases as E_f is increased until heavy fermion feature has practically disappeared when $E_f = -4.2$. In order to explain the unusual increase in C_v/T, we have plotted in Figure 2.5 the energy-level diagram of the first few many-body states for each of these E_f as well as for $E_f = -3.0$.

In Figure 2.5, for $E_f = -5.0$, the ground state is a singlet but the next two higher energy states are a triplet and a quintet which are nearly degenerate with the ground state. The low-temperature rise in C_v is determined by these three levels. As E_f increases, the

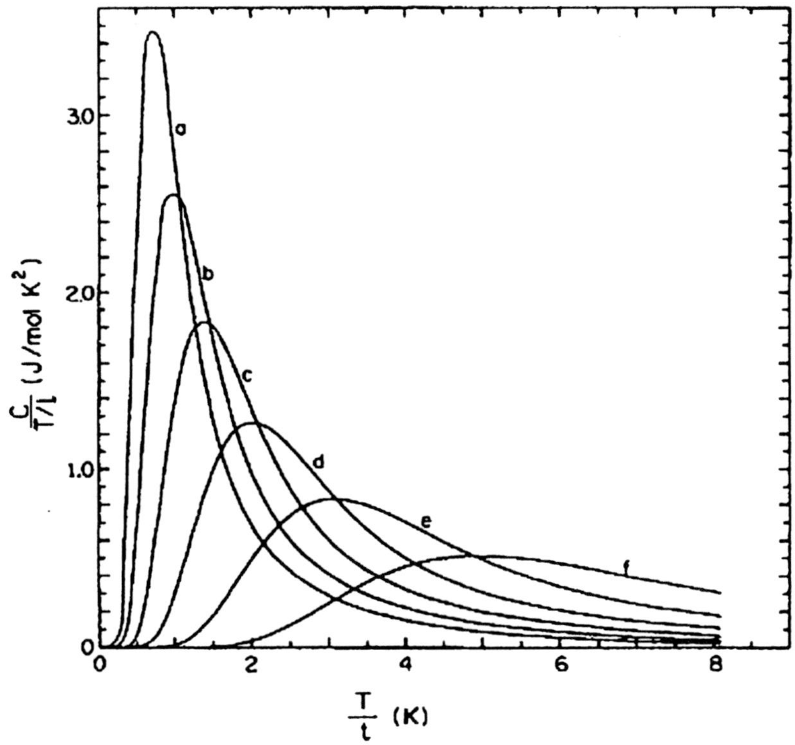

Figure 2.4. $C_v/T/|t|$ versus $T/|t|$ for various E_f for t negative, $U = 50$ and $V = 0.1$ for a tetrahedron. Curve (a) $E_f = -5.0$; curve (b) $E_f = -4.8$; curve (c) $E_f = -4.6$; curve (d) $E_f = -4.4$; curve (e) $E_f = -4.2$; curve (f) $E_f = -4.0$. (All parameters in units of $|t|$.) Reproduced with the permission of the American Physical Society from Ref. [58].

separation between the lowest three levels increases and the rise in C_v/T correspondingly decreases. Thus, the heavy fermion behavior is obtained when the many-body ground state is a singlet but nearly degenerate to two other magnetically ordered states. The same pattern is repeated for a tetrahedron for $t = 1$, except that in some cases the ground state is a magnetically ordered triplet state. In such cases, the ground state of the heavy fermion system would be magnetically ordered.

In Figure 2.6, we have plotted $k_B\chi_f T/(g\mu_B)^2(\equiv\chi_f' T)$ versus $\ln T/|t|$ in order to compare our results with the benchmark results for the single-impurity Anderson model [8]. Following Ref. [8], we define "frozen-impurity" regime ($\chi_f' T = 0$), a free-orbital regime ($\chi_f' T \approx 0.125$), a valence-fluctuation regime ($\chi_f' T \approx 0.167$), and a local-moment regime ($\chi_f' T = 0.25$). In addition, we define an "intermediate regime" for which $0 < \chi_f' T < 0.125$ but $\chi_f' T$ essentially remains a constant in this regime.

We note when $E_f = -5$ ($n_f = 0.994$), there is a transition from the frozen-impurity to the local-moment regime. For $E_f = -3$ ($n_f = 0.725$), we get a transition from the frozen-impurity to the valence-fluctuation regime. When E_f is further increased, the transition is from the frozen-impurity to the free-orbital regime. The high-temperature results are in

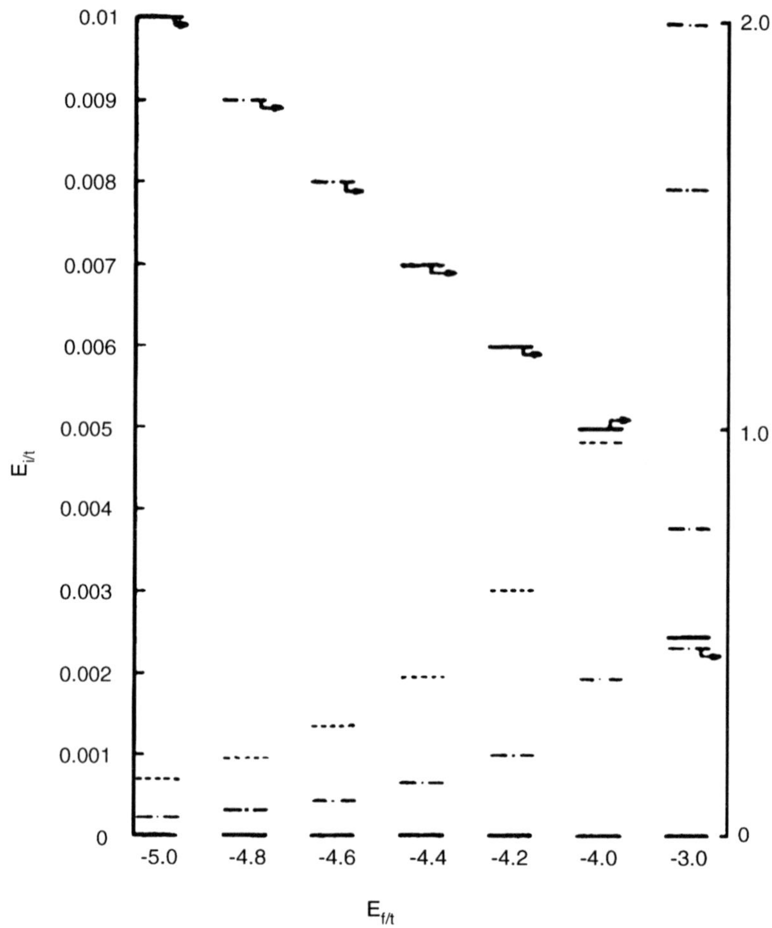

Figure 2.5. Energy level diagram of the first few many-body states for various E_f for t negative, $U = 50$ and $V = 0.1$ for a tetrahedron. (All parameters in units of $|t|$.) Reproduced with the permission of the American Physical Society from Ref. [58].

excellent agreement with the single-impurity "benchmark" results (Figure 8 in Ref. [8]) as expected while at low temperatures, impurity–impurity interactions mediated by the conduction electrons are important.

When we compare the specific heat curves with the χT curves for the same parameters, the specific heat maxima generally occur below the temperature at which χT reaches its high-temperature value (i.e., the crossover temperature from enhanced Pauli- to Curie-like susceptibility). The main reason is that at low temperatures where C_v is maximum, the many-body states with magnetic moments are still just becoming thermally populated. The same broad features have also been observed experimentally [65].

Heretofore, it has been believed that as E_f is increased from far below E'_d (Figure 2.2), there will be a transition from the magnetic to the Kondo-lattice regime. However, we

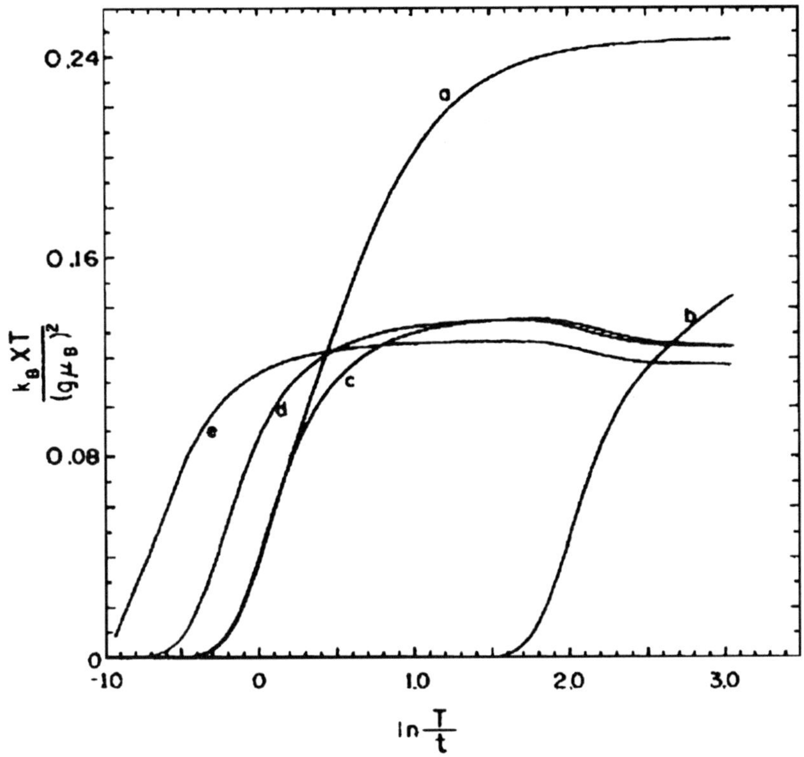

Figure 2.6. $k_B\chi_f T/(g\mu_B)^2$ versus $\ln T/|t|$ for $t = -1$, $U = 50$, and $V = 0.1$ for various E_f for a tetrahedron. Curve (a) $E_f = -5.0$; curve (b) $E_f = -3.0$; curve (c) $E_f = -1.0$; curve (d) $E_f = -0.5$; curve (e) $E_f = 0.5$. (All parameters in units of $|t|$.) Reproduced with the permission of the American Physical Society from Ref. [58].

find that for some choice of parameters, the system undergoes a transition from a Kondo-lattice to a magnetic regime as E_f is increased. Subsequently it reenters a Kondo-lattice regime for higher values of E_f. This unusual feature of reentry to the Kondo-lattice regime is very sensitive to the hybridization parameter and occurs only for low $V/|t|$ values which are the most important parameter in determining n_f as well as the thermodynamic properties.

References

[1] A.C. Hewson, *The Kondo Problem to Heavy Fermions* (Cambridge University Press, Cambridge, 1993).
[2] C.M. Verma and Y. Yafet, Phys. Rev. B **13**, 295 (1975).
[3] A. Bringer and H. Lustfeldt, Z. Phys. B **28**, 213 (1977).
[4] F.D. Haldane, Phys. Rev. Lett. **40**, 416 (1978).
[5] K.G. Wilson, Rev. Mod. Phys. **47**, 773 (1975).
[6] P. Nozieres, J. Low Temp. Phys. **17**, 31 (1974).

[7] P.W. Anderson, Phys. Rev. B **124**, 41 (1961).
[8] (a) H.R. Krishnamurthy, J.W. Wilkins, and K.G. Wilson, Phys. Rev. B **21**, 1003 (1980); (b) **21**, 1044 (1980).
[9] P.B. Wiegmann, V.M. Filyov, and A.M. Tsvelick, JETP Lett. **35**, 92 (1982).
[10] P. Schlottmann, Z. Phys. B **51**, 49 (1983), Phys. Rev. B **50**, 1697 (1983).
[11] A.C. Hewson and J.W. Rasul, J. Phys. C **16**, 6799 (1983).
[12] V.T. Rajan, Phys. Rev. Lett. **51**, 308 (1983).
[13] N. Andrei, K. Furuya, and J.H. Lowenstein, Rev. Mod. Phys. **55**, 331 (1983).
[14] P.B. Wiegmann and A.M. Tsvelick, J. Phys. C **16**, 2281 (1983).
[15] A. Okiji and N. Kawakami, Phys. Rev. Lett. **50**, 1157 (1983).
[16] A.M. Tsvelick and P.B. Wiegmann, Adv. Phys. **32**, (1983); Z. Phys. B **54**, 201 (1984).
[17] P. Schlottmann, Z. Phys. **56**, 127 (1984); Z. Phys. **57**, 23 (1984); J. Phys. C **17**, L267 (1984); Phys. Lett. **100A**, 509 (1984).
[18] A.M. Tsvelick, J. Phys. C **18**, 159 (1985).
[19] P. Schlottmann, J. Magn. Magn. Mater. **47–48**, 367 (1985).
[20] T. Mihalisin, P. Scoroboria, and J. Ward, Phys. Rev. Lett. **46**, 862 (1981).
[21] R. Selim and T. Mihalisin, J. Magn. Magn. Mater. **54–57**, 407 (1986).
[22] V.V. Moschalkov, J. Magn. Magn. Mater. **47–48**, 7 (1985).
[23] J.R. Schrieffer and P.A. Wolff, Phys. Rev. **149**, 491 (1966).
[24] B. Coqblin and J.R. Schrieffer, Phys. Rev. **185**, 847 (1969).
[25] T.V. Ramakrishnan and K. Sur, Phys. Rev. B **26**, 1798 (1982).
[26] F.C. Zhang and T.K. Lee, Phys. Rev. B **28**, 33 (1983).
[27] O. Gunnarson and K. Schonhammer, Phys. Rev. Lett. **50**, 604 (1983); Phys. Rev. B **28**, 4315 (1983); Phys. Rev. B **31**, 4815 (1985).
[28] P. Coleman, Phys. Rev. B **28**, 5255 (1983); **29**, 3035 (1984).
[29] N. Read and D.M. Newns, J. Phys. C **16**, 3273 (1983); J. Phys. C **16L**, 1055 (1983); Solid State Commun. **52**, 993 (1984).
[30] N. Read, D.M. Newns, and S. Doniach, Phys. Rev. B **30**, 3841 (1984).
[31] J.W. Rasul and A.C. Hewson, J. Phys. C **17**, 2555 (1984); J. Phys. C **17**, 3337 (1984).
[32] N. Read, J. Phys. C **18**, 2651 (1985).
[33] N.E. Bickers, D.L. Cox, and J.W. Wilkins, J. Magn. Magn. Mater. **47–48**, 335 (1985); Phys. Rev. Lett. **54**, 230 (1985).
[34] A. Sumiyama, Y. Oda, H. Nagano, Y. Onuki, and T. Komatsubara, J. Phys. Soc. Jpn. **54**, 877 (1985).
[35] T. Penney, J. Stankiewicz, S. von Molnar, Z. Fisk, J.L. Smith, and H.R. Ott, J. Magn. Magn. Mater. **54–57**, 370 (1986).
[36] T.M. Rice and K.M. Ueda, Phys. Rev. Lett. **55**, 995 (1985).
[37] T.M. Rice, K. Ueda, H.R. Ott, and H. Rudigier, Phys. Rev. B **31**, 594 (1985).
[38] C.M. Varma, W. Weber, and J.L. Randall, Phys. Rev. **B 33**, 1015 (1986); M.C. Gutzwiller, Phys. Rev. 137, A1726 (1963).
[39] H. Tsunetsugu, M. Sigrist, and K. Ueda, Rev. Mod. Phys. **69**, 809 (1997).
[40] S.E. Barnes, J. Phys. F **6**, 1375 (1976).
[41] P. Coleman, Phys. Rev. B **29**, 3035 (1984).
[42] G. Kotliar and A.E. Ruckenstein, Phys. Rev. Lett. **57**, 1362 (1986).
[43] C. Lacroix and M. Cyrot, Phys. Rev. B **20**, 1969 (1979).

[44] M. Lavagna, C. Lacroix, and M. Cyrot, J. Phys. F **12**, 745 (1982); Phys. Lett. **89A**, 154 (1982); Phys. Lett. **90A**, 210 (1982).
[45] N. Read, D.M. Newns, and S. Doniach, Phys. Rev. B **30**, 384 (1984).
[46] N. Read and D.M. Newns, Solid State Commun. **52**, 993 (1984).
[47] P. Coleman, J. Magn. Magn. Mater. **47–48**, 323 (1985).
[48] K. Yamada and K. Yosida, J. Magn. Magn. Mater. **31–34**, 461 (1983).
[49] R. Julien, J. Fields, and S. Doniach, Phys. Rev. Lett. **38**, 1500 (1977); Phys. Rev. B **16**, 4489 (1977).
[50] R. Julien and R.M. Martin, J. Appl. Phys. **53**, 2137 (1982); Phys. Rev. B **26**, 6173 (1982).
[51] H.R. Ott, H. Rudigier, K. Ueda, Z. Fisk, and J.L. Smith, Phys. Rev. Lett. **52**, 1915 (1984).
[52] O. Vallis and Z. Tesanovich, Phys. Rev. Lett. **53**, 1497 (1984).
[53] P.W. Anderson, Phys. Rev. B **30**, 1549 (1984).
[54] H. Razahfimandimby, P. Fulde, and J. Keller, Z. Phys. **54**, 111 (1984).
[55] N. d'Ambrumenil and P. Fulde, J. Magn. Magn. Mater. **47–48**, 1 (1985).
[56] C.M. Varma, Phys. Rev. Lett. **55**, 2723 (1985).
[57] B.H. Brandow, Phys. Rev. **33**, 215 (1986).
[58] P.K. Misra, D.G. Kanhere, and J. Callaway, Phys. Rev. B **35**, 5013 (1987).
[59] R.G. Arnold and K.W.H. Stevens, J. Phys. C **12**, 5037 (1979).
[60] J.C. Parlebas, R.H. Victoria, and L.H. Falicov, J. Magn. Magn. Mater. **54–57**, 405 (1986).
[61] J.E. Hirsh and R.M. Fye, Phys. Rev. Lett. **56**, 2521 (1986).
[62] L.M. Falicov and R.H. Victoria, Phys. Rev. B **30**, 1695 (1984); Phys. Rev. Lett. **55**, 1140 (1985).
[63] J. Callaway, D.P. Chen, D.G. Kanhere, and P.K. Misra, Phys. Rev. B **38**, 2583 (1988).
[64] J. Callaway, D.P. Chen, Q. Li, D.G. Kanhere, and P.K. Misra, Studies of model Hamiltonian in finite cluster, in *Quantum Simulations of Condensed Matter Phenomena*, eds. J.D. Doll and J.E. Gubernatis, pp. 19–30 (World Scientific, Singapore, 1990).
[65] K. Satoh, T. Fujita, Y. Maeno, Y. Onuki, T. Komatsubara, and T. Onuki, Solid State Commun. **56**, 327 (1985).

Chapter 3

Dynamical, Extended Dynamical, and Cluster Dynamical Mean-Field Theories: (DMFT, EDMFT, and Cluster DMFT)

3.1. The local impurity self-consistent approximation (LISA)

The dynamical mean-field theory (DMFT) is a very powerful tool to study the strongly correlated systems (for an extensive review, see Ref. [1]). In this approach, a lattice model is replaced by a single-site quantum impurity problem embedded in an effective medium determined self-consistently. This leads to an intuitive picture of the local dynamics of a quantum many-body problem. Since the impurity problem has been extensively studied, the self-consistency condition incorporates the translation invariance and the coherence effects of the lattice. This approach is now popularly known as the "local impurity self-consistent approximation" or LISA. The LISA freezes spatial fluctuations but includes local quantum fluctuations and is therefore characterized as a "dynamical mean-field theory" (DMFT). The on-site quantum problem is still a many-body problem, which can be addressed by using a variety of techniques. The DMFT becomes exact in the limit of large spatial dimensions $d \to \infty$ or in the limit of large lattice coordination.

Metzner and Vollhardt [2] first showed that even in $d = \infty$ dimensions, the Hubbard model, when scaled properly, describes non-trivial correlations among fermions. They found that diagrammatic treatments are much simpler than in finite dimensions. They also showed that the weak-coupling correlation energy is a good approximation for $d = 3$. This is demonstrated by an elegant plot (Figure 3.1) of second-order correlation energy versus density for lattice dimensions $d = 1$, 3, and ∞. There is no significant difference between $n = 3$ and $n = \infty$.

In their pioneering work, which contributed to the evolution of the LISA approach, Metzner and Vollhardt [2] also showed that the approximations based on slave boson technique are recovered by the exact evaluation of explicit variational wave function in $d = \infty$. Muller-Hartmann [3–5] proved the locality of many-body Green's function perturbation theory. He used it to derive self-consistent equations for the self-energy in terms of the Luttinger–Ward functional and obtained results by applying weak-coupling perturbation theory to various orders. Brandt and Mielsch [6–8] solved for the Falicov–Kimball model [9] by deriving self-consistent functional equations.

Ohkawa [10, 11] and Georges and Kotliar [12] derived the main essence of the LISA approach by demonstrating that the functional equations can be interpreted as an Anderson impurity model subject to a self-consistent impurity bath. Georges and Kotliar [12]

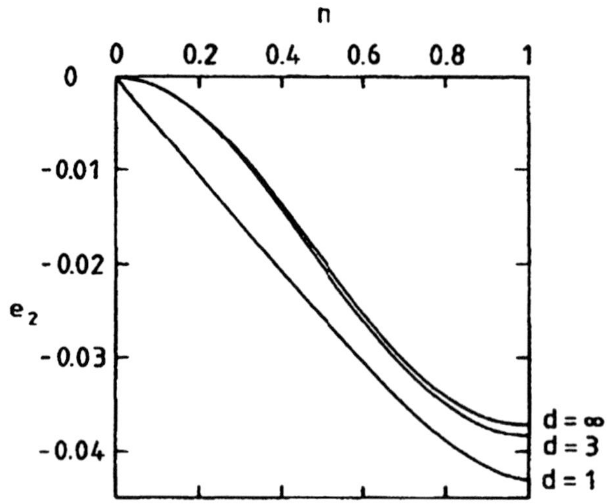

Figure 3.1. Second-order correlation energy versus density for lattice dimensions $d = 1$, 3, and ∞. Reproduced with the permission of the American Physical Society from Ref. [2].

obtained a correspondence with the classical mean-field theory and the analog of the Weiss effective field. They also showed, by using the self-consistency condition with the single-impurity Anderson model, that the metallic phase of $d = \infty$ Hubbard model is a Fermi liquid for arbitrary interaction strength and doping. This result led them to conclude that the methods used to solve the Anderson impurity model can also be used to study correlated electrons in large dimensions. Georges et al. [13] have used the LISA method to a large number of strongly correlated systems as well as to phases with broken symmetry.

As mentioned earlier, in the mean-field theory, a lattice problem with many degrees of freedom is approximated by a single-site effective problem. The dynamics at a given site is the interaction of the degrees of freedom at this site with an external bath created by the degrees of freedom on the other sites. A simple example is the application to Hubbard model in which the Hamiltonian is

$$H = -\sum_{\langle ij\rangle,\sigma} t_{ij}(C^{\dagger}_{i\sigma}C_{j\sigma} + C^{\dagger}_{j\sigma}C_{i\sigma}) + U\sum n_{i\uparrow}n_{i\downarrow}. \tag{3.1}$$

An imaginary-time action (the local effective action) for the fermionic degrees of freedom $(C_{o\sigma}, C^{\dagger}_{o\sigma})$ at site o is

$$S_{\text{eff}} = -\int_0^{\beta} d\tau \int_0^{\beta} d\tau' \sum_{\sigma} C^{\dagger}_{o\sigma}(\tau) g_0^{-1}(\tau - \tau') C_{o\sigma}(\tau') + U\int_0^{\beta} d\tau n_{\sigma\uparrow}(\tau) n_{\sigma\downarrow}(\tau). \tag{3.2}$$

Here $g_0(\tau-\tau')$, the generalized "Weiss function", is the effective amplitude for a fermion to be created on the isolated site at time τ (coming from the "external bath") and destroyed at time τ' (going back to the bath). Since g_0 is a function of time, it accounts for local quantum fluctuations. It can be shown [1] that

$$g_0(i\omega_n)^{-1} = i\omega_n + \mu + G(i\omega_n)^{-1} - R[G(i\omega_n)^{-1}]. \tag{3.3}$$

$G(i\omega_n)$, the on-site interacting Green's function, is calculated from

$$G(\tau - \tau') = -\langle TC(\tau)C^\dagger(\tau')\rangle_{S_{\text{eff}}}, \tag{3.4}$$

$$G(i\omega_n) = \int_0^\beta d\tau G(\tau)e^{iw_n\tau}, \ \omega_n \equiv \frac{(2n+1)\pi}{\beta}. \tag{3.5}$$

Here $R(G)$ is the reciprocal function of the Hilbert transform of the density of states corresponding to the lattice [1]. As an example, in the Hubbard model

$$D(\varepsilon) = \sum_k \delta(\varepsilon - \varepsilon_k), \ \varepsilon_k = \sum_{ij} t_{ij} e^{i\vec{k}\cdot(\vec{R_i} - \vec{R_j})}. \tag{3.6}$$

The Hilbert transform $\overline{D}(\xi)$ and its reciprocal function R are defined by

$$\overline{D}(\xi) \equiv \int_{-\infty}^{\infty} d\varepsilon \frac{D(\varepsilon)}{\xi - \varepsilon}, \ R[\overline{D}(\xi)] = \xi. \tag{3.7}$$

Equations (3.2–3.4) are the basic equations of the LISA method. However, the major difficulty lies in the solution of S_{eff}. It can be shown [1] that solving these equations yields the local quantities and all the $\vec{k}$-dependent correlation functions of the original lattice Hubbard model can be obtained.

It may be noted that the LISA approach freezes spatial fluctuations but retains local quantum fluctuations. Each site undergoes transition between the four possible quantum states $|0\rangle, |\uparrow\rangle, |\downarrow\rangle, |\uparrow, \downarrow\rangle$ by exchanging electrons with the rest of the lattice or "the external bath". As an example, one can consider $(C_{o\sigma}, C^\dagger_{o\sigma})$ as an impurity orbital. The bath can be described as a "conduction band" described by the operators $(a_{l\sigma}, a^\dagger_{l\sigma})$ and the Hamiltonian is the well-known single-impurity Anderson Hamiltonian [14]:

$$H_{\text{AM}} = \sum_{l\sigma} \tilde{\epsilon}_l a^\dagger_{l\sigma} a_{l\sigma} + \sum_{l\sigma} V_l (a^\dagger_{l\sigma} C_{o\sigma} + C^\dagger_{o\sigma} a_{l\sigma}) - \mu \sum_\sigma C^\dagger_{o\sigma} C_{o\sigma} + U n_{\sigma\uparrow} n_{\sigma\downarrow}. \tag{3.8}$$

Equation (3.8) is quadratic in $a^\dagger_{l\sigma}, a_{l\sigma}$ and integrating these gives rise to S_{eff} of the form given in equation (3.2), provided

$$g_0^{-1}(i\omega_n)^{\text{AM}} = i\omega_n + \mu - \int_{-\infty}^{\infty} d\omega \frac{\Delta(\omega)}{i\omega_n - \omega} \tag{3.9}$$

and

$$\Delta(\omega) = \sum_{l\sigma} V_l^2 \delta(\omega - \tilde{\varepsilon}_l). \tag{3.10}$$

If the parameters V_l, $\tilde{\varepsilon}_l$ are chosen to obtain g_0, the solution of the mean-field equations, H_{AM} becomes the Hamiltonian representation of S_{eff}. Here $\tilde{\varepsilon}_l$'s are effective parameters and not $\varepsilon_{\mathbf{k}}$, the single-particle energy. In addition, $\Delta(\omega)$, the conduction bath density of states, is obtained when the self-consistent problem is solved.

Thus, by using the LISA approach, one obtains the Anderson impurity embedded in a self-consistent medium from the Hubbard model [10–13, 15]. The dynamical mean-field equations are solved such that the proper g_0 is obtained. When this g_0 is inserted into the Anderson model, the resulting Green's function should obey the self-consistency condition in equation (3.3). The mapping onto impurity models, which have been studied by a variety of analytical and numerical techniques, are thus used to study the strongly correlated lattice models in large dimensions. However, it is important to solve S_{eff} by using reliable methods.

3.2. Brief discussions of the dynamical mean-field equations

The correct equations for a given model with on-site equations can be guessed by postulating that the self-energy can be computed from a single-site effective interaction and an arbitrary retarded quadratic term. One can obtain the self-consistency equation by writing that the interacting Green's function of the single-site action is the same as the site-diagonal Green's function of the lattice model. These approximations are obviously valid in large dimensions. We briefly discuss several derivations.

3.2.1. *The cavity method*

In Figure 3.2, we show a cavity created in the full lattice by removing a single site and its adjacent bonds. This method can be easily generalized to several models. The main technique used in this method is to concentrate on one site, say $i = 0$, and then integrate the

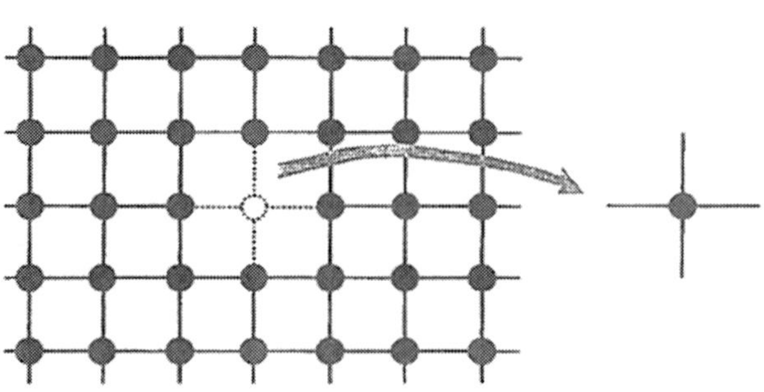

Figure 3.2. Cavity created in the full lattice by removing a single site and its adjacent bonds. Reproduced with the permission of the American Physical Society from Ref. [1].

degrees of freedom on the rest of the lattice sites. Thus, one obtains an effective dynamics for the particular site.

In the Hubbard model, one can show [1] that the local effective action S_{eff} in the "cavity method" is given by

$$S_{\text{eff}} = \sum_{n=1}^{\infty} \sum_{i_1 \ldots j_n} \int \eta^\dagger_{i_1}(\tau_{i_1}) \ldots \eta^\dagger_{i_n}(\tau_{i_n}) \eta_{j_1}(\tau_{j_1}) \ldots \eta_{j_n}(\tau_{j_n}) G^{(o)}_{i_1 \ldots j_n} \times (\tau_{i_1} \ldots \tau_{i_n}, \tau_{j_1} \ldots \tau_{j_n}) + S_o + \text{const.}, \tag{3.11}$$

where

$$S_o = \int_o^\beta d\tau \left(\sum_\sigma C^\dagger_{o\sigma}(\partial_\tau - \mu) C_{o\sigma} + U n_{o\uparrow} n_{o\downarrow} \right), \tag{3.12}$$

$\eta_i \equiv t_{i\sigma} C_{o\sigma}$ plays the role of a source coupled to $C^\dagger_{i\sigma}$, $G^{(o)}$ is the Green's function of the cavity Hamiltonian (Hubbard model with one site removed) and the integration is over fermions $i \neq o$. Here $C^\dagger$ and C are the creation and annihilation operators of the electrons. It can also be shown that in the limit $d \to \infty$ [1]

$$g_0^{-1}(i\omega_n) = i\omega_n + \mu - \sum_{ij} t_{oi} t_{oj} G^{(o)}_{ij}(i\omega_n). \tag{3.13}$$

Equation (3.13) relates $G^{(o)}_{ij}$ to the Weiss function g_0.
It can also be shown that [1] for a general lattice,

$$G^{(o)}_{ij} = G_{ij} - \frac{G_{io} G_{oj}}{G_{oo}}, \tag{3.14}$$

where $G^{(o)}$ is the cavity Green's function and G is the full Green's function.

3.2.2. Perturbation theory in infinite dimensions

It was observed by Metzner and Vollhardt [2], Muller-Hartmann [3], and Metzner [16] that many-body diagrammatics, while using perturbation theory in the interaction strength U, becomes simple in infinite dimensions. The main reason is that the local Green's function obeys a closed set of equations. This is shown in Figure 3.3, in which the interaction term $U n_{i\uparrow} n_{i\downarrow}$ is shown as a four-leg vertex at site I and each line stands for a free-fermion propagator. It can be shown that in infinite dimensions, when two internal vertices (i, j) can be connected by at least three paths, they correspond to identical sites.

In the momentum-space formulation of perturbation theory, whenever two vertices can be "collapsed", the fermion propagators $G^{(0)}(\mathbf{k}, i\omega_n)$ connecting them can be replaced by their local $\mathbf{k}$-independent counterpart $\Sigma_{\mathrm{k}} G^{(0)}(\mathbf{k}, i\omega_n)$ ignoring momentum conservation at the vertices. Since evaluation of momentum sums poses a real problem in going to higher orders, this method simplifies the weak-coupling expansions.

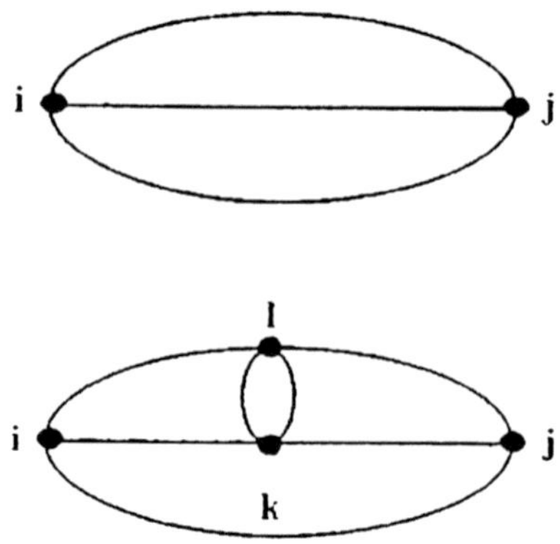

Figure 3.3. Examples of skeleton diagrams contributing to the self-energy at second and fourth order which can be collapsed to a single site. Reproduced with the permission of the American Physical Society from Ref. [1].

Figure 3.4. First two contributions to the Luttinger–Ward functional. Reproduced with the permission of the American Physical Society from Ref. [1].

An example of application of this method is the Luttinger–Ward [17] free-energy functional $\Phi[\{G_{ij}\}]$, which is the sum of all vacuum-to-vacuum skeleton graphs as shown in Figure 3.4.

It can be shown that this functional is related to the self-energy [18]

$$\Sigma_{ij}(i\omega_n) = \frac{\delta\Phi}{\delta G_{ij}(i\omega_n)}. \tag{3.15}$$

When $d \to \infty$, Φ and $\Sigma_{ij}^{\text{skel}}$ depend on the local Green's functions G_{ii}:

$$\Phi = \sum_i \phi[G_{ii}], \quad d \to \infty. \tag{3.16}$$

Thus, the self-energy is site-diagonal.

There are also other methods such as derivation based on an expansion around the atomic limit [19, 20] and the effective medium (coherent potential) approximation [21, 22]. However, we shall not discuss them here since they are not often used for heavy fermions.

3.3. Methods of solution

The essence of DMFT is to map lattice models of correlated fermions onto a single-impurity model. The self-consistence condition specifies the relation between the Weiss function g_0 (in the impurity model effective action) and the local Green's function G. However, G can be obtained by solving the effective impurity model.

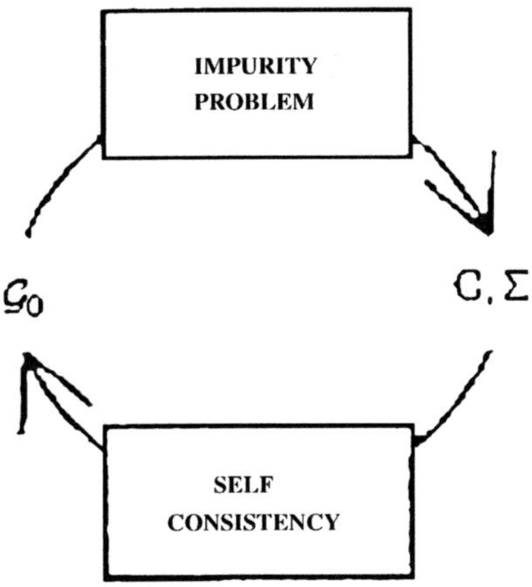

Figure 3.5. For a given g_0, the local impurity model is solved to obtain the interacting Green's function G which is used in the self-consistency condition to produce anew g_0. The loop is reiterated until a converged set (G, g_0) is obtained. Reproduced with the permission of the American Physical Society from Ref. [1].

Therefore an iterative process is followed by first guessing g_0, then calculating G and Σ which are used as an input into the self-consistency condition to produce a new g_0.

This process is reiterated until a converged solution (G, g_0) is obtained. This process is shown in Figure 3.5. There are both numerical and analytical approximate methods available to solve the problem. We shall briefly describe the numerical methods.

The two numerical techniques used to solve the LISA equations are the quantum Monte Carlo (QMC) and the exact diagonalization methods. In the QMC method, one uses the Hirsch–Fye [23] algorithm in which the single-impurity problem is considered in discretized imaginary time. The effective bath enters through g_0 so that it is not necessary to discretize the conduction band. Thus, the method is primarily used to calculate the local Green's function at finite temperature, which was introduced in equation (3.4). The QMC method was first independently used by Jarrell [15], Rozenberg et al. [24], and Georges and Krauth [25].

In the exact diagonalization method [26–28], the single-impurity problem is solved exactly with an effective bath which is approximated by only a few orbitals. Thus, the effective bath is parameterized. These parameters are functions of the site energies and the hopping amplitudes of the fictitious electrons. They are also dependent on the choice of the geometry of their connections. Since it is possible to have several geometries of the electronic bath, an appropriate choice can be made only through the physical insight of the relevant problem. This would lead to the correct parameterization. The main limitation of this method is the size of the exponentially growing Hilbert space.

There are several analytic methods available to solve quantum impurity models. These can be divided into two broad categories. There are several analytic tools to study the low-energy universal properties which are important in the LISA context. Another class of methods are designed to obtain quantitative solution of impurity models but these methods are mostly approximate. One useful analytic method is the weak-coupling perturbation theory which leads to the iterated perturbation theory in the LISA context. The non-crossing approximation and the high-temperature equation of motion method have also been proven useful [29, 30].

3.4. Application of LISA to periodic Anderson model

We shall now briefly describe the application of LISA method to heavy fermion systems and the Kondo insulators. This is done by using the periodic Anderson model (PAM). This model describes a band of conduction electrons which hybridize with localized f electrons at each lattice site. The PAM Hamiltonian can be written as

$$\begin{aligned} H = & \sum_{\mathbf{k}\sigma} \epsilon_{\mathbf{k}} C^{\dagger}_{\mathbf{k}\sigma} C_{\mathbf{k}\sigma} + V \sum_{i\sigma} (C^{\dagger}_{i\sigma} f_{i\sigma} + f^{\dagger}_{i\sigma} C_{i\sigma}) \\ & + \epsilon_{\mathrm{f}} \sum_{i\sigma} f^{\dagger}_{i\sigma} f_{i\sigma} + U \sum_{i} \left(n_{\mathrm{f}i\uparrow} - \frac{1}{2} \right) \left(n_{\mathrm{f}i\downarrow} - \frac{1}{2} \right), \end{aligned} \tag{3.17}$$

and the terms have been defined in Chapter 2. In the $d \to \infty$ limit, the local interaction gives rise to **k**-independent self-energy and the various Green's functions are obtained in the following forms [1]:

$$\begin{aligned} G_{\mathrm{c}}(i\omega_n, \mathbf{k})^{-1} &= i\omega_n - \epsilon_{\mathbf{k}} - \frac{V^2}{i\omega_n - \epsilon_{\mathrm{f}} - \Sigma_{\mathrm{f}}(i\omega_n)}, \\ G_{\mathrm{f}}(i\omega_n, \mathbf{k})^{-1} &= i\omega_n - \epsilon_{\mathrm{f}} - \Sigma_{\mathrm{f}}(i\omega_n) - \frac{V^2}{i\omega_n - \epsilon_{\mathbf{k}}}, \\ G_{\mathrm{cf}}(i\omega_n, \mathbf{k})^{-1} &= \frac{1}{V} \{ [(i\omega_n - \epsilon_{\mathbf{k}})(i\omega_n - \epsilon_{\mathrm{f}} - \Sigma_{\mathrm{f}}(i\omega_n)] - V^2 \}, \end{aligned} \tag{3.18}$$

where $\Sigma_{\mathrm{f}}(i\omega_n)$ is the self-energy of the f electrons and μ, the chemical potential, has been absorbed in the definitions of $\epsilon_{\mathbf{k}}$ and ϵ_{f}. It can be shown by reducing to a self-consistent single-site model that the effective action is [1]

$$\begin{aligned} S_{\mathrm{eff}} = & -\int_0^{\beta} d\tau \int_0^{\beta} d\tau' \sum_{\sigma} f^{\dagger}_{\sigma}(\tau) g_0^{-1}(\tau - \tau') f_{\sigma}(\tau') \\ & + U \int_0^{\beta} d\tau \left[n_{\mathrm{f}\uparrow}(\tau) - \frac{1}{2} \right] \left[n_{\mathrm{f}\downarrow}(\tau) - \frac{1}{2} \right]. \end{aligned} \tag{3.19}$$

The f self-energy is obtained from

$$\Sigma_{\mathrm{f}} = g_0 - G_{\mathrm{f}}^{-1}, \; G_{\mathrm{f}} \equiv - \langle \mathrm{T f f}^{\dagger} \rangle_{S_{\mathrm{eff}}}. \tag{3.20}$$

Since the self-consistency condition requires that the Green's function of the impurity problem must be equal to the local f Green's function of the lattice model, one obtains

$$G_{\mathrm{f}}(i\omega_n) = \int_{-\infty}^{\infty} \frac{d\epsilon \, D(\epsilon)}{i\omega_n - \epsilon_{\mathrm{f}} - \Sigma_{\mathrm{f}}(i\omega_n) - V^2/(i\omega_n - \epsilon)}. \tag{3.21}$$

Here $D(\epsilon)$ is the density of states (non-interacting) of the conduction electrons.

The quantitative studies of PAM ($d = \infty$) have been mostly done on the half-filled case (Kondo insulators). However, Schweitzer and Czycholl [31] have shown, by using a self-consistent second-order perturbation theory (SOPT) in terms of the Coulomb repulsions U, that the temperature dependence of the electronic transport properties of heavy fermion systems can be calculated. The Luttinger sum rules are fulfilled in the limit of a large spatial dimension d for correlated lattice electrons. In fact, their calculations of temperature dependence of resistivity (Figure 3.6) and thermopower (Figure 3.7), by using the $d = 3$ result for $d = \infty$, agree with the experimental results.

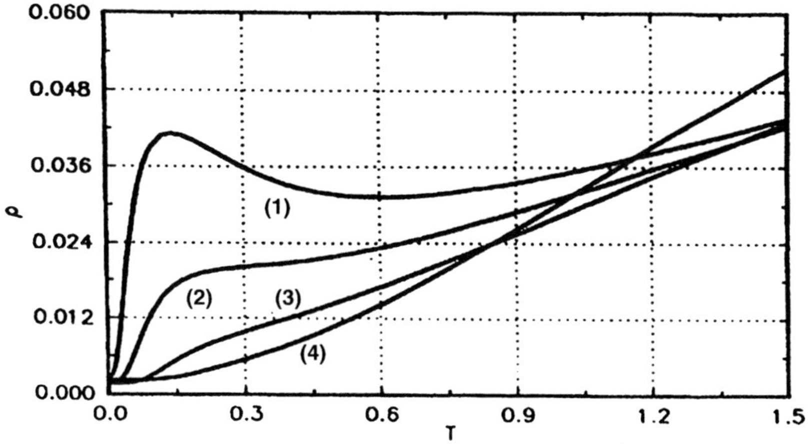

Figure 3.6. Temperature dependence of the resistivity obtained for the PAM within the SOPT for $d = \infty$ and the parameters $E_f = -0.5$, $V = 0.4$, $U = 1$. (1) $n_{tot} = 0.8$, (2) $n_{tot} = 0.6$, (3) $n_{tot} = 0.4$, (4) $n_{tot} = 0.2$. Reproduced with the permission of the American Physical Society from Ref. [31].

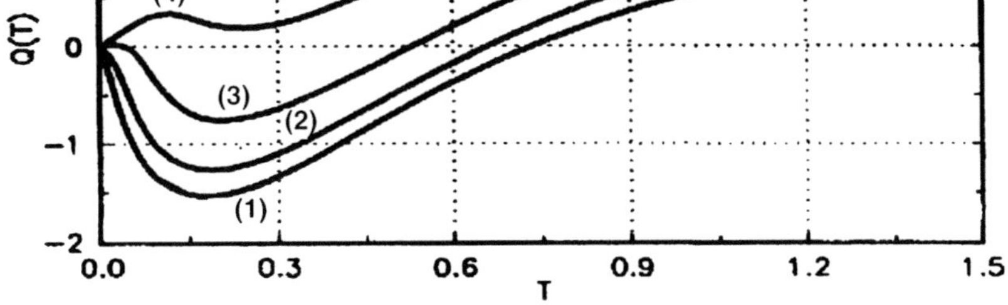

Figure 3.7. Temperature dependence of thermopower $Q(T)$; the parameters are the same as in Figure 3.6. Reproduced with the permission of the American Physical Society from Ref. [31].

3.5. Kondo insulators

The Kondo insulators such as CeNiSn, $Ce_3Bi_4Pt_3$, and SmB_6 are the half-filled case $\langle n_c \rangle + \langle n_f \rangle = 2$ ($\varepsilon_f = 0$ in this notation). Jarrell et al. [32] have used the QMC method to solve the associated impurity problem and calculate the density of states and various response functions. One typical result is shown in Figure 3.8. The spectral density displays a narrow insulating gap with two sharp peaks on each side.

The results for density of states for c and f electrons for the half-filled PAM, obtained by using the site-excluded perturbation theory (SEPT) (in U) method in which the self-energy is subtracted off from the site propagator during the self-consistency step, is shown in Figure 3.9. As mentioned above, the results are very similar to that of QMC method.

3.6. The multichannel Kondo lattice

The charge fluctuations can be ignored for large U and close to half-filling. As mentioned earlier, by using the Schrieffer–Wolff transformation [33], the PAM can be formulated as Kondo lattice model. The conduction electrons interact with only the localized spins. One obtains

$$H = \sum_{\mathbf{k}\sigma} \epsilon_{\mathbf{k}} C^{\dagger}_{\mathbf{k}\sigma} C_{\mathbf{k}\sigma} + J_{\mathrm{K}} \sum_{i\sigma\sigma'} \sum_{a=x,y,z} S^a_i \frac{1}{2} C^{\dagger}_{i\sigma} \sigma^a_{\sigma\sigma'} C_{i\sigma'}. \tag{3.22}$$

The conduction electron Green's function is

$$G_{\mathrm{c}}(i\omega_n, \mathbf{k}) = \frac{1}{i\omega_n - \epsilon_{\mathbf{k}} - \Sigma_{\mathrm{c}}(i\omega_n)}, \tag{3.23}$$

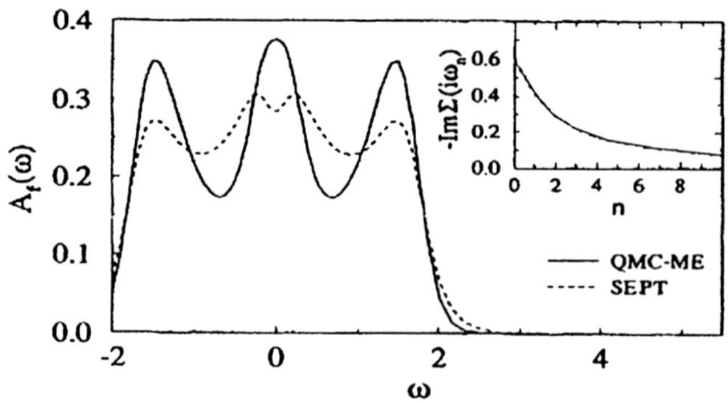

Figure 3.8. The f-electron DOS obtained from QMC and maximum entropy (QMC-ME) method, and that obtained from second-order SEPT when $U = 2.0$, $V = 0.5$, $\varepsilon_d = \varepsilon_f = 0$, and $T = 0.2$. The corresponding imaginary frequency self-energies are shown in the inset. Reproduced with the permission of the American Physical Society from Ref. [32].

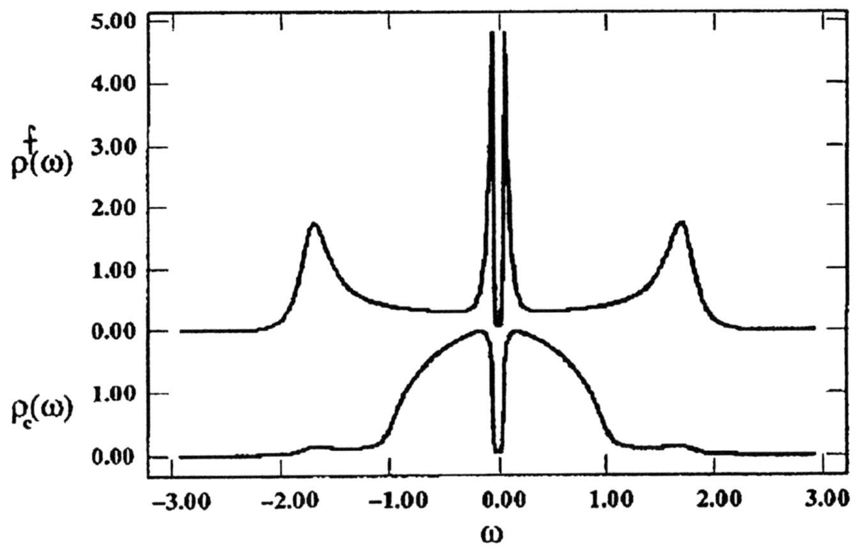

Figure 3.9. Density of states for c and f electrons (bottom and top) in the half-filled PAM for $U/D = 2.5$ and $V/D = 0.4$ at $T/D = 0.01$. Reproduced with the permission of the American Physical Society from Ref. [1].

where Σ_c is the conduction electron self-energy. The single-site effective action is [13]

$$S_{\text{eff}} = -\int_0^\beta d\tau \int_0^\beta d\tau' \sum_\sigma C_\sigma^\dagger(\tau) g_0^{-1}(\tau - \tau') C_\sigma(\tau')$$
$$+ J_K \sum_a \int_0^\beta d\tau S^a \frac{1}{2} \sum_{\sigma\sigma'} C_\sigma^\dagger \sigma_{\sigma\sigma'}^a C_{\sigma'} + L_s[S]. \tag{3.24}$$

Here S_{eff} describes a single-impurity Kondo model in an effective conduction bath g_0 and $L_s[S]$ is a spin-1/2 Lagrangian.

The self-consistency condition is obtained from

$$G_c(i\omega_n) \equiv -\langle C^\dagger(i\omega_n) C(i\omega_n)\rangle_{S_{\text{eff}}} = \int_{-\infty}^{\infty} \frac{D(\epsilon)\, d\epsilon}{i\omega_n - \epsilon - \Sigma_c(i\omega_n)}, \tag{3.25}$$

and

$$\Sigma_c \equiv g_0^{-1} - G_c^{-1}. \tag{3.26}$$

The multichannel Kondo lattice Hamiltonian [1] is obtained by generalizing equation (3.22) by including "channels" for the conduction electrons $C_{\mathbf{k}\sigma}^\dagger \to C_{\mathbf{k}\sigma\alpha}^\dagger$, with $\alpha = 1, \ldots, M$:

$$H = \sum_{\mathbf{k}\sigma\alpha} \epsilon_{\mathbf{k}} C_{\mathbf{k}\sigma\alpha}^\dagger C_{\mathbf{k}\sigma\alpha} + J_K \sum_{\alpha=1}^{M} \sum_{i\sigma\sigma' a} S_i^a C_{i\alpha\sigma}^\dagger \frac{\sigma_{\sigma\sigma'}^a}{2} C_{i\alpha\sigma'}. \tag{3.27}$$

The single-impurity multichannel Kondo is the associated impurity model of H.

The single-site ($d = \infty$) DMFT has been applied extensively to many problems by using the techniques briefly described earlier. However, it essentially ignores the feedback of interimpurity interactions into single-particle quantities. Since it fails to capture the effects of intersite interactions in either the charge channel (nearest-neighbor interactions) or the spin channel (exchange), it does not take into account the dynamical effects of fluctuations in the high-temperature disordered phase. In heavy fermion systems, the competition of the Kondo effect with the Rudderman–Kittel–Kasuya–Yosida (RKKY) interaction [34] plays a very important role but it is impossible to solve this by using the DMFT. There are also other crucial problems in strongly correlated systems which cannot be understood in the LISA framework.

3.7. RKKY interaction

Ruderman and Kittel [34] considered the problem of nuclear-spin ordering in a metal and used SOPT to derive an expression for the indirect nuclear spin–spin interaction,

$$H_{\mathrm{RKKY}} = -\frac{9\pi}{8} n_{\mathrm{c}}^2 \frac{J^2}{\varepsilon_{\mathrm{F}}} \sum_{\langle ij \rangle} \frac{\mathbf{S}_i \cdot \mathbf{S}_j}{r_{ij}^3} \left[2k_{\mathrm{F}} \cos(2k_{\mathrm{F}} r_{ij}) - \frac{\sin(2k_{\mathrm{F}} r_{ij})}{r_{ij}} \right], \tag{3.28}$$

where k_{F} is the Fermi wave vector and n_{c} is the density of conduction electrons. The spin–spin interaction is long ranged and changes its sign depending on the distance between the pair of spins. Kasuya [34] discussed the magnetic properties of rare earth metals based on equation (3.28). Yosida [34] showed that the oscillatory behavior originates from the Friedel oscillation of the spin polarization of conduction electrons induced by a localized spin. Therefore equation (3.28) is known as the RKKY interaction.

For rare earth metals, the Fourier transform of the RKKY interaction is given by $\chi(\mathbf{q})$, the susceptibility of the conduction electrons for wave number $\mathbf{q}$. The ground state is usually ferromagnetic if $\chi(\mathbf{q})$ is maximum at $\mathbf{q} = 0$. If the maximum of $\chi(\mathbf{q})$ occurs at $\mathbf{q} = \mathbf{Q}$, the antiferromagnetic wave vector, the ground state becomes antiferromagnetic. The ground state may have a spiral spin ordering if $\chi(\mathbf{q})$ becomes maximum at a general wave vector. The Kondo effect is suppressed whenever there is any type of magnetic ordering. The low-energy physics of the Kondo effect is given by the Kondo temperature

$$T_{\mathrm{K}} = \varepsilon_{\mathrm{F}} e^{-1/J\rho(\varepsilon_{\mathrm{F}})}. \tag{3.29}$$

However, the characteristic energy of the RKKY interaction is given by $J^2/\varepsilon_{\mathrm{F}}$. This energy dominates over the Kondo temperature in the weak-coupling regime.

In the strong-coupling regime, the local moments are quenched because of the formation of local singlets. The Kondo effect or the effect of singlet formation is not considered for the derivation of the RKKY interaction. The relation between RKKY interaction and the Kondo effect depends on the conduction electron density, dimensionality, and the exchange coupling. As an example, we consider two localized spins S_1 and S_2. The direct exchange coupling between the two spins can be expressed as

$$H = J_{\mathrm{RKKY}} \mathbf{S}_1 \cdot \mathbf{S}_2, \tag{3.30}$$

where J_{RKKY}, the intersite coupling constant, is arbitrary. For $J > 0$, the Kondo coupling is antiferromagnetic and the ground state is a singlet. When $J \gg J_{RKKY}$, each of the two localized spins forms a singlet with conduction electrons and hence the interaction between the singlets is weak. When $J_{RKKY} \ll J$, the two localized spins form a singlet by themselves and J is no more important. There have been several theories as to whether the change between the two regimes is smooth or sharp.

3.8. Extended dynamical mean-field theory (EDMFT)

3.8.1. Overview

We have discussed in the previous section that there are two competing interactions in the heavy fermion systems, the indirect exchange between the moments mediated by the RKKY interaction and the Kondo exchange between the conduction electrons and the moments. At high temperatures, the moments and conducting electrons retain their identities and interact weakly. As the temperature is lowered, when the RKKY interaction is much larger than the Kondo energy, the moments order magnetically. The conduction electrons follow the magnetization and there is no strong renormalization. Since the local moments are ordered, the spin entropy is mostly quenched. When the Kondo energy is much larger than the RKKY interaction, there is formation of heavy Fermi-liquid quasiparticles which are composites of local-moment spins bound to the conduction electrons. However, at low temperatures, the Kondo and RKKY interactions are comparable and the problem remains unresolved. At least two materials [35], $YbRh_2Si_2$ [36] and $CeCu_{6-x}Au_x$ [37], can be driven continuously from the paramagnetic (PM) to the antiferromagnetic (AF) phase by application of alloying, magnetic field, and pressure. We show the experimental results for $YbRh_2Si_2$ in Figures 3.10–3.12.

In Figure 3.10, χ_{ac} is plotted as a function of temperature (on a logarithmic scale). The data down to very low temperature indicates a weak antiferromagnetic transition $T_N \approx 65\,\text{mK}$ which is one of the lowest reported for the onset of magnetism in strongly correlated systems. The inset of Figure 3.10 shows the $p = 0$ electrical resistivity of $YbRh_2Si_2$ is almost constant above 100 K, but decreases toward lower temperatures indicating the development of coherence effects. This temperature dependence is typical of heavy fermion compounds with a strong interaction between the localized 4f and the conduction electrons and resembles some non-Fermi-liquid systems.

In Figure 3.11(a), the existence of NFL effects in $\rho(T)$ of $YbRh_2Si_2$ at low temperatures is shown. At low temperatures and at $p = 0$, $\rho(T) = \rho(0) + bT^{\varepsilon}$, with $\varepsilon \approx 1$. $\partial \ln \Delta\rho/\partial \ln T$ remains within the value (1 ± 0.05) up to almost 10 K. Figure 3.11(b) indicates that the application of magnetic field leads to a gradual recovery of $\Delta\rho(T) = T^2$ below a crossover temperature that increases upon increasing B.

In Figure 3.12(a), two anomalies in $\rho(T)$ of $YbRh_2Si_2$ at $p = 1.5\,\text{GPa}$ are shown. These are denoted as T_m and T_l. Figure 3.12(b) suggests an evolution toward complex magnetic ordering for $p > 0$.

Thus, theoretical studies are needed to understand the microscopic mechanisms leading to the experimental results observed in $YbRh_2Si_2$ and other similar heavy fermion compounds exhibiting quantum-critical phenomena as well as non-Fermi-liquid behavior.

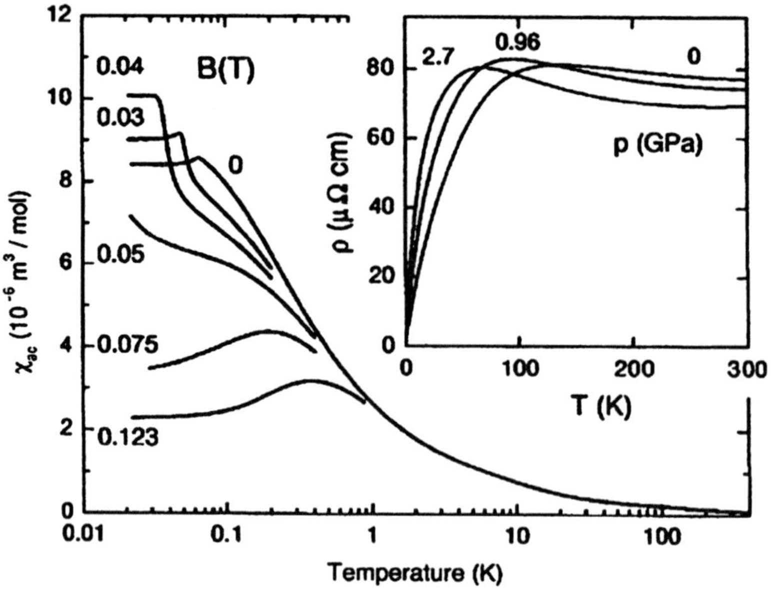

Figure 3.10. ac magnetic susceptibility of $YbRh_2Si_2$ as χ_{ac} versus T (on a logarithmic scale), measured on a tetragonal plane at varying fields. Inset: high-temperature electrical resistivity at different pressures. Reproduced with the permission of the American Physical Society from Ref. [36].

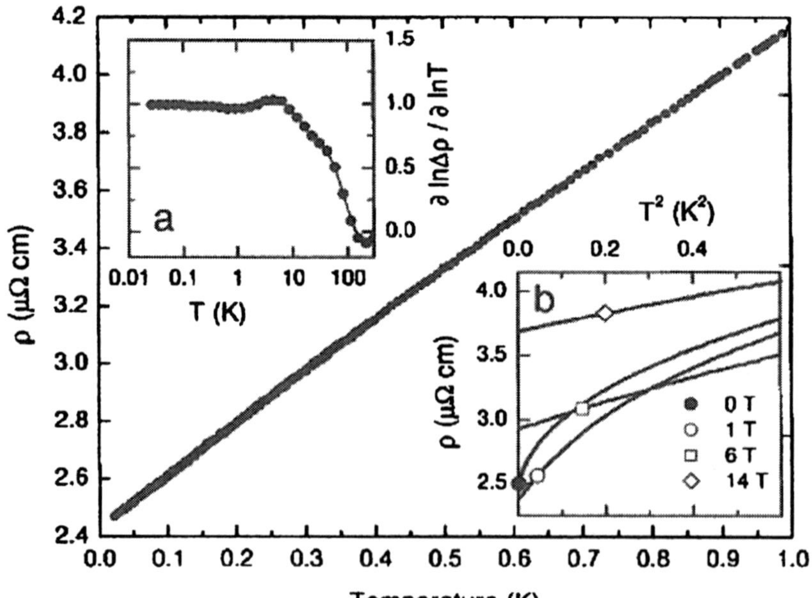

Figure 3.11. Low-temperature electrical resistivity of $YbRh_2Si_2$ at $p = 0$ measured along the a axis as a function of temperature, obeying $\rho(T) = \rho(0) + bT^{\varepsilon}$, with $\varepsilon \approx 1$. (a) Temperature dependence of the effective exponent ε, defined as the logarithmic derivative of $\Delta\rho = \rho - \rho_0$ with respect to T. (b) $\rho(T)$, plotted as ρ versus T^2, for $B \leq 14T$ applied along the C axis. The position of the symbols indicates the crossover temperature below which a T^2 law is recovered. Reproduced with the permission of the American Physical Society from Ref. [36].

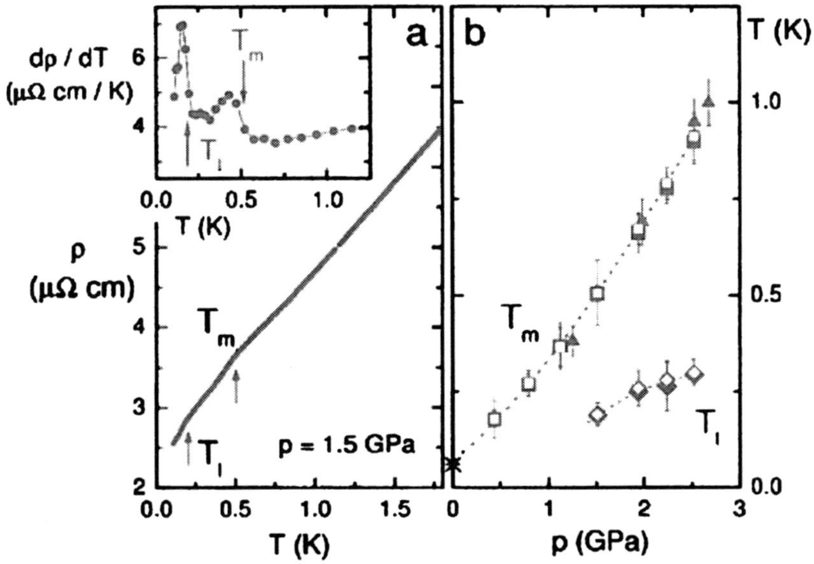

Figure 3.12. (a) Low-temperature electrical resistivity of $YbRh_2Si_2$ for $p = 1.5$ GPa. Inset: temperature derivative of $\rho(T)$ showing well-defined, second-order-type transitions at T_m and T_l. (b) Pressure dependence of T_m and T_l measured in three different samples. Reproduced with the permission of the American Physical Society from Ref. [36].

The extended dynamical mean-field theory (EDMFT) [38–41], which is an extension of DMFT, is particularly suitable to solve such problems as the competition between the exchange interaction and kinetic energy, since conventional many-body theory and DMFT are inadequate. As we have noted, the DMFT reduces the lattice problem to a self-consistent Anderson impurity model. The quantum impurity is coupled to a self-consistent fermionic bath. The local quantum fluctuations are retained and the fermionic bath of the impurity problem reflects the influence of the rest of the lattice on the impurity site. In this process the non-local quantum fluctuations are lost since the intersite correlations of the lattice are neglected.

In the EDMFT, the local quantum fluctuations are treated on the same level as the intersite quantum fluctuations. This is achieved by reducing the correlated lattice problem to a novel effective impurity problem corresponding to an Anderson impurity model with additional self-consistent bosonic baths. These bosonic baths reflect the influence of the rest of the lattice on the impurity site. As an example, they represent the fluctuating magnetic fields induced by the intersite spin-exchange interactions in the magnetic case. The intersite quantum fluctuations are included through self-consistency. The mean-field equations were derived using the "cavity method" in the earlier work but later "diagrammatic perturbation method" [40, 41] has been used. Parcollet and Georges [42] have applied EDMFT in the context of a one-band model. Haule et al. [43] have shown that the increasing exchange reduces the coherence temperature.

3.8.2. Application to Kondo lattice model

Smith and Si [40] have applied EDMFT method to the two-band Kondo lattice model:

$$H = \sum_{\langle ij\rangle,\sigma} t_{ij} C_{i\sigma} C_{j\sigma} + \sum_i J_{\mathrm{K}} \vec{S_i} \cdot \vec{s_{c_i}} - \sum_{\langle ij\rangle} J_{ij} \vec{S_i} \cdot \vec{S_j}, \tag{3.31}$$

where $\vec{S_i}$ is the impurity spin at site i, and $\vec{S_{c_i}}$ the spin of conduction (c-) electrons at site i, t_{ij} the hopping integral, and J_{ij} the spin-exchange interaction. In the large D (dimension) limit, with $t_0 = t_{ij}\sqrt{D}$ and $J_0 = J_{ij}\sqrt{D}$, they have derived an expression for the impurity action

$$\begin{aligned} S^{\mathrm{MF}} &= S_{\mathrm{top}} + \int_0^\beta d\tau J_{\mathrm{K}} \vec{S}_i \cdot \vec{s}_{\mathrm{c}} - \int_0^\beta d\tau \int_0^\beta d\tau' \\ &\times \left[\sum_\sigma C_\sigma^\dagger(\tau) G_0^{-1}(\tau - \tau') C_\sigma(\tau') + \vec{S}(\tau) \cdot \chi_{s,0}^{-1}(\tau - \tau') \vec{S}(\tau') \right], \end{aligned} \tag{3.32}$$

where S_{top} is the Berry-phase of the impurity spin. The Weiss fields G_0^{-1} and $\chi_{\mathrm{s},0}^{-1}$ are determined by the self-consistency equations,

$$G_0^{-1}(i\omega_n) = i\omega_n + \mu - \sum_{ij} t_{i0} t_{0j} \left[G_{ij}(i\omega_n) - G_{i0}(i\omega_n) G_{0j}(i\omega_n) / G_{\mathrm{loc}}(i\omega_n) \right] \tag{3.33}$$

and

$$\chi_{s,0}^{-1} = \sum_{ij} J_{i0} J_{0j} \left(\chi_{s,ij} - \chi_{s,i0} \chi_{s,0j} / \chi_{s,\mathrm{loc}} \right). \tag{3.34}$$

Here χ_{s} is the spin susceptibility. Smith and Si [40] have also shown that the effective action can be written in terms of the impurity problem:

$$\begin{aligned} H_{\mathrm{imp}} &= \sum_{k\sigma} E_k \eta_{k\sigma}^\dagger \eta_{k\sigma} + \sum_q w_q \vec{\phi}_q^\dagger \cdot \vec{\phi}_q - \mu \sum_\sigma C_\sigma^\dagger C_\sigma \\ &+ t \sum_{k\sigma} (C_\sigma^\dagger \eta_{k\sigma} + \mathrm{H.C.}) + J_{\mathrm{K}} \vec{S} \cdot \vec{s}_{\mathrm{c}} + g \sum_q \vec{S} \cdot (\vec{\phi}_q + \vec{\phi}_{-q}^\dagger), \end{aligned} \tag{3.35}$$

where E_{k}, t, w_q and g are determined from the Weiss fields G_0^{-1} and $\chi_{\mathrm{s},0}^{-1}$ specified by

$$i\omega_n + \mu - t^2 \sum_k \frac{1}{iw_n - E_k} = G_0^{-1}(i\omega_n) \tag{3.36}$$

and

$$g^2 \sum_q \frac{w_q}{(iv_n)^2 - w_q^2} = \chi_{s,0}^{-1}(iv_n). \tag{3.37}$$

In a later paper, Si et al. [44] described the SU(2) Kondo lattice model in terms of the criticality of an impurity model in a self-consistent medium. They conclude that in 2D and within EDMFT this model has a quantum-critical point (QCP) which has non-universal exponents. We shall discuss their theory in more detail in Chapter 4.

In a subsequent paper, Si and co-workers [45] have studied the magnetic quantum phase transition in an anisotropic Kondo lattice model. They have used EDMFT to treat the competition between the Kondo and RKKY interactions and have used a QMC method to solve at very low temperatures. They have also used the Hamiltonian

$$H = \sum_{ij\sigma} t_{ij} C_{i\sigma}^{\dagger} C_{j\sigma} + \sum_i J_K \mathbf{S}_i \cdot \mathbf{s}_{c,i} + \sum_{ij} \left(\frac{I_{ij}}{2}\right) S_i^z S_j^z. \tag{3.38}$$

We note that equation (3.38) is similar to equation (3.31) except that I_{ij} is the Ising-exchange interactions between the local moments. The chemical potential μ has been chosen by them so that the system is half-filled and all the phases are metallic. At the antiferromagnetic wave vector $\mathbf{Q}$, $I_{\mathbf{q}}$, the Fourier transform of I_{ij}, is most negative ($I_{\mathbf{q}} = -I$).

In order to access local quantum-critical point, they have considered two-dimensional magnetism and the RKKY density of states are of the form

$$\rho_I(\epsilon) \equiv \sum_{\mathbf{q}} \delta(\epsilon - I_{\mathbf{q}}) = \left(\frac{1}{2I}\right) \Theta(I - |\epsilon|), \tag{3.39}$$

where Θ is Heaviside function. In addition, they have considered $\rho_0(\epsilon) \equiv \sum_{\mathbf{k}} \delta(\epsilon - \epsilon_{\mathbf{k}})$. In order to obtain an order parameter for the transition into an antiferromagnetic state, the last term in equation (3.38) is separated into its Hartree and normal-ordered parts [39, 40],

$$H_I = \sum_{ij} I_{ij} \left[\left(\frac{1}{2}\right) :S_i^z::S_j^z: + \langle S_j^z \rangle S_i - \left(\frac{1}{2}\right) \langle S_i^z \rangle \langle S_j^z \rangle \right], \tag{3.40}$$

where $:S_i^z: \equiv S_i^z - \langle S_i^z \rangle$. After some algebra which involves the EDMFT mapping to a self-consistent model, the first term of H_I leads to a coupling of $:S^z:$ at a selected site to a fluctuating magnetic field. Zhu et al. [45] obtained an expression for the effective impurity action in the form

$$\begin{aligned} S_{\text{imp}} = S_{\text{top}} &+ \int_0^\beta d\tau [h_{\text{loc}} S_z(\tau) + J_K \mathbf{S}(\tau) \cdot \mathbf{s}_c(\tau) \\ &- \iint_0^\beta d\tau d\tau' \sum_\sigma C_\sigma^{\dagger}(\tau) G_{0,\sigma}^{-1}(\tau - \tau') C_\sigma(\tau') \\ &- \iint_0^\beta d\tau d\tau' S^z(\tau) \chi_0^{-1}(\tau - \tau') S^z(\tau'). \end{aligned} \tag{3.41}$$

Here S_{top} describes the Berry phase of the local moment, and h_{loc}, $G_{0,\sigma}^{-1}$, and χ_0^{-1} are the static and dynamic Weiss fields. Their self-consistency conditions are

$$\begin{aligned} h_{loc} &= -\left[I - \chi_0^{-1}(\omega = 0)\right] m_{AF}, \\ \chi_{loc}(\omega) &= \int_{-I}^{I} d\epsilon \frac{\rho_I(\epsilon)}{M(\omega) + \epsilon}, \\ G_{loc,\sigma}(\omega) &= \int_{-D}^{D} d\epsilon \frac{\rho_0(\epsilon)}{\omega + \mu - \epsilon - \Sigma_\sigma(\omega)}. \end{aligned} \tag{3.42}$$

Here $M(\omega)$ and $\Sigma_\sigma(\omega)$ are the spin- and conduction-electron self-energies:

$$\begin{aligned} &M(\omega) = \chi_0^{-1}(\omega) + \left[\frac{1}{\chi_{loc}(\omega)}\right] \text{ and } \Sigma_\sigma(\omega) = G_{0,\sigma}^{-1}(\omega) - \left[\frac{1}{G_{loc,\sigma}(\omega)}\right], \\ &\chi_{loc}(\omega) \text{ is the Fourier transform of } \langle T_\tau[:S^z(\tau)::S^z(0):]\rangle_{imp}, \\ &G_{loc,\sigma}(\omega) \text{ is the Fourier transform of } -\langle T_\tau[C_\sigma(\tau)C_\sigma^\dagger(0)]\rangle_{imp}, \end{aligned} \tag{3.43}$$

and $m_{AF} = \langle S_z \rangle_{imp}$ is the staggered magnetization. The lattice spin susceptibility is [39, 40]

$$\chi(\mathbf{q}, \omega) = \frac{1}{I_\mathbf{q} + M(\omega)}. \tag{3.44}$$

Equation (3.41) can also be written in a Hamiltonian form,

$$\begin{aligned} H_{imp} &= h_{loc} S^z + J_K \mathbf{S} \cdot \mathbf{s}_c + \sum_{p\sigma} E_{p,\sigma} C_{p\sigma}^\dagger C_{p\sigma} \\ &\quad + g \sum_p S^z(\phi_p + \phi_{-p}^\dagger) + \sum_p w_p \phi_p^\dagger \phi_p. \end{aligned} \tag{3.45}$$

Here $E_{p,\sigma}$, w_p, and g are defined by

$$\chi_0^{-1}(\omega) = -\sum_p \frac{2w_p g^2}{\omega^2 - \omega_p^2} \quad \text{and} \quad G_{0,\sigma} = \sum_p \frac{1}{\omega - E_{p\sigma}}. \tag{3.46}$$

Equation (3.45) describes a local moment coupled to a fermionic bath ($C_{p\sigma}$), a dissipative bosonic bath (ϕ_p), and a static field.

Zhu et al. [45] have solved the effective impurity model by using QMC method. The details of their calculation including canonical transformations of H_{imp} is described in Ref. [45]. We will briefly discuss a few of their results. Figure 3.13(a) shows the temperature dependence of the order parameter m_{AF} as a function of temperature at $I = 3T_K^0$ and $I = 1.5T_K^0$, and antiferromagnetism develops at large RKKY interactions. m_{AF} is non-zero at $T_N \approx 0.29T_K^0$ and $T_N \approx 0.05T_K^0$. The jump in m_{AF} at T_N suggests that the finite-temperature magnetic transition is of first order. In Figure 3.13(b), it is seen that $\chi_{loc}(T_N)$ is finite,

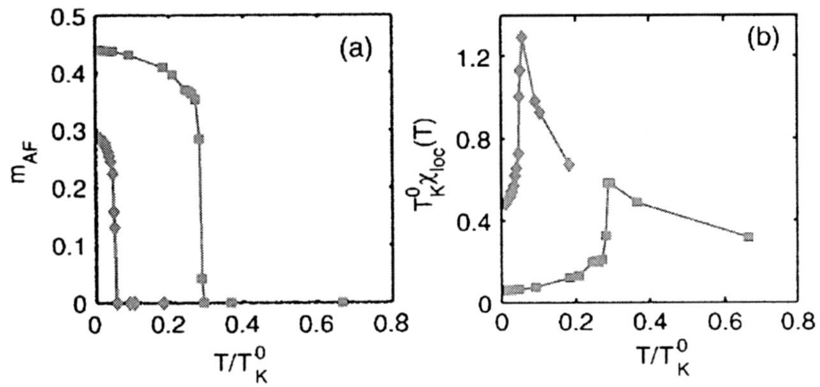

Figure 3.13. Temperature dependence of m_{AF} (a) and χ_{loc} (b) for the values of RKKY interaction $I/T^0_K = 3.0$ (squared curves) and 1.5 (diamond curves). Reproduced with the permission of the American Physical Society from Ref. [45].

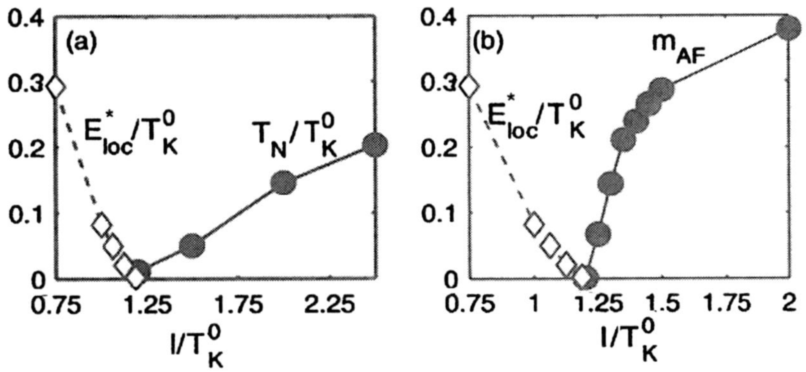

Figure 3.14. (a) The Neel temperature (solid dots) and E^*_{loc} (open diamonds) as a function of I/T^0_K. (b) The staggered magnetization m_{AF} as a function of I/T^0_K at $T = 0.0117T^0_K$. Reproduced with the permission of the American Physical Society from Ref. [45].

which confirms the first-order transition since, with density of states as in equation (3.39), χ would have been divergent had the transition been of second order.

Figure 3.14(a) shows the Neel temperature (solid dots) and the coherence scale of the Kondo lattice E^*_{loc} (open diamonds) as a function of I/T^0_K, the ratio of the RKKY interaction to the bare Kondo scale. Figure 3.14(b) shows the staggered magnetization m_{AF} (solid dots) as a function of I/T^0_K, at $T = 0.011\ T^0_K$. Since E^*_{loc} characterizes the coherence scale of the paramagnetic solution, Figure 3.14(a) represents the overall phase diagram which shows that the finite-temperature transition is of first order.

3.8.3. Application to periodic Anderson model

Sun and Kotliar [46] have used a PAM to study the competition between RKKY and Kondo interactions in heavy fermion systems. They have focused on the evolution of the electronic structure at finite temperature and not too close to the phase transition.

They have solved the problem using EDMFT and a continuous field Hubbard–Stratonovich QMC method as an impurity solver [47, 48].

The periodic Anderson Hamiltonian is given by

$$H = \sum_{\vec{k}\sigma} \epsilon_{\vec{k}} n_{\vec{k}\sigma} + V \sum_{i\sigma} (C^{\dagger}_{i\sigma} f_{i\sigma} + f^{\dagger}_{i\sigma} C_{i\sigma}) + E_{\rm f} \sum_{i\sigma} n^{f}_{i\sigma}$$
$$+ U \sum_{i} \left(n^{f}_{i\uparrow} - \frac{1}{2} \right) \left(n^{f}_{i\downarrow} - \frac{1}{2} \right) + \frac{J}{2} \sum_{\langle ij \rangle} S^{f}_{i,z} S^{f}_{j,z}, \tag{3.47}$$

where $C_{i\sigma}(f_{i\sigma})$ is a conduction (localized f) electron of spin σ at site i. $n^{a}_{i\sigma} = a^{\dagger}_{i\sigma} a_{i\sigma}$, with $a = c, f$, and $S^{f}_{i,z} = n^{f}_{i\uparrow} - n^{f}_{i\downarrow}$.

In order to simplify the solution, Sun and Kotliar [46] have introduced an independent RKKY interaction which can be induced by hybridization of the f electron with either the c electron or the electrons in the other orbitals which are not included explicitly. They have considered a short-range exchange of the form $J_{\vec{q}} = J_{\rm RKKY} (\cos q_x + \cos q_y)/2$. They have taken the dispersion, $\epsilon_{\vec{k}} = (\cos k_x + \cos k_y)/2$ for the c band. In order to obtain the EDMFT equations in the AF phase [1], they have introduced two effective impurity models. They have used the symmetry that electrons at one impurity site are equivalent to the electrons on the other with opposite spins. The self-energies of both spins were included in the self-consistent condition. In the PM phase, the EDMFT equations are obtained when the self-energies are independent of spin. The Green's functions and the self-energies were obtained from the quantum impurity model:

$$S_0 = -\int_0^{\beta} d\tau \int_0^{\beta} d\tau' \sum_{\sigma} f^{\dagger}_{0\sigma}(\tau) G^{-1}_{0,\sigma}(\tau - \tau') f_{0,\sigma}(\tau')$$
$$- \frac{1}{2} \int_0^{\beta} d\tau \int_0^{\beta} d\tau' \phi_0(\tau) D_0^{-1}(\tau - \tau') \phi_0(\tau')$$
$$+ U \int_0^{\beta} d\tau \left[n^{f}_{0,\uparrow}(\tau) - \frac{1}{2} \right] \left[n^{f}_{0,\downarrow}(\tau) - \frac{1}{2} \right] - \int_0^{\beta} d\tau \phi_0(\tau) [n^{f}_{0,\uparrow}(\tau) - n^{f}_{0,\downarrow}(\tau)]. \tag{3.48}$$

Here G_0 and D_0 are the dynamical Weiss functions which describe the environment on the f electrons and their spin. These were obtained from a self-consistent solution of the EDMFT equations described in detail in Ref. [46]. Sun and Kotliar [46] have used QMC to solve the impurity problem described in equation (3.48) and then applied the self-consistent solution. The new dynamical Weiss functions are obtained by successive iteration until convergence. They have used $U = 3.0$, $V = 0.6$, $E_{\rm f} = -0.5$, and $\mu = 0.0$. They have measured the energies in terms of the Kondo temperature at zero $J_{\rm RKKY}$ coupling, $T^0_{\rm K} = 1/\beta^0_{\rm K} \simeq 1/8.0$, which is determined from the position of the peak in the local susceptibility versus temperature plot (Figure 3.17).

The phase diagram of the PAM as $J_{\rm RKKY}$ and temperature are varied is shown in Figure 3.15. There are two lines, J_{c1} and J_{c2}, between which there are two solutions. There is a first-order phase transition, the strength of which is reflected in the size of the coexistence region, and is large at high (region I) and low (region II) temperatures. As the temperature is lowered, the transition becomes weaker along J_{c1} and stronger along J_{c2}.

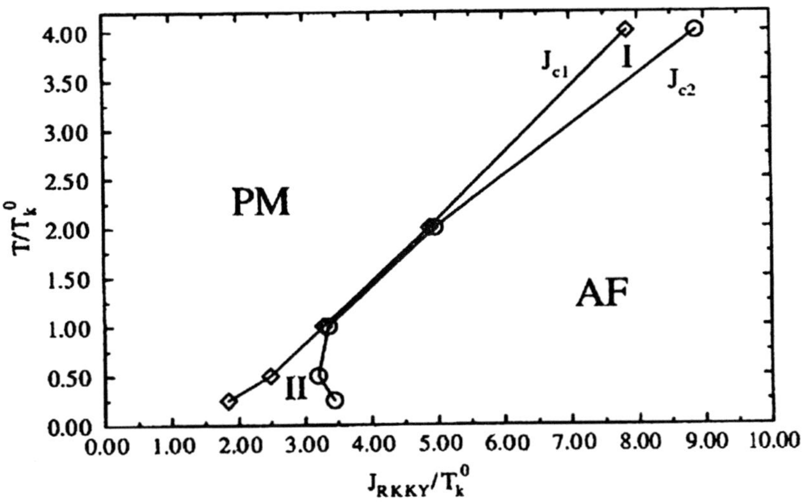

Figure 3.15. EDMFT phase diagram of the PAM. The diamonds and circles are the EDMFT results of the phase boundaries. In between the two solid lines connecting J_{c1} and J_{c2}, both the PM and AF solutions are obtained. Reproduced with the permission of the American Physical Society from Ref. [46].

In Figure 3.16, the inverse quasiparticle residue Z^{-1}, which is proportional to the effective electron mass, is plotted as a function of J_{RKKY} at $T/T_{\mathrm{K}}^{0} = 1.0$. It may be noted that the effective mass is enhanced as the transition is approached from both sides. The mass enhancement of the majority carriers is larger than the minority carriers in the AF phase. In the inset, the extrapolation of the self-energy to zero Matsubara frequency (which is a measure of the quasiparticle lifetime) is displayed.

In Figure 3.17, the evolution of the magnetic properties as a function of temperature and J_{RKKY} is studied by plotting the local susceptibility. Following the PM solution, Sun and Kotliar [46] have made a sweep of the phase diagram from $J_{\mathrm{RKKY}} = 0$ up to the line J_{c2} in Figure 3.15. The position of the Kondo peak moves toward lower temperature while its height increases by approximately 37%. However, if the EDMFT equations are solved without forcing any magnetic order, either the AF or the PM solution can appear. When $J_{\mathrm{RKKY}}/T_{\mathrm{K}}^{0} \geq 2.00$, the Neel cusp occurs below the Kondo peak and disappears when J_{RKKY} further increases, which happens before the Kondo temperature goes to zero.

The EDMFT is a mean-field treatment of the spatial degrees of freedom and breaks down in the immediate vicinity of the transition. It has proved to be unreliable in the coexistence region of the region I of the phase diagram in Figure 3.15. In finite dimensions, the transition is of second order while EDMFT induces a spurious first-order transition. The non-Gaussian thermal fluctuations are important in the vicinity of the transition but they are absent in the EDMFT theory. The existence of EDMFT solution between J_{c1} and J_{c2} lines at low temperatures (region II in Figure 3.15) indicates a crossover to a different regime where non-local effects are important. In addition, the spins in the bath are treated classically instead of through quantum mechanics. In order to address all the deficiencies mentioned above, it became necessary to introduce a different extension of DMFT to treat a cluster of spins in a self-consistent medium [49].

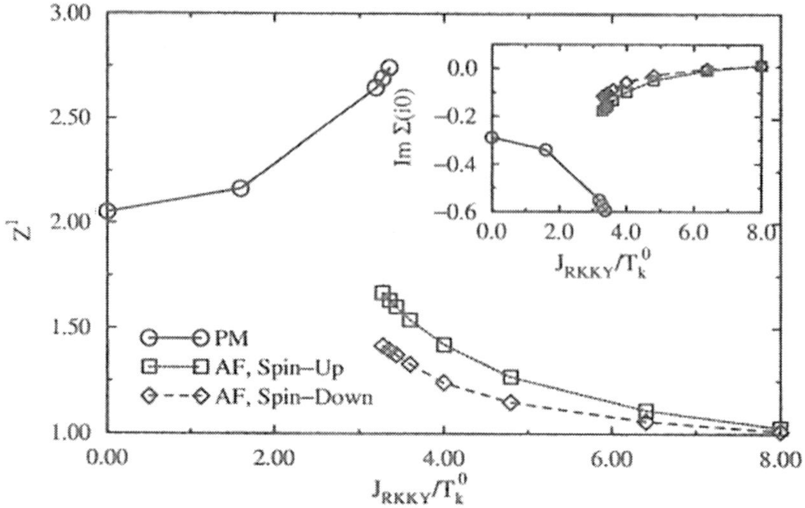

Figure 3.16. Evolution of the inverse quasiparticle residue Z^{-1} as a function of J_{RKKY} at $T/T_K^0 = 1.0$. In the calculation, the majority spin is spin-up. The majority spin band in the AF phase is more strongly renormalized. The extrapolation of the f-electron self-energy to zero Matsubara frequency is plotted in the inset. Reproduced with the permission of the American Physical Society from Ref. [46].

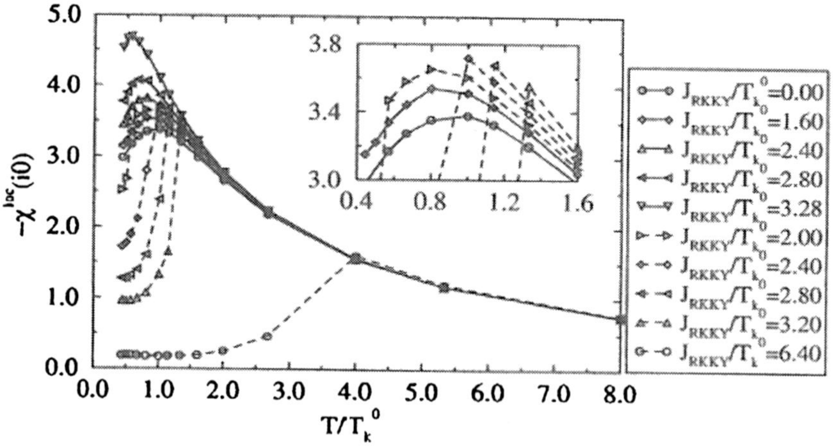

Figure 3.17. Evolution of Kondo peak into the Neel cusp as J_{RKKY} is increased. The EDMFT equations are solved by forcing the PM order where the symbols are connected by the solid lines. The results with dashed lines are obtained without any constraint. The inset shows the behavior around the Kondo peak with the same symbol scheme. Reproduced with the permission of the American Physical Society from Ref. [46].

3.8.4. Two-impurity cluster DMFT

The two-impurity problem, which has the dynamics of both the RKKY and Kondo interactions, had been used to study heavy fermions by Jones and Varma [50] and later by Affleck et al. [51]. In order to study a lattice of impurities, Sun and Kotliar [52] have combined DMFT with the two-impurity model. In the process, the RKKY and Kondo interactions are treated in a balanced way.

They have considered the PAM in three dimensions,

$$H = \sum_{\vec{k},\sigma}(\epsilon_{\vec{k}} - \mu)C^{\dagger}_{\vec{k},\sigma}C_{\vec{k},\sigma} + (E_{\rm f} - \mu)\sum_{j,\sigma} n^{f}_{j,\sigma} + U\sum_{j}\left(n^{f}_{j,\uparrow} - \frac{1}{2}\right)\left(n^{f}_{j,\downarrow} - \frac{1}{2}\right)$$
$$+ V\sum_{j,\sigma}(f^{\dagger}_{j,\sigma}C_{j,\sigma} + C^{\dagger}_{j,\sigma}f_{j,\sigma}). \tag{3.49}$$

They have taken $\epsilon_{\vec{k}} = -1/3\sum_{i=1}^{3}\cos k_i$ and divided the lattice into two interpenetrating sublattices, A and B, so that the unit cell is doubled. By applying the cavity method [1], they have derived an expression for the local action:

$$S^0 = -\int_0^{\beta} d\tau \int_0^{\beta} d\tau' \sum_{\sigma}\sum_{X,Y={\rm A,B}} f^{\dagger}_{X,0\sigma}(\tau)[G^{f}_{0\sigma}]^{-1}_{XY}(\tau - \tau')f_{Y,0\sigma}(\tau')$$
$$+ U\int_0^{\beta} d\tau \sum_{X={\rm A,B}}\left(n^{f}_{X,0,\uparrow} - \frac{1}{2}\right)\left(n^{f}_{X,0,\downarrow} - \frac{1}{2}\right). \tag{3.50}$$

The impurity Green's function has been obtained by using QMC method and the Weiss function $G^{f}_{0\sigma}$ is determined self-consistently, which has been described in detail in Ref. [52].

Sun and Kotliar [52] have studied the phase diagram of temperature versus V at $E_{\rm f} = -0.15$ and $U = 1.2$. In order to avoid crossing the band gap, they have changed the chemical potential μ along with V such that the free ($U = 0$) particle density per site at $T = 0$ is fixed at $N^{\rm Free}_{\rm tot} = 2.5423$. They have studied the PM and AFM phases and the transition between them. In Figure 3.18, the phase diagram has been plotted. The crossover temperature T_0 was obtained by them by using the saturated static spin susceptibility $\chi(\vec{k}= 0, i0) \rightarrow C/T_0$ at low temperatures (they have used $C = 1/2$). The inset of Figure 3.18 shows the saturation behavior.

In Figure 3.19, the local, $\chi_{\rm AA}(i0)$ and the nearest-neighbor, $\chi_{\rm AB}(i0)$, spin susceptibilities have been plotted as a function of V. $\chi_{\rm AB}(i0)$ changes sign as V increases due to the evolution of the RKKY correlation from being AFM to FM. In this process, the mediating conduction electrons change from those near the Fermi surface to the band bottom. $\chi_{\rm AA}(i0)$ continues to be positive and becomes strong in the crossover regime. Extrapolating this to $T = 0$, at the QCP, $\chi_{\rm AB}(i0) = 0$ and $\chi_{\rm AA}(i0)$ becomes critical.

In Figure 3.20, the evolution of the spatial correlation in the spin responses is plotted which indicates that the spin fluctuations are local in space up to $T \sim 0.1$ for $V = 0.26$ and $T \sim 0.2$ for $V = 0.50$. The FM susceptibility becomes dominant at lower temperatures, which has been observed experimentally in $YbRh_2Si_2$ [53, 54]. In addition, the logarithmic temperature dependence of χ as shown in Figure 3.20 has been observed in many heavy fermion systems [35].

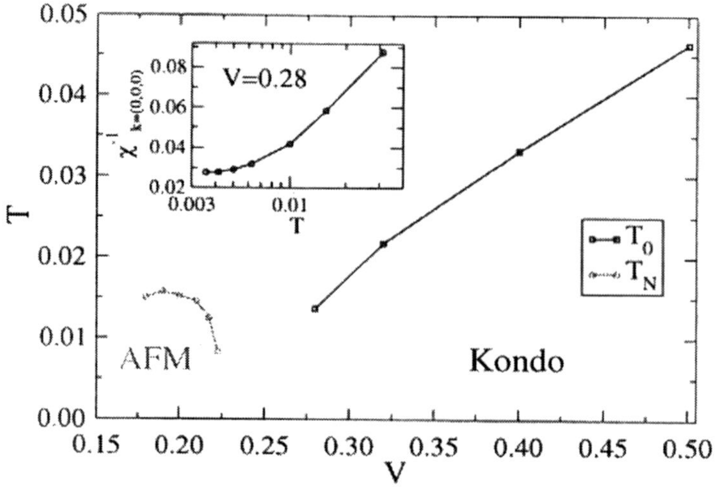

Figure 3.18. The phase diagram calculated by Sun and Kotliar. $T_N(V)$ and $T_0(V)$ do not cross at the lowest temperatures reached by them. The inset shows the low-temperature saturation of static spin susceptibility χ at $V = 0.28$. Reproduced with the permission of the American Physical Society from Ref. [52].

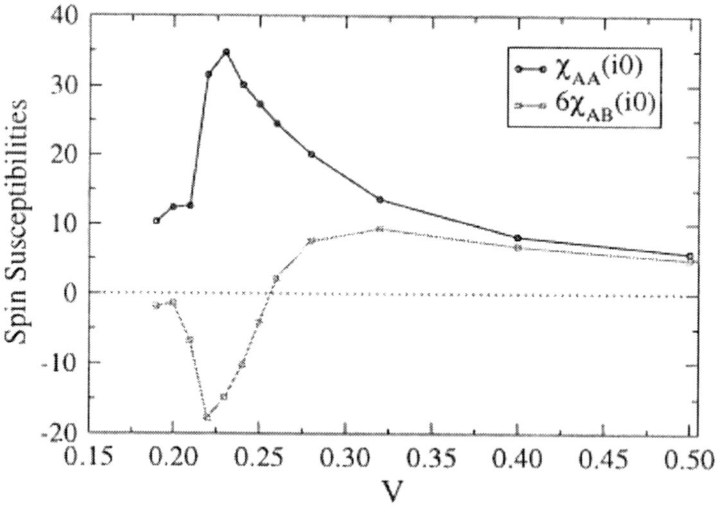

Figure 3.19. $\chi_{AA}(i0)$ and $\chi_{AB}(i0)$ versus V at $\beta = 120$. χ_{AB} has been multiplied by the local coordination number to reflect its contribution to the lattice spin susceptibility. Reproduced with the permission of the American Physical Society from Ref. [52].

3.9. Quantum cluster theories

We shall briefly discuss the quantum cluster theories which reduce the complexity of the infinite lattice problem by mapping it to a cluster with a few spatial degrees of freedom. These approaches might be applied to heavy fermion systems in the future. The quantum

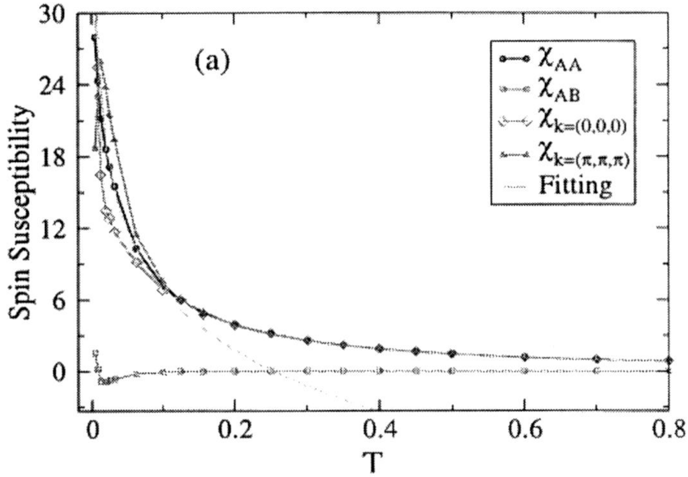

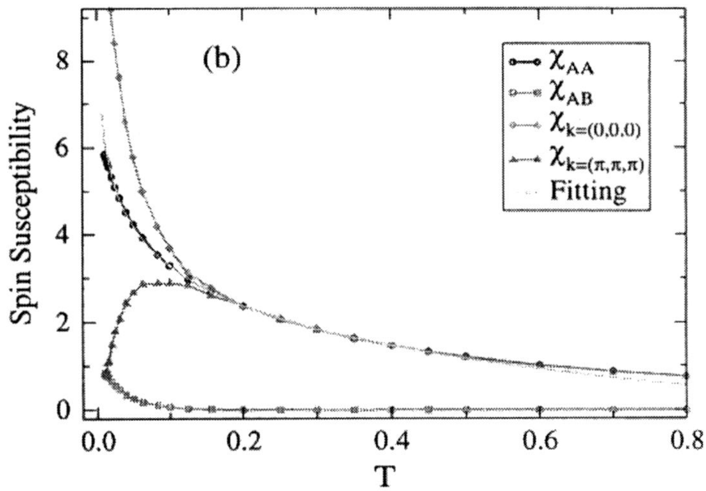

Figure 3.20. The static spin susceptibilities as functions of temperature at (a) $V = 0.26$ and (b) $V = 0.5$. The fittings are given by $\chi(T) = 7.399\ln(0.253/T)$ in (a) and $\chi(T) = 1.314\ln(1.218/T)$ in (b). Reproduced with the permission of the American Physical Society from Ref. [52].

cluster theories are non-perturbative and by increasing the cluster size, they provide additional information on finite-size simulations. Since the lattice problem is mapped to a finite cluster, the short-ranged correlations within the cluster are described accurately while the long-ranged physics is obtained in a mean-field level. An excellent review of the quantum cluster theories have recently been published in Ref. [55].

There are three quantum cluster approaches used to solve many-particle systems. The cluster perturbation theory (CPT) maps the problem to an isolated finite-size cluster. By combining it with the self-energy functional approach, the CPT can be used to study the instabilities of broken-symmetry phases. The dynamical cluster approximation (DCA) maps the lattice to an embedded cluster in reciprocal space. One obtains a self-consistent theory with a single-particle coupling between the cluster and the host. The DCA cluster preserves the translational symmetries of the lattice since it has periodic boundary conditions. The cellular dynamical mean-field theory (CDMFT) is a direct generalization of DMFT to a cluster in real space. This approach leads to a cluster with open boundary conditions which violate translational symmetries. An excellent review of CDMFT approach to Mottness has recently been published in Ref. [56].

A large number of methods used to solve DMFT problems (briefly discussed in the beginning of this chapter) are, in principle, available for the study of effective cluster models. However, with the increase of cluster size, the complexity of the solution increases rapidly.

References

[1] A. Georges, G. Kotliar, W. Krauth, and M.J. Rozenberg, Rev. Mod. Phys. **68**, 13 (1996).
[2] W. Metzner and D. Vollhardt, Phys. Rev. Lett. **62**, 324 (1989).
[3] E. Muller-Hartmann, Z. Phys. B **74**, 507 (1989).
[4] E. Muller-Hartmann, Z. Phys. B **76**, 211 (1989).
[5] E. Muller-Hartmann, Int. J. Mod. Phys. B **3**, 2169 (1989).
[6] U. Brandt and C. Mielsch, Z. Phys. **75**, 365 (1989).
[7] U. Brandt and C. Mielsch, Z. Phys. **79**, 295 (1990).
[8] U. Brandt and C. Mielsch, Z. Phys. **82**, 37 (1991).
[9] L.M. Falicov and J.C. Kimball, Phys. Rev. Lett. **22**, 997 (1967).
[10] F.J. Ohkawa, J. Phys. Soc. Jpn. **60**, 3218 (1991).
[11] F.J. Ohkawa, Prog. Theor. Phys. (Suppl.) **106**, (1991) 95.
[12] A. Georges and G. Kotliar, Phys. Rev. B **45**, 6479 (1992).
[13] A. Georges, G. Kotliar, and Q. Si, Int. J. Mod. Phys. B **6**, 705 (1992).
[14] P.W. Anderson, Phys. Rev. **124**, 41 (1961).
[15] M. Jarrell, Phys. Rev. Lett. **69**, 168 (1992).
[16] W. Metzner, Z. Phys. B **77**, 253 (1889).
[17] J.M. Luttinger and J.C. Ward, Phys. Rev. **118**, 1417 (1960).
[18] A.A. Abrikosov, L.P. Gorkov, and I.E. Dzialoshinski, *Methods of Quantum Field Theory in Statistical Physics* (Pergamon, Elmsford, NY, 1965).
[19] W. Metzner, Phys. Rev. B **43**, 8549 (1991).
[20] G. Hulsenbeck and F. Dtephen, Z. Phys. B **94**, 281 (1994).
[21] V. Janis, Z. Phys. B **83**, 227 (1991).
[22] V. Janis and D. Volhardt, Int. J. Mod. Phys. B **6**, 731 (1992).
[23] J.E. Hirsch and R.M. Fye, Phys. Rev. Lett. **56**, 2521 (1986).
[24] M.J. Rozenberg, X.Y. Zhang, and G. Kotliar, Phys. Rev. Lett. **69**, 1236 (1992).
[25] A. Georges and W. Krauth, Phys. Rev. Lett. **69**, 1240 (1992).
[26] M. Caffarel and W. Krauth, Phys. Rev. Lett. **72**, 1545 (1994).
[27] M.J. Rozenberg, G. Moeller, and G. Kotliar, Mod. Phys. Lett. B **8**, 535 (1994).
[28] Q.M. Si, M.J. Rozenberg, K. Kotliar, and A.E. Ruckenstein, Phys. Rev. Lett. **72**, 2761 (1994).
[29] A.M. Tsvelick and P.B. Wiegmann, Adv. Phys. **32**, 453 (1989).

[30] A.C. Hewson, *The Kondo Problem to Heavy Fermions* (Cambridge University Press, Cambridge, England, 1993).
[31] H. Schweitzer and G. Czycholl, Phys. Rev. Lett. **67**, 3724 (1991).
[32] M. Jarrell, H. Akhlaghphour, and Th. Pruschke, Phys. Rev. Lett. **70**, 1670 (1993).
[33] J.R. Schrieffer and P.A. Wolff, Phys. Rev. **149**, 491 (1966).
[34] M.A. Ruderman and C. Kittel, Phys. Rev. **96**, 99 (1954); T. Kasuya, Prog. Theor. Phys. **16**, 45 (1956); K. Yosida, Phys. Rev. **106**, 893 (1957).
[35] G.R. Stewart, Rev. Mod. Phys. **48**, 797 (2001).
[36] O. Trovarelli, C. Geibel, S. Mederle, C. Langhammer, F.M. Grosche, P. Gegenwart, M. Lang, G. Sparn, and F. Steglich, Phys. Rev. Lett. **85**, 626 (2000); Physica B **281–282**, 372 (2000).
[37] H.V. Lohneysen, J. Phys. Condens. Matter **8**, 9689 (1996).
[38] S. Sachdev and J. Ye, Phys. Rev. Lett. **70**, 3339 (1993).
[39] Q. Smith and J.L. Smith, Phys. Rev. Lett. **77**, 3391 (1996).
[40] J.L. Smith and Q. Si, Phys. Rev. B **61**, 5184 (2000).
[41] R. Chitra and G. Kotliar, Phys. Rev. B **63**, 115110 (2001).
[42] O. Parcollet and A. Georges, Phys. Rev. B **59**, 5341 (1999).
[43] K. Haule, A. Rosch, J. Kroha, and P. Wolfle, Phys. Rev. Lett. **89**, 236402 (2002).
[44] Q. Si, S. Rabello, K. Ingersent, and J. Lleweilun Smith, Nature **43**, 804 (2001).
[45] J.-X. Zhu, D.R. Grempel, and Q. Si, Phys. Rev. Lett. **91**, 156404 (2003).
[46] P. Sun and G. Kotliar, Phys. Rev. Lett. **91**, 037209 (2003).
[47] Y. Mototome and G. Kotliar, Phys. Rev. B **62**, 12800 (2000).
[48] P. Sun and G. Kotliar, Phys. Rev. B **66**, 85120 (2002).
[49] P. Sun and G. Kotliar, Phys. Rev. B **71**, 245104 (2005).
[50] B.A. Jones and C. M. Varma, Phys. Rev. B **40**, 324 (1989).
[51] I. Affleck, A.W.W. Ludwig, and B.A. Jones, Phys. Rev. B **52**, 9528 (1995).
[52] P. Sun and G. Kotliar, Phys. Rev. Lett. **95**, 016402 (2005).
[53] K. Ishida, K. Okamoto, Y. Kawasaki, Y. Kitaoka, O. Trovarelli, C. Geibel, and F. Steglich, Phys. Rev. Lett. **89**, 107202 (2002).
[54] P. Gegenwart, J. Custers, Y. Tokiwa, C. Geibel, and F. Steglich, Phys. Rev. Lett. **94**, 076402 (2005).
[55] T. Maier, M. Jarrell, T. Pruschke, and M.H. Heitler, Rev. Mod. Phys. **77**, 1027 (2005).
[56] T.D. Stanescu, M. Civelli, K. Haule, and G. Kotliar, Ann. Phys. **321**, 1682 (2006).

Chapter 4

Fermi-Liquid, Heavy Fermi-Liquid, and Non-Fermi-Liquid Models

4.1. Fermi-liquid theory of Landau

Landau's Fermi-liquid theory [1] was originally developed to explain the properties of liquid ^{3}He but has been applied extensively to explain the ground-state properties of electrons in metals. Since the electron–electron interactions are strong in metals, Landau referred to these interacting electrons as quasiparticles. He referred as "normal Fermi system" to those systems of interacting particles obeying Fermi–Dirac statistics, for which the quasiparticle representation is valid. In fact, by using Green's function methods, Landau showed that to all orders of perturbation theory in the interaction, every interacting Fermi system is normal. Thus, in a sense, there is a one-to-one correspondence between the free electrons (Fermi gas) and the strongly interacting electrons in metals (Fermi liquid).

Varma et al. [27b] have recently summarized the essential difference between non-interacting fermions and an interacting Fermi liquid from a microscopic perspective. The momentum states $|\mathbf{k}\rangle$ are eigenstates of the Hamiltonian of the free fermions with eigenvalue $\varepsilon_{\mathbf{k}} = h^2k^2/8\pi^2m$. The thermal distribution function $n^0_{\mathbf{k}\sigma}$ is given by the Fermi–Dirac distribution function, where σ denotes the Fermi level. At $T = 0$, all states are occupied within the Fermi sphere and the distribution is 1 which jumps to zero at $|\mathbf{k}| = k_F$, and the energy is equal to the chemical potential. This is shown in Figure 4.1.

When a particle is added to an interacting system of fermions, the added particle (assuming a repulsive interaction) will kick particles from below the Fermi surface to above. In Figure 4.2, a schematic illustration of the perturbative expansion of the change of wave function as a result of the addition of an electron to the Fermi sea due to interactions with the particles in the Fermi sea is shown.

The temperature dependence of some physically observable quantities at very low temperatures is predicted by Landau's Fermi-liquid model. For example, the ratio of specific heat C divided by temperature T, (C/T), is approximately a constant. The magnetic susceptibility χ becomes independent of temperature, and the electrical resistivity $\rho \approx \rho_0 + AT^2$. As $T \to 0$, if the electron interactions are short ranged in both space and time and become temperature independent, the correct description of the low-temperature parameters of metals can be done through the Fermi-liquid model.

Landau's Fermi-liquid theory can essentially be summarized as follows.

One defines

$$\delta n(\mathbf{k},\sigma) = n(\mathbf{k},\sigma) - n_0(\mathbf{k},\sigma), \tag{4.1}$$

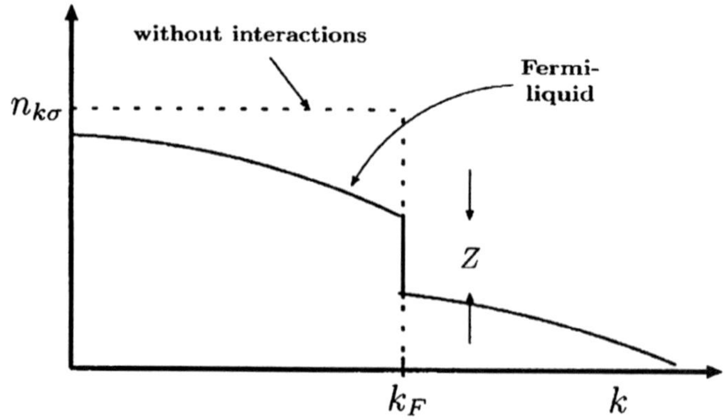

Figure 4.1. Bare-particle distribution at $T = 0$ for a spin direction in a translationally invariant Fermi system with interactions (full line) and without interactions (dashed line). Reproduced with the permission of Elsevier from Ref. [27b].

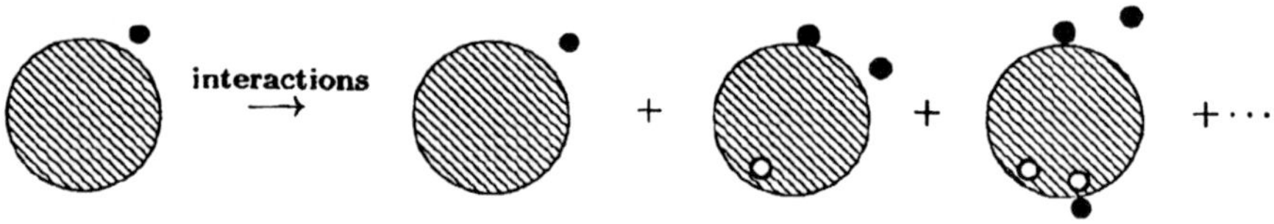

Figure 4.2. Schematic illustration of the perturbative expansion of the change of wavefunction as a result of the addition of an electron to the Fermi sea due to interactions with the particles in the Fermi sea. Reproduced with the permission of Elsevier from Ref. [27b].

where $n_0(\mathbf{k}, \sigma)$ is the equilibrium distribution function and $n(\mathbf{k}, \sigma)$ the quasiparticle distribution function. The deviation of the Gibbs free energy from its ground-state value is written as a function of $\delta n(\mathbf{k}, \sigma)$,

$$G = G_0 + \frac{1}{V}\sum_{\mathbf{k},\sigma}(\tilde{\epsilon}_k - \mu)\delta n_{\mathbf{k}\sigma} + \frac{1}{2V^2}\sum_{\mathbf{kk}',\sigma\sigma'} f_{\mathbf{kk}',\sigma\sigma'}\delta n_{\mathbf{k}\sigma}\delta_{\mathbf{k}'\sigma'} + \cdots. \tag{4.2}$$

The function $f_{\mathbf{kk}',\sigma\sigma'}$ has symmetric (f^{s}) and antisymmetric (f^{a}) parts since it depends on spin. In addition, $f_{\mathbf{kk}',\sigma\sigma'}$ depends on momenta and for the low-energy and long-wavelength phenomena, only $\mathbf{k} \approx \mathbf{k}_{\mathrm{F}}$ has a role. f would depend on the relative angle between $\mathbf{k}$ and $\mathbf{k}'$ and can be expanded in terms of the Legendre polynomials $P_l(x)$. Thus, one obtains

$$N(0)f^{\mathrm{s,a}}_{\mathbf{kk}',\sigma\sigma'} \xrightarrow{\mathbf{k}\approx\mathbf{k}'\approx\mathbf{k}_{\mathrm{F}}} \sum_{l=0}^{\infty} F_l^{\mathrm{s,a}} P_l(\hat{\mathbf{k}}\cdot\hat{\mathbf{k}}'), \tag{4.3}$$

where $N(0)$ is the density of states (DOS) at the Fermi energy. One can relate the lowest order Landau coefficients to the thermodynamic quantities C_{v} (specific heat), κ (compressibility), and χ (susceptibility) as follows:

$$\frac{C_\mathrm{v}}{C_{\mathrm{v}0}} = \frac{m^*}{m}, \quad \frac{\kappa}{\kappa_0} = (1 + F_0^\mathrm{s})\frac{m^*}{m}, \quad \frac{\chi}{\chi_0} = (1 + F_0^\mathrm{a})\frac{m^*}{m}, \tag{4.4}$$

where m^* is the effective mass.

4.2. Fermi-liquid model for Kondo-lattice systems

I shall first describe a Fermi-liquid model for Kondo-lattice systems derived by Misra and Callaway [17]. The ground state of a Kondo ion is a singlet, and renormalization group analysis and exact solution [2–7] suggest that a renormalized resonant level (RL) is a good description of the low-energy behavior of such a system. The RL is characterized by an effective width Δ^* and an effective f-level position E_f^*. It describes a renormalized f state with $2J + 1 = N$ scattering channels, each with a specified azimuthal quantum number [8] m_j. In the Kondo regime, $\Delta^* \approx D\exp(-\pi|E_\mathrm{f}|/N\Delta)$ for $-E_\mathrm{f} \gg N\Delta$. Here Δ and E_f are the bare hybridization width and f-level position and $2D$ the bandwidth.

The energy scale for lattice coherence effects in a Kondo lattice is smaller than the Kondo temperature. In a Kondo lattice, the contribution to the free energy from intersite coherence effects are of higher order in $1/N$ than intrasite effects [9–11]. Razahfimandimby et al. [12] and d'Ambrumenil and Fulde [13] have assumed that the scattering at each Kondo site for a Kondo lattice can be described by a phase shift

$$\delta_\nu(\varepsilon) = \delta(\mu) + \frac{\varepsilon - \mu}{T_\mathrm{K}} + \Sigma_{\nu'\varepsilon'}\varphi_{\nu\nu'}\delta n_{\nu'}(\varepsilon'), \tag{4.5}$$

where ν and ν' denote the spin-orbital states of the localized f level, T_K is the Kondo temperature, and $\delta(\mu)$ is fixed by the valence of the Kondo ion and by using the Friedel sum rule. Requiring that the scattering of the lattice at $T = 0$ ($\delta n_{\nu'} = 0$) is coherent leads to the KKRZ equations [14] whose solutions near μ and E_K correspond to energies of single quasiparticle excitations. Two quasiparticles interact when one "senses" the virtual polarization of a Kondo ion induced by another. This interaction is characterized by the parameters $\varphi_{\nu\nu'}$. Pettifor [15] and Razahfimandimby et al. [12] cast their results in a hybridization model by using the well-known results from the theory of transition metals developed by Heine [16]. However, they assumed that the additional phase shift caused by the Kondo ions is of s electrons instead of f electrons, thereby neglecting spin-orbit effects which are important for heavy fermions.

I present an outline of a Fermi-liquid theory derived by us [17] for some Kondo-lattice systems in which spin-orbit effects are specifically included. Zou and Anderson [18] have also formulated a Fermi-liquid model by using a similar technique. However, our theory is more general and has several advantages, which are discussed in Ref. [17]. We have assumed the ground state of the Kondo lattice to be single-ion Kondo singlets built coherently into the lattice ($CeCu_2Si_2$, $CeAl_3$, etc.). In cerium systems, only the $l = 3$ phase shift is important. The spin-orbit splitting between $J = 5/2$ and $7/2$ is much larger than Δ^*, and therefore $\delta_3(\varepsilon)$ is large in the $J = 5/2$ state. It is easy to show, by using a procedure similar to the KKRZ method and including spin-orbit effects, that the band structure is obtained from

$$\det \| (k_n^2 - E)\delta_{k_n s, k_{n'} s'} + \Gamma_{k_n s, k_{n'} s'} \| = 0. \tag{4.6}$$

Here

$$\Gamma_{k_n s, k_{n'} s'} = -\frac{4\pi^2}{K\Omega} \Sigma_{\lambda\mu} \tan \eta_{l(\lambda)} \frac{j_{l(\lambda)}(k_n R) j_{l(\lambda)}(k_{n'} R)}{j^2_{l(\lambda)}(KR)} C(l\tfrac{1}{2} j, \mu - m_s, m_s)$$
$$\times C(l\tfrac{1}{2} j, \mu - m_{s'}, m_{s'}) Y_{l(\lambda)}^{\mu - m_s}(\hat{\mathbf{k}}_n) Y_{l(\lambda)}^{*\mu - m_{s'}}(\hat{\mathbf{k}}_{n'}) \tag{4.7}$$

and $k_n = k + G_n$, $K = (2mE)^{1/2}$, Ω is the volume of the unit cell, s is the spin index, the C is Clebsch-Gordon coefficient, det$\|U\|$ means the determinant of matrix $\mathbf{U}$. Also,

$$l = \lambda \qquad j = l - \tfrac{1}{2} \qquad (\lambda > 0)$$

$$l = -\lambda - 1 \qquad j = l + \tfrac{1}{2} \qquad (\lambda < 0)$$

$$\cot \eta_{l(\lambda)} = \cot \delta_{l(\lambda)} - \frac{\eta_{l(\lambda)}(KR)}{j_{l(\lambda)}(KR)},$$

where $j_l(x)$ and $\eta_l(x)$ are the spherical Bessel functions. We redefine

$$E_{\mathrm{f}} = E_{\mathrm{F}}^* - \left(\frac{\eta_3(KR)}{j_3(KR)} \right) \Delta^*. \tag{4.8}$$

It may be noted that Γ is proportional to $\Delta^*/(E_{\mathrm{f}} - E)$ and is strongly energy-dependent. There is a singularity at resonance which can be tamed by a transformation in which RLs are added as basis states.

The Bloch states are constructed as

$$\Psi_{ks}(r) = \frac{1}{\sqrt{N}} \Sigma_{i\mu} \exp(i\mathbf{k}\cdot\mathbf{R}_i) A_{\lambda\mu}^{ks} \Phi_\lambda(|\mathbf{r} - \mathbf{R}_i|)\, \chi_{\lambda\mu}(i), \tag{4.9}$$

where

$$A_{\lambda\mu}^{ks} = \left(\frac{4\pi}{3}\right)^{1/2} i^l C\left(l\frac{1}{2} j,\, \mu - m_s,\, m_s\right) Y_l^{*\mu - m_s}(\hat{k}) \tag{4.10}$$

and

$$\chi_{\lambda\mu}(i) = C(l\tfrac{1}{2} j,\, \mu - m_s,\, m_s)\, Y_l^{\mu - m_s}(\mathbf{r} - \mathbf{R}_i) \chi^{m_s}. \tag{4.11}$$

In equation (4.9), $\Phi_\lambda(r)$ is the radial part of the exact solution of the Schrodinger equation within one Wigner-size cell for a given scattering potential with eigenvalue E but for $r > R$, the Φ decays into plane wave states. One can easily show that

$$\Sigma_\mu A^{ks}_{\lambda\mu} A^{ks'}_{\lambda\mu} = \delta_{ss'}. \tag{4.12}$$

The determinant in equation (4.6) can be transformed by using equation (4.9) and the integral formula for phase shift by adapting a technique used for transition metals by Heine [16]. One obtains

$$\| U \| \cdot \left\| \begin{array}{c|c} (k_n^2 - E)\delta_{\mathbf{k}_n s'',\mathbf{k}_{n'} s'''} + W_{\mathbf{k}_n s'',\mathbf{k}_{n'} s'''} & \gamma_{ks,k_{n'} s''} \\ \hline \gamma^*_{k_n s'',ks'} & (E_\mathrm{f} - E)\delta_{ks,ks'} \end{array} \right\| = 0, \tag{4.13}$$

where $\mathbf{U}$ is a matrix with a non-vanishing determinant, $W_{\mathbf{k}_n s''\mathbf{k}_{n'} s'''}$ the pseudopotential, and s the spin index. The hybridization form factors $\gamma^{3,5/2}_{\mathbf{k}s,\mathbf{k}_n s''}$ and $\gamma^{3,7/2}_{\mathbf{k}s,\mathbf{k}_n s''}$, are given by

$$\begin{aligned} \gamma^{3,5/2}_{\mathbf{k}s,\mathbf{k}_n s''} = & \left(\frac{3B_{3,5/2}}{8\pi}\right)[5(\mathbf{k}\cdot\mathbf{k}_\mathbf{n})^3 - 3(\mathbf{k}\cdot\mathbf{k}_\mathbf{n})k^2k_n^2]\mathbf{I}_{ss'} \\ & + \left(\frac{3iB_{3,5/2}}{8\pi}\right)[5(\mathbf{k}\cdot\mathbf{k}_\mathbf{n})^2 - k^2k_n^2]\mathbf{k}\times\mathbf{k}_\mathbf{n}\cdot\sigma_{ss''} \end{aligned} \tag{4.14}$$

and

$$\begin{aligned} \gamma^{3,7/2}_{\mathbf{k}s,\mathbf{k}_n s''} = & \left(\frac{B_{3,7/2}}{2\pi}\right)[5(\mathbf{k}\cdot\mathbf{k}_\mathbf{n})^3 - 3(\mathbf{k}\cdot\mathbf{k}_\mathbf{n})k^2k_n^2]\mathbf{I}_{ss'} \\ & + \left(\frac{3iB_{3,7/2}}{8\pi}\right)[5(\mathbf{k}\cdot\mathbf{k}_\mathbf{n})^2 - k^2k_n^2]\mathbf{k}\times\mathbf{k}_\mathbf{n}\cdot\sigma_{ss''}, \end{aligned} \tag{4.15}$$

where

$$B_{3,5/2} = -\left(\frac{(4\pi)^3}{3}\right)^{1/2}\left(\frac{\Delta_{5/2}}{K'\Omega}\right)^{1/2}\left(\frac{1}{k^3K^3}\right), \tag{4.16}$$

$$B_{3,7/2} = -\left(\frac{(4\pi)^3}{3}\right)^{1/2}\left(\frac{\Delta_{7/2}}{K'\Omega}\right)^{1/2}\left(\frac{1}{k^3K^3}\right), \tag{4.17}$$

$\mathbf{I}$ is an identity matrix and $\boldsymbol{\sigma}$ the Pauli spin matrix. It may be noted that the hybridization potential, which is obtained by multiplying the structure factor by the form factor, includes the spin-orbit coupling and mixes the dispersionless resonance states with extended states of all Brillouin zones.

Equation (4.13) can be considered as the secular equation of a hybridization Hamiltonian of the form

$$H = \Sigma_{k,n,s}\varepsilon_n(\mathbf{k})C^{\dagger}_{kns}C_{kns} + \Sigma_{k,s,J}E_{fJ}f^{\dagger}_{ksJ}f_{ksJ} + \Sigma_{k,n,s,s',J}\gamma^{J}_{\mathbf{ks},\mathbf{k}_n s'}(f^{\dagger}_{ksJ}C_{kns'} + C^{\dagger}_{kns'}, f_{ksJ}). \quad (4.18)$$

Here $\varepsilon_n(\mathbf{k})$ is the energy of the nth Brillouin zone obtained by diagonalizing the Hamiltonian containing the kinetic energy term and the corresponding spin-orbit pseudopotential term (including $l = 3$), $\gamma^{J}_{\mathbf{ks},\mathbf{k}_n s'}$ are the hybridization potentials derived by us, and C_{kns} ($C^{\dagger}_{kns}$) and f_{ksJ} ($f^{\dagger}_{ksJ}$) are the usual fermion operators. The energy E_{fJ} is dispersionless but the hybridization term mixes the localized states with the extended states of all Brillouin zones. The Hamiltonian (4.18) can be transformed into one with one extended band only by a projection method suggested by Heine [16]. If this procedure is adopted, one obtains the hybridization of one conduction band with localized states which have dispersion in their energy and one obtains a periodic Anderson Hamiltonian in which spin-orbit effects are explicitly included.

One can use degenerate perturbation theory to construct Bloch functions which are eigenfunctions of the Hamiltonian in equation (4.18). These functions have been used to evaluate the momentum and spin-matrix elements which occur in the general expressions for spin magnetic susceptibility (χ_s) of Misra et al. [19]. One obtains

$$\chi_s = -\mu_0^2\,\Sigma_k\left\{\frac{1}{3}g_j^2 j(j+1)(2j+1)a_1^2(k) + 2[(a_2(k)+b_2(k))^2 + (a_3(k)+b_3(k))^2 + (a_4(k)+b_4(k))^2]\right\}\frac{f'(E_k)}{1-\alpha(k)}. \quad (4.19)$$

The momentum and spin-matrix elements have also been used to calculate the Knight shift (K_s) [20]. One obtains

$$K_s = -\Sigma_k\left[\frac{2}{3}\mu_0 g_j^2(2j+1)\left(\langle 1/r_0^3\rangle_{nl}\, l(l+1) + \frac{4}{3\pi}\langle\delta(r_0)\rangle_{nl}[j(j+1) - l(l+1) + s(s+1)]\right)a_1^2(k) + \frac{16}{3}\pi\bar{\Omega}\langle|\Psi_{kf}(0)|^2\rangle_{av}[a_2(k)(a_2(k) + b_2(k)) + a_3(k)(a_3(k)+b_3(k)) + a_4(k)(a_4(k) + b_4(k))]\right]\frac{f'(E_k)}{1-\alpha(k)}. \quad (4.20)$$

In equations (4.19) and (4.20), $a_n(k)$, $b_n(k)$, and $\alpha(k)$ are complicated functions of k and the other symbols have their usual meanings. It is evident that the Knight shift K_s is not directly proportional to the spin susceptibility χ_s, a result which is well known for Kondo-lattice systems. In view of the fact that for metals the Knight shift is proportional to the Pauli spin susceptibility, such deviation for Kondo-lattice systems had baffled physicists.

4.3. Heavy Fermi liquids

In fact, there are a number of universal features associated with the coherent Fermi-liquid state in heavy fermion systems. They can be summarized as follows.

(i) The dimensionless Wilson ratio $R = \left[\chi(0)/g_J^2 J(J+1)\mu_B^2\right] / \left[\gamma(0)/\pi^2 k_B^2\right]$ is close to the value of unity [21].
(ii) The specific heat [22] C_v has a rapid downturn with increasing temperature which has been fit to a function of the form $T^3 \ln T$.
(iii) The resistivity [23, 24] ρ is proportional to T^2.
(iv) The low-T susceptibility [24, 25] χ_T also appears to vary as T^2.

In addition, the evidence of universal behavior of heavy fermions is the observation of Kadowaki and Woods [23] that $\rho/T^2 \equiv A$ is the same multiple of γ^2 for essentially all materials.

A large number of the heavy fermion systems become heavy Fermi liquids at low temperatures in the sense that Landau's Fermi-liquid theory is still adequate to describe the physics provided the large effective mass is included. In the Fermi-liquid theory [26], the specific heat enhancement C_v/C_{v0} is related, at low temperatures, to the quasiparticle DOS at the Fermi surface. This is equivalent to an average Fermi velocity or to that of an average mass. In view of the above, the effective mass is defined as [27]

$$\frac{C_v}{C_{v0}} = \frac{m^*}{m}. \tag{4.21}$$

Here, it is important to comment on the physical interpretation of the quasiparticles. The f electrons are supposed to be hopping from site to site. There are a large number of f electrons and the Luttinger theorem requires the Fermi surface to contain the total number of states and not a volume containing the mobile holes in the f band. The f electrons have very large mass due to the weak effective hybridization. Thus, the quasiparticles are essentially f electrons and the quasiparticle bands are f bands which have moved upto the Fermi energy and have been narrowed by correlation. The heavy Fermi liquid arises due to the Kondo screening of the localized moments at each lattice site. In a sense, the localized moments "dissolve" into the Fermi sea.

Several authors [27–31] have contributed to the development of the heavy Fermi-liquid theory. In this review, we shall summarize the heavy Fermi-liquid model of Auerbach and Levin [30]. The objective of their work was to derive a consistent Fermi-liquid description of the Kondo lattice with $1/N$ Kondo-boson (KB) approach as the basis. They have computed the Landau parameters and associated vertex functions and related these to R, χ, C_v, and ρ. They have studied the Coqblin–Schrieffer lattice (CSL) which is the Kondo limit of the large-U Fano-Anderson lattice in which the f electrons have vanishingly small charge fluctuations. It has already been shown [32–35] that the same Lagrangian describes the lattice model away from that limit provided one includes the f-charge susceptibility. The partition function is given by ($h = 1$)

$$Z_{\mathrm{CSL}} = \int D\Pi_{i,m} \lambda C^* C f^* f \exp\left\{ -\int_0^\beta d\tau \left[L(\tau) + i \sum_{i,m} \lambda_i (f_{i,m}^* f_{i,m} - Q_0) \right] \right\}, \tag{4.22}$$

$$L = \sum_{k,m} C^*_{km}(\partial_\tau + \epsilon_k)C_{km} + f^*_{km}\partial_\tau f_{km}\,(J/N)\sum_{i,m,m'} C^*_{i,m} f^*_{i,m'} f_{i,m} C_{i,m'}\,. \tag{4.23}$$

Here, $C_{i,m}$ and $f_{i,m}$ are Grassmann variables of an electron at site r_i and magnetic quantum number m, $|m| \leq (N-1)/2$, in the conduction band (of dispersion in $\in_k$) and the dispersionless f band, respectively. The antiferromagnetic Kondo-interaction energy is given by J. The integration over $\lambda_i(\tau)$ imposes the local constraint of the f charge conservation, $n_f = Q_0$, at all times and sites. Q_0 has been kept as a fixed parameter in order to define a true N-independent mean-field theory [34, 35]. In the radial gauge, one defines the magnitude r_i and phase θ_i of the boson field at lattice site i as $b_i(\tau) = r_i e^{i\theta_i}$. This can be introduced by a Hubbard–Stratonovich transformation which allows one to integrate out the fermions and evaluate the Z_{CSL} by steepest descents [32–35]. We note that the radial gauge suffers from the disadvantage that a functional integral formulation of the theory must be used.

After considerable algebra, Auerbach and Levin [30] have derived expressions for the quasiparticle self-energy Σ and vertex function Γ to leading order in $1/N$ by functionally differentiating Z_{CSL} with respect to source currents. The quasiparticles interact via the exchange of a single Kondo-boson propagator as shown in Figure 4.3. The details of the derivations are described in Ref. [30]. Their expression for Γ can be written as $\Gamma = \Gamma^{dir} - \Gamma^{exch}$.

Using these expressions, considering the $\omega \to 0$ limit of Γ evaluated on the Fermi surface ($E_k = E_{k'} = \mu$) and projecting it onto Legendre polynomials, P_l, they have derived expressions for the Landau scattering amplitudes $\{A_l^{s,a}\}$.

$$A_l^s = T_K^{-1} N \delta_{l,0} \lim_{|q|\omega/|q| \to 0} \Gamma^{dir} + A_l^a, \tag{4.24}$$

$$A_l^a = -T_K^{-1} \frac{2l+1}{2k_F^2} \int_0^{2k_F} d\kappa \Gamma^{exch}(\kappa, 0) P_l\left(1 - \frac{\kappa^2}{2k_F^2}\right)\kappa, \tag{4.25}$$

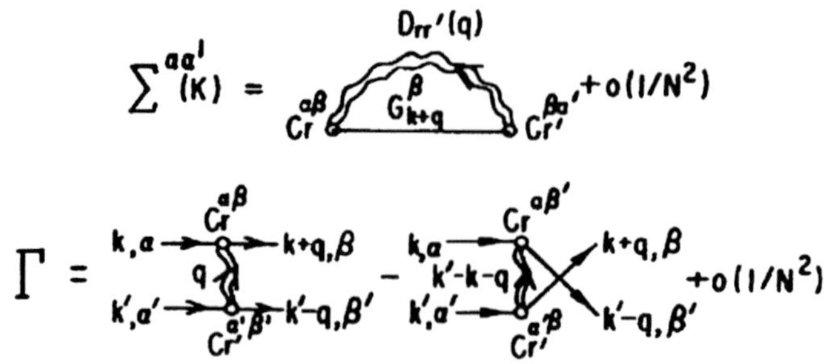

Figure 4.3. The self-energy Σ and vertex function Γ to leading order in $1/N$, which involves a single Kondo-boson (wavy line) exchange. The two terms Γ^{dir} and Γ^{exch} are shown in the diagram. Reproduced with the permission of the American Physical Society from Ref. [30].

where T_K^{-1} is the Fermi-level DOS and k_F the Fermi wave vector ($\epsilon_F = \mu_0$). In this Fermi-liquid theory, the "bare particles" are the heavy mean-field quasiparticles. Auerbach and Levin [30] have numerically evaluated $\{A^{s,a}\}$ by using a parabolic band structure $(\varepsilon(k) \propto |k^2|)$ and obtained the following results for $l = 0,1$ parameters.

$$A_0^a = \frac{-1.0}{N} + \frac{0.08}{N}\frac{Q_0}{\mu_0}, \quad A_0^s = 1.0 + A_0^a, \quad A_1^s = A_1^a = -\frac{0.12}{N}\frac{Q_0}{\mu_0}. \tag{4.26}$$

In the Fermi-liquid theory of Landau, the Wilson ratio $R = 1 - A_0^a$ which can be calculated from equation (4.22). Auerbach and Levin have also obtained expressions for the susceptibility χ and for γ, the coefficient of the linear term in specific heat.

$$\chi = \chi^0\left[1 + \frac{\delta m}{m} - A_0^a + O\left(\frac{1}{N^2}\right)\right] \quad \text{and} \quad \gamma = \gamma^0\left[1 + \frac{\delta m}{m} + O\left(\frac{1}{N^2}\right)\right], \tag{4.27}$$

where

$$\chi^0 = \frac{g^2 N(N^2-1)}{12T_K} \quad \text{and} \quad \gamma^0 = \frac{\pi^2 N}{3T_K}. \tag{4.28}$$

Here, g is the electron magnetic moment. It may be noted that the terms in equation (4.27) are known as Fermi-liquid identities related to Ward identities of spin and charge conservation.

By following an analysis similar to that of Engelsberg and co-workers and Riedel [38–40], Auerbach and Levin [30] have shown that, in addition to the correction to γ, there exists a specific heat correction ΔC_v analogous to the paramagnon $T^3 \ln T$ contribution to liquid ^{3}He. They have derived

$$\Delta C_v = 0.85\left(\frac{T}{\bar{T}_K}\right)^3 \ln\left(\frac{T}{\bar{T}_K}\right) + O(T^3), \tag{4.29}$$

where $\bar{T}_K = Q_0 T_K$.

Finally, Auerbach and Levin have calculated the expression for resistivity, ρ, by following an analogous paramagnon calculation [41]. They have derived

$$\rho = \rho_{max}\left(\frac{T}{\lambda \bar{T}_K}\right)^2 + O(T^3), \tag{4.30}$$

where $\rho_{max} = h/e^2 k_F N^2$. Here, $\bar{T}_K$ is the coherence energy scale and Q_0 a key parameter which determines the $l > 0$ Landau parameters, and λ is of the order of unity.

Nozieres [42] had pointed out that while magnetic moments can still be screened locally for temperatures lower than the single-impurity Kondo scale T_K, the formation of a Fermi-liquid regime is a global phenomenon requiring coherence over the whole lattice. This could be associated with a much lower coherence temperature T^*. However, in the limit of low concentration ($n_c \ll 1$) the coherence problem becomes important since only a few conduction electrons are available to screen the local spins. In this "exhaustion" regime, either magnetic ordering wins over Kondo screening or a paramagnetic Fermi-liquid state is formed with a much suppressed coherence scale $T^* \ll T_K$. Nozieres [43] had later suggested that $T^* \sim n_c D$ (where D is half the bandwidth) for strong coupling ($J_K/D \gg 1$, where $T_K \sim J_K$), while $T^* \sim T_K^2/D$ for weak coupling ($J_K \ll D$), but the latter assumption was considered unlikely by other authors.

Burdin et al. [44] have derived expressions for T_K and T^* using the slave boson approach in the form of a controlled large-N solution of the Kondo-lattice model in the weak-coupling regime. They have obtained

$$T_K = De^{-1/J_k \rho_0(\epsilon_F)} \sqrt{1 - \left(\frac{\epsilon_F}{D}\right)^2} F_K(n_c), \tag{4.31}$$

where

$$F_K(n_c) = \exp\left(\int_{-(D+\epsilon_F)}^{D-\epsilon_F} \frac{d\omega}{|\omega|} \frac{\rho_0(\epsilon_F + \omega) - \rho_0(\epsilon_F)}{2\rho_0(\epsilon_F)} \right), \tag{4.32}$$

and ϵ_F is the non-interacting Fermi level given by

$$\frac{n_c}{2} = \int_{-D}^{\epsilon_F} d\epsilon \rho_0(\epsilon). \tag{4.33}$$

This expression is valid for an even DOS $\rho_0(-\epsilon) = \rho_0(\epsilon)$ which vanishes outside the interval $-D < \epsilon < D$.

At the low-temperature limit $T \ll T_K$, Burdin et al. [44] have derived an analytical expression for $T^* = r^2(T = 0)/D$ (where r is the boson condensation amplitude) in the weak-coupling limit, which is valid for a general DOS and arbitrary density n_c:

$$T^* = De^{-1/J_K \rho_0(\epsilon_F)} \left(1 + \frac{\epsilon_F}{D}\right) \frac{\Delta\epsilon_F}{D} F^*(n_c), \tag{4.34}$$

where

$$F^*(n_c) = \exp\left(\int_{-(D+\epsilon_F)}^{\Delta\epsilon_F} \frac{d\omega}{|\omega|} \frac{\rho_0(\epsilon_F + \omega) - \rho_0(\epsilon_F)}{\rho_0(\epsilon_F)} \right), \tag{4.35}$$

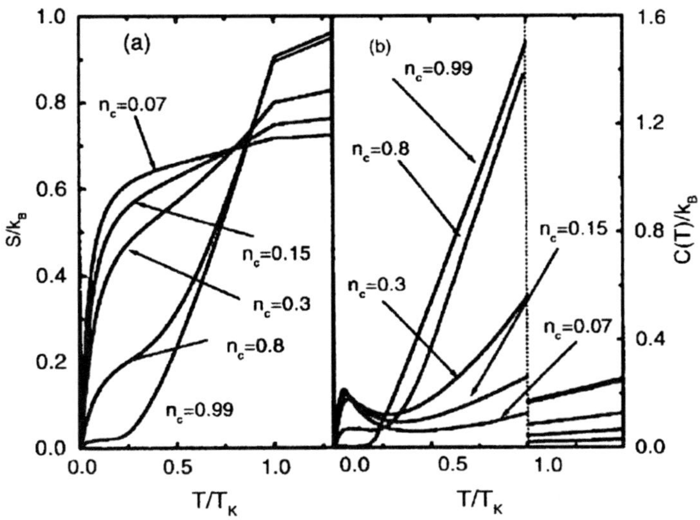

Figure 4.4. Entropy (a) and specific heat (b) for the lattice model for several values of density. $J_K/D = 0.75$. Reproduced with the permission of the American Physical Society from Ref. [44].

$\Delta\epsilon_F = \epsilon_F^> - \epsilon_F$, and $\epsilon_F^>$ the non-interacting value of the Fermi level corresponding to $(n_c + 1)/2$ fermions per spin color. Thus, two different energy scales are relevant for the Kondo-lattice model: T_K is associated with the local Kondo screening and T^* is associated with Fermi-liquid coherence and the properties of physical quantities at $T = 0$. It can be shown that in the exhaustion limit ($n_c \ll 1$), $T^* \ll T_K$, a result which is in qualitative agreement with Nozieres [43], but not his estimate $T^*/T_K \propto T_K/D$.

In Figure 4.4(a), Burdin et al. [44] have plotted the temperature dependence of the entropy. At low temperatures, for $n_c \neq 1$, $S(T) = \gamma T$. The slope $\gamma \propto \rho(0)$ decreases with increasing density as does the temperature T_F^* scale up to which $S(T)$ is linear. T_F^* is of the order of $T^*(\ll T_K)$ at low n_c, while $T_F^* \approx T_K/|\ln(1 - n_c)| \ll T_K \simeq T^*$ for $n_c \to 1$. For $n_c \approx 1$, T_F^* is a better estimate of the coherence scale than T^*. The variation of specific heat $C(T)$ with temperature is shown in Figure 4.4(b). $C(T)$ has a two-peak structure: the peak at T_K shows the onset of Kondo screening. The second peak, at T_F^*, signals the Fermi-liquid heavy fermion regime. As n_c increases, weight is gradually transferred to the high-temperature peak until the low-temperature peak disappears in the Kondo-insulator limit.

We shall now summarize the concepts of the various heavy Fermi-liquid models [27–31] by following the elegant but brief review of Senthil et al. [45]. The Kondo-lattice model can be written as

$$H_K = \sum_k \varepsilon_k C_{k\alpha}^\dagger C_{k\alpha} + \frac{J_k}{2} \sum_r \mathbf{S}_r \cdot C_{r\alpha}^\dagger \sigma_{\alpha\alpha'} C_{r\alpha'}. \tag{4.36}$$

Here n_c is the density of conduction electrons with dispersion ε_k, $C_{k\alpha}^\dagger$ and $C_{k\alpha}$ are the creation and annihilation operators of conduction states, k the momentum, and $\alpha = \uparrow, \downarrow$ is a spin index. The conduction electrons interact with f electron spins $\mathbf{S}_r$

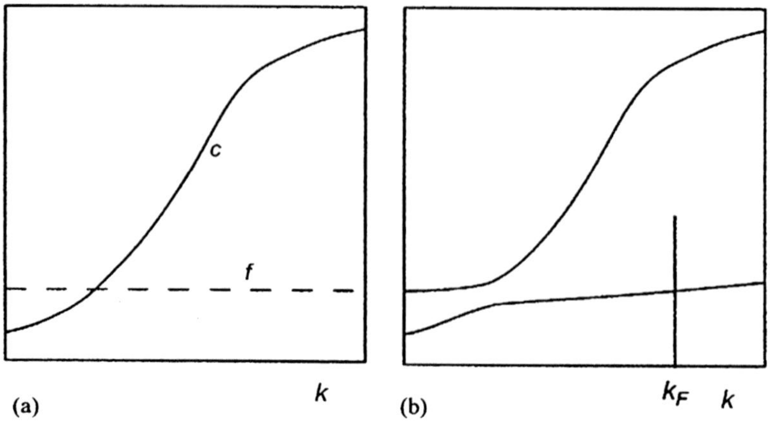

Figure 4.5. A completely flat f electron "band" (dashed line in (a)) mixes with the conduction electrons to obtain the renormalized bands in (b). The single Fermi surface at k_F in the Fermi liquid state contains states whose wave number equals the sum of the c and f electrons. Reproduced with the permission of Elsevier from Ref. [45].

via antiferromagnetic Kondo-exchange coupling constant J_K. Here r is a lattice position and $\boldsymbol{\sigma}$ are the Pauli spin matrices.

In the heavy fermion liquid models [27–31], the charge of the $f_{r\alpha}$ electrons is fully localized on the rare earth sites. These electrons occupy a flat dispersionless band shown in Figure 4.5(a). Since this band is half-filled, it is placed at the Fermi level. The $C_{r\alpha}$ electrons occupy their own conduction band. The Kondo exchange turns on a small hybridization between these two bands. The hybridization can be represented by a bosonic operator:

$$b_r \sim \sum_{\alpha} C^{\dagger}_{r\alpha} f_{r\alpha}. \tag{4.37}$$

Since $\langle b_r \rangle$ is non-zero, renormalized bands are formed (Figure 4.5(b)) due to the mixture of the two bands.

Since the f band was initially dispersionless, the renormalized bands do not overlap. One now applies the Fermi surface sum rule by Luttinger [46], also known as the Luttinger theorem. According to Luttinger, the volume enclosed by the Fermi surface is entirely determined by only the electron density. The volume is independent of the type and strength of an interaction, if the system remains a Fermi liquid and no phase transition occurs. Using this theorem leads to the conclusion that the occupied states are entirely within the lower band, and a single Fermi surface is obtained within wave vector k_F. The volume within k_F is obtained by the total density of f and c electrons. The Fermi surface is in a region (Figure 4.5) where the electrons primarily have an f character and the band is flat. According to this model, this accounts for the large effective mass of the fermionic quasiparticles.

Since the charge fluctuations are quenched at the f electron sites, every rare earth site has a constraint

$$\sum_{\alpha} f^{\dagger}_{r\alpha} f_{r\alpha} = 1, \tag{4.38}$$

which is obeyed at each rare earth site. This implies that the theory is invariant under the space-time dependent $U(1)$ gauge transformation

$$f^{\dagger}_{r\alpha} \rightarrow f_{r\alpha} e^{i\phi_r(\tau)}, \tag{4.39}$$

where τ is imaginary time. This invariance leads to the emergence of a dynamical compact $U(1)$ gauge field A_μ, where μ is space-time index and $A_\mu \rightarrow A_\mu + 2\pi$. It may be noted that the bosonic field b in equation (4.19) also carries a $U(1)$ gauge charge. The b field is condensed in the heavy Fermi-liquid state. In this state, the $U(1)$ gauge theory can be considered to be in a "Higgs phase". The A_μ fluctuations are quenched and are relatively innocuous in the heavy Fermi-liquid state.

4.4. Non-Fermi-liquid behavior in f electron metals

In 1991, Seaman et al. [47] discovered non-Fermi-liquid (NFL) behavior in the Kondo alloy $Y_{1-x}U_xPd_3$. Their experimental result, particularly for $x = 0.2$, has led to intense theoretical and experimental activity in the general area of NFL behavior in d- and f-electron metals. Many alloys, which exhibit NFL behavior, have been experimentally fabricated. In fact, there are more than 50 materials which exhibit NFL behavior. Several theories have been proposed but there is yet no consensus about whether any particular theory explains the wide range of NFL properties. The divergent theories include disorder causing incomplete screening of local moments, nearness to a magnetic instability in the phase diagram, or overcompensation of local moments. The only consensus on the $Y_{1-x}U_xPd_3$ results is that it is a system with electron–electron interactions that are too strong to permit entry into the Fermi-liquid ground state at low temperatures. The seminal experimental results of Seaman et al. [47], particularly for $Y_{0.8}U_{0.2}Pd_3$, are shown in Figures 4.6–4.8. An excellent and extensive review in this rapidly expanding field has recently been published by Stewart [48].

Figure 4.6 shows ρ versus $\ln(T)$ below 300 K for $Y_{0.8}U_{0.2}Pd_3$. The background contribution of YPd_3 is negligible below 100 K. Above 80 K, $\rho(T)$ is linear in $-\ln(T)$ which is consistent with single-ion Kondo scattering of conduction states off localized U 5f states. Seaman et al. [47] have plotted the data as $\ln(1 - \rho/\rho_0)$ versus $\ln(T)$ which, using $\rho_0 = 357.70 \pm 0.05$ µΩ cm, yields a straight line below ~20 K (inset). Linear least squares fit for $0.2 \leq T \leq 20$ K yielded $\rho/\rho_0 = 1 - (T/T_0)^n$, where $n = 1.13 \pm 0.04$ and $T_0 = 180 \pm 20$ K. In case of single channel, spin-1/2 magnetic Kondo effect, one would expect exclusively T^2 resistivity saturation below 20 K.

In Figure 4.7, $\Delta C(T)/T$ versus $\ln(T)$ has been plotted for $Y_{0.8}U_{0.2}Pd_3$. The data are linear in $-\ln(T)$ from about 0.8 to 20 K. The solid line represents a linear least squares fit of the $\Delta C(T)/T$ versus $\ln(T)$ data which gives a slope of $-0.0500(2)$ J/mol U K^2 and an intercept

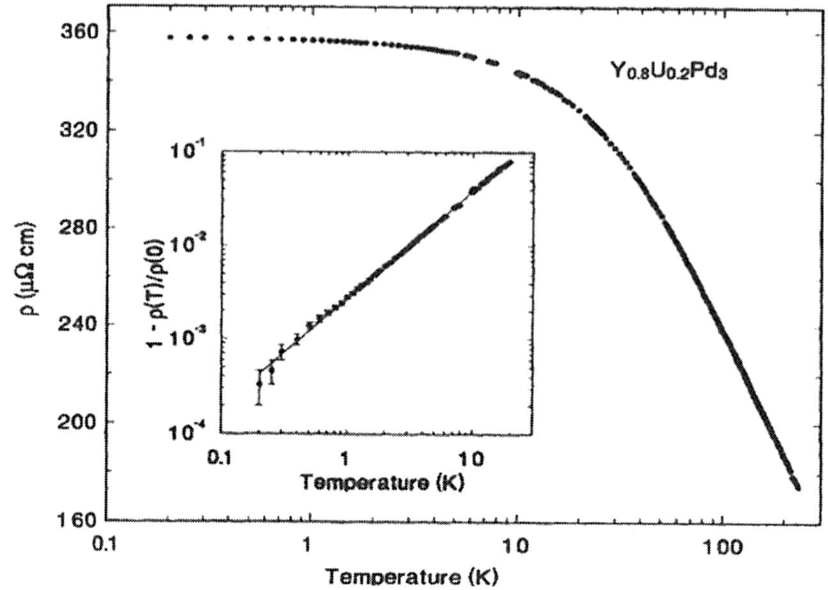

Figure 4.6. Temperature dependence of the electrical resistivity $\rho(T)$ of $Y_{0.8}U_{0.2}Pd_3$. The data follow $-\ln T$ behavior, indicative of the Kondo effect, above about 80 K and saturate below ~20 K (inset) with power-law behavior $\rho/\rho_0 = 1-(T/T_0)^n$, with best fit using $\rho(0) = 357.7\ \mu\Omega$ cm, $n = 1.13$, and $T_0 = 180$ K. Reproduced with the permission of the American Physical Society from Ref. [47].

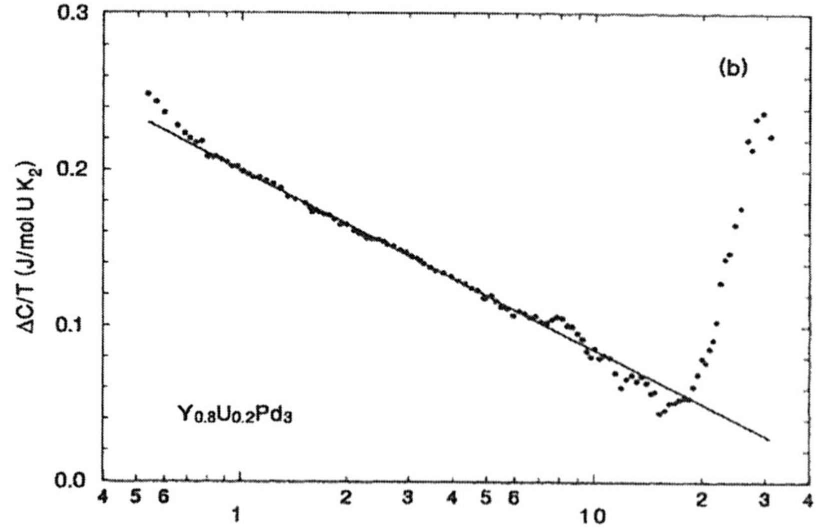

Figure 4.7. $\Delta C(T)/T$ versus $\ln T$ for $Y_{0.8}U_{0.2}Pd_3$. The solid line represents a least-square fit of the data by the form $-(0.25/T_K)\ln(T/(0.41T_K)) + b$. From the slope, one obtains $T_K = 42$ K and $b = 61$ mJ/mol K^2 which likely arises from a Schottky anomaly. Reproduced with the permission of the American Physical Society from Ref. [47].

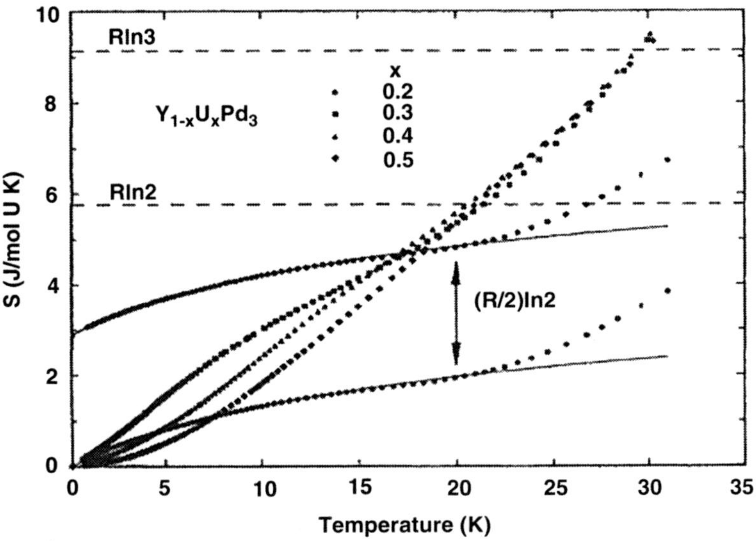

Figure 4.8. Temperature dependence of the added electronic entropy per U, $\Delta S(T) - \Delta S(0)$, versus T for $Y_{1-x}U_xPd_3$. For $0.3 \leq x \leq 0.5$, $\Delta S(T)$ converges to $\sim R\ln(2)$ near $\sim$20 K, suggestive of freezing out of a doublet ground state. For $x = 0.2$, $\Delta S(T) - \Delta S(0)$ is greatly reduced with apparent saturation to $(R/2)\ln(2)$, suggestive of a ground state entropy of the same value (offset curve). The upturn above 20 K is likely due to an excited state Schottky anomaly. The solid line through the $x = 0.2$ data was obtained from the fit of Figure 4.4. Reproduced with the permission of the American Physical Society from Ref. [47].

of about 55 K. Below 0.8 K, the samples exhibit an upturn which is responsible for the deviation from the $\ln(T)$ behavior. The above data seem to be consistent with the properties of a two-channel $S = 1/2$ Kondo model [49]. The above results show that the low-temperature slope with respect to $\ln(T)$ has the universal value [50] $(1/R)d(\Delta C/T)/d\ln T = -0.251/T_K$. Applying this formula to Figure 4.7 yields $T_K = 42$ K. Using this value of Kondo temperature, the low-temperature power-law behavior of resistivity of Figure 4.6 has the form $\rho(T)/\rho(0) = 1 - [T/(4.3 \pm 0.5)\, T_K]^{1.13\pm0.04}$.

The entropy $\Delta S(T) - \Delta S(0)$ versus T is shown in Figure 4.8. For $0.3 \leq x \leq 0.5$, the $\Delta S(T) - \Delta S(0)$ curves converge for entropy values $\geq R\ln(2)$ and temperatures $\geq$20 K, above T_{SG} in each case. The spin-glass freezing removes entropy from doublet ground states and the rise above 20 K is due to a Schottky anomaly arising from excited crystal-field levels. It is obvious that for $x = 0.2$, $\Delta S(T) - \Delta S(0)$ has a different character than the higher concentration data. The entropy below 20 K is much lower than $R\ln(2)$ which suggests that the ground-state degeneracy is not lifted. The solid line through the $x = 0.2$ data represents the integrated fit of the $\Delta C(T)/T$ data plus $(R/2)\ln(2)$ for the offset curve. The extrapolated curve appears to saturate to $R\ln(2)$.

The NFL behavior in d- and f-electron systems was originally discovered for doped systems. However, there are a variety of ways in which such behavior can be achieved. In some doped systems, doping can cause NFL behavior far from antiferromagnetism. In some other systems, doping either suppresses the Neel temperature becoming zero ($T_N \to 0$) or is about to induce antiferromagnetism. Surprisingly, in some other doped systems, antiferromagnetism coexists with NFL behavior. Finally, in some systems, doping either suppresses a ferromagnetic ordering temperature, $T_c \to 0$, or at a higher concentration causes ferromagnetic ordering. Subsequently, such NFL behavior was discovered for some undoped systems.

4.5. The quadrupolar Kondo model

The first theoretical explanation of NFL behavior of $Y_{0.8}U_{0.2}Pd_3$ was given by Cox based on the two-channel Kondo model [49]. This is also known as the quadrupolar Kondo model [50]. The multichannel Kondo model for a single impurity was introduced by Nozieres and Blandin [51]. The Hamiltonian for the model is described by

$$H_K = \sum_{k,m,\sigma} \varepsilon_k a^t_{km\sigma} a_{km\sigma} + J \sum_{k,k',m,\sigma,\sigma'} \vec{S} \cdot a^t_{km\sigma} \vec{\sigma}_{\sigma\sigma'} a_{k'm\sigma'}, \tag{4.40}$$

where J is the antiferromagnetic coupling constant which couples the conduction states to the impurity spin $\vec{S}$, $\vec{\sigma}$ are the Pauli matrices, and m denotes the orbital "channels" or the degrees of freedom. The spins of the conduction electrons near the impurity are bound together to compensate the impurity spin.

The multichannel Kondo model has been solved exactly [52–54] for dilute impurities. The following three relations between n, the orbital degrees of freedom, and the impurity spin S describe the physics of the system. When $n = 2S$, the conduction electron channels compensate the impurity spin which gives rise to normal Fermi-liquid behavior and for $n = 1$ and $S = 1/2$, this is the Kondo problem. When $n < 2S$, the impurity spin is not compensated as there are not enough conduction electrons degrees of freedom to form a singlet ground state. When $n > 2S$, the impurity spin is overcompensated, and when the temperature and external field > 0, critical behavior sets in and the system become a non-Fermi liquid. The two-channel model describes the case when $n = 2$ and $S = 1/2$.

Cox [49] had originally used the two-channel Kondo model to explain the difference in magnetic response between UBe_{13} and $CeCu_2Si_2$. His argument was that both UBe_{13} and $CeCu_2Si_2$ are superconductors with identical C_{el} curves from 1 to 10 K, and corresponding effective Fermi temperatures of about 10 K. The resistivities of the two materials are very similar, but they differ substantially in their magnetic response. Up to 10 T, $C_{el}(T)$ is nearly field independent for UBe_{13} [55] while it drops by 20% for $CeCu_2Si_2$ [48]. The magnetic neutron-scattering cross section of UBe_{13} has a quasielastic peak at 15 meV [56], while the corresponding peak for $CeCu_2Si_2$ is only 1 meV [57]. Cox assumed that the model U site has the following features. (i) A stable $5f^2$, $J = 4$ Hund's rule ground state within the LS coupling scheme is assumed at a site of cubic symmetry (later he extended this argument to tetragonal and hexagonal symmetry [58]). (ii) The crystal-field split $J = 4$ multiplet has a ground-state Γ_3 non-magnetic doublet at energy ε_f. (iii) Only a Γ_4 triplet excited level is retained within the $J = 4$ multiplet, at energy $\varepsilon_f + \Delta$. (iv) A $5f^1$ configuration lies above the Γ_3 level by $|\varepsilon_f|$ and all other configurations are neglected.

The Kondo effect is incorporated into this model by hybridizing the 5f electrons with the conduction electron states and considering virtual excitations into the $5f^1$ configuration.

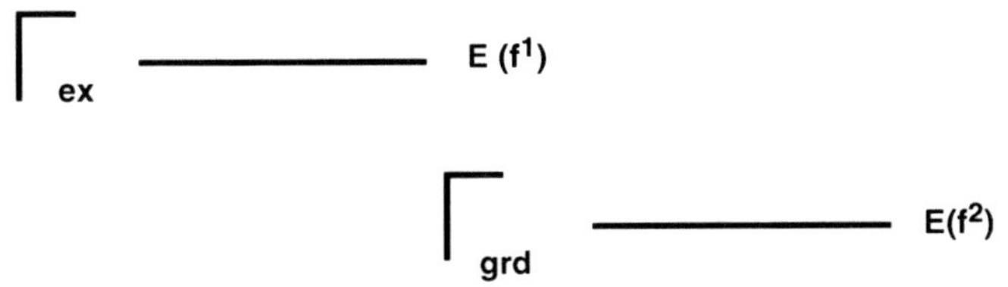

Figure 4.9. Energy level scheme for the U^{4+} ions (energy runs vertically). Two configurations are kept for simplicity. The same physics will ensue so long as f^3 has the lowest doublet level. Reproduced with the permission of Elsevier from Ref. [58].

The non-magnetic ground state has an electric crystalline field quadrupolar moment from which the name "quadrupolar Kondo effect" is derived. Figure 4.9 shows the basic picture of the states of an impurity lanthanide/actinide ion and conduction electrons required to achieve a two-channel Kondo model description of low-energy-scale physics.

The two-channel (or quadrupolar) Kondo model developed by Cox and co-workers is a very simple model which explains some but not all of the properties of the NFL systems. For example, the non-linear susceptibility measurements in UBe_{13} [59] contradict the results of quadrupolar Kondo model. However, this theory can be easily applied to various systems and has the ability to predict some results. Yet there is a need for microscopic theories which has led to the development of some elegant theories such as quantum-critical point (QCP) or quantum phase transitions (QPT). We shall briefly discuss these elegant models.

4.6. Quantum-critical point theories

The NFL state in many systems is linked to a magnetic instability which arises at $T = 0$ since NFL behavior in those systems is found near a magnetically ordered phase. Many authors have theoretically considered the transport and thermodynamics properties in the special case of a QPT (for an overview of QPT, see Ref. [60]). A QPT is driven by a control parameter other than temperature (in contrast to classical phase transitions), i.e., magnetic field, pressure, or doping at $T = 0$. In the path-integral formulation, a d-dimensional quantum system can be viewed as a $(d + z)$-dimensional classical system, where z is the scaling exponent for the dynamics. The QCP in the phase diagram relies on the findings of the scaling behavior since the experimental results can be obtained only for $T \neq 0$. There are several QCP models for NFL behavior of heavy fermion systems. These models can be grouped into two categories, elegantly shown in Figure 4.10 by Coleman [61]. In the "weak-coupling approach", the instability is approached from the Fermi-liquid side and the QCP is treated as a magnetic instability of the Fermi surface. The magnetic instability develops in momentum space; NFL behavior is driven by the infinitely long-range and retarded interactions which develop between the quasiparticles at the QCP. In the "strong-coupling approach", the instability is approached from the magnetic side, where local magnetic moments exist. The local-moment system loses its magnetism once the single-ion Kondo temperature is large

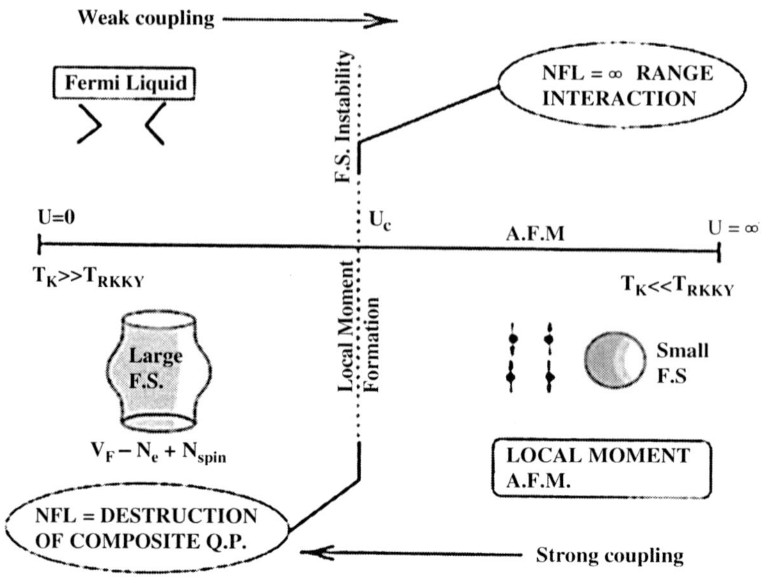

Figure 4.10. Contrasting the weak- and strong-coupling picture of an antiferromagnetic QCP. Reproduced with the permission of Elsevier from Ref. [61].

enough to develop a dense Kondo effect. In this process, quasiparticles in the Fermi liquid are composite bound states formed in real space between conduction electrons and local moments. At the critical point, the bound states which characterize the Kondo lattice disintegrate to reveal an underlying lattice of ordered magnetic moments.

4.7. Weak-coupling theories

The weak-coupling theory was pioneered by Hertz [62]. He showed that in quantum systems at $T = 0$, the value of the dynamic exponent, z, affects the static critical behavior, in contrast to $T > 0$ phase transitions. The scaling dimension of a quartic interaction in a model with dynamic exponent z in spatial dimensionality d was $[4 - (d + z)]$. Hertz also showed that in Fermi-liquid-like systems, the dynamic exponent $z > 1$ since the fermion excitations will often overdamp the mode associated with the ordering field. Hertz also noted that at a QCP, the correlation time τ diverges more rapidly than the correlation length ξ, according to the relation

$$\tau \propto \xi^z, \tag{4.41}$$

where z is the dynamical critical exponent mentioned above. Hertz identified z as an anomalous scaling dimension of time $[\tau\,] = [x]^z$.

Millis [63] has used renormalization group techniques and used Hertz's model to study the Gaussian or marginal fixed points occurring when itinerant fermion systems in two or three spatial dimensions undergo $T = 0$ phase transition at wave vector Q which do not span the Fermi surface. He has assumed that all systems arise from an underlying Fermi-liquid state and except for the two-dimensional antiferromagnet ($d = z = 2$) are above their upper critical dimension, d_c^+, i.e., hyperscaling does not apply. The results of the various models depend on the dimension d, the critical exponent z, the reduced temperature ($t = T/T^*$), where T^* is a characteristic temperature, and δ, a control parameter which is related to a Hamiltonian parameter such as magnetic field, doping, or pressure [64]. From Millis' theory, one can draw a qualitative phase diagram for the different cases. In the models considered by him, the particle-hole continuum Fermi excitations overdamps the mode associated with the ordering transition and implies to $z = 3$ for the clean ferromagnet and $z = 2$ for antiferromagnet.

Stewart [48] has drawn such a phase diagram for the three-dimensional antiferromagnetic scenario of a QCP, which is shown in Figure 4.11. In Figure 4.11, the different regions are indicated as I–III. $T_{\rm I}$ and $T_{\rm II}$ are the crossover temperatures between these regions. Fluctuations on the scale of ξ have energies much greater than k_BT in region I. Thus, the quantum nature and the Fermi-liquid behavior appears in this region and the specific heat $C/T = \gamma + a_1T^2\log T + a_2T^2 + \cdots$. $T_{\rm I} \propto (\delta - \delta_c)^{z/2}$ which means that the crossover line to region II is linear for an antiferromagnet (since $z = 2$). However, region II is a quantum classical crossover regime, where the energy of the modes becomes less than k_BT. In the classical region III, $T_{\rm II} > (\delta - \delta_c)^{z/d+z-2}$ the correlation length is controlled by T rather than $(\delta - \delta_c)$. The quantum problem is transferred into an effective classical problem, thereby neglecting interactions. This is the cause of the presence of classical Gaussian behavior in a large range of the phase diagram.

Moriya and Takimoto [65] have made a self-consistent renormalization study of the spin fluctuations near magnetic phase transitions and have made several predictions about NFL behavior. Their model gives a more systematic treatment of mode–mode coupling

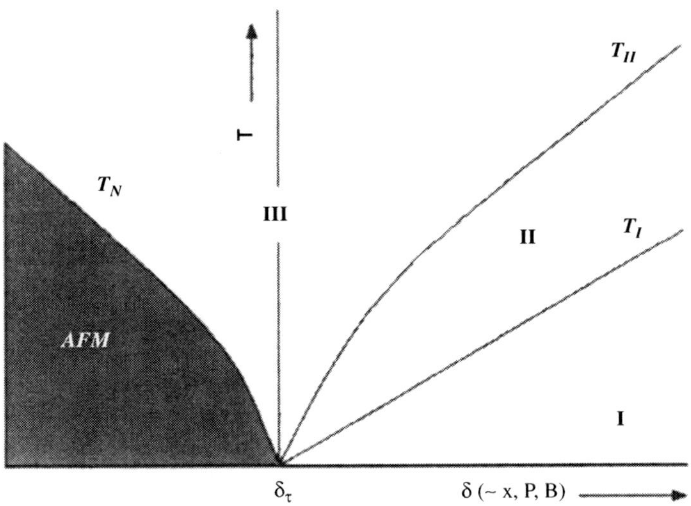

Figure 4.11. A qualitative phase diagram for a QCP for three-dimensional antiferromagnetic interactions (from the theory of Millis [63]). Antiferromagnetism (AFM) and T_N are suppressed by the tuning parameter δ_c (either doping (x), pressure (P), or magnetic field (B)). The NFL behavior occurs in both regions II and III. Reprinted with the permission of the American Physical Society from Ref [48].

of spin fluctuation at $q = 0$ as well as at $q = Q$ (antiferromagnetic ordering). Here equation for the dynamical susceptibility for a three-dimensional antiferromagnetic metal has two characteristic energy parameters T_A and $T_{0,m}$ which are related to the spin-fluctuation energies in ω and q space. They have also considered additional coupling between spin-fluctuation modes within the self-consistent renormalization model. The details of their results can be obtained in Ref. [65].

In an elegant but brief review, Coleman [61] has pointed out the inadequacy of the weak-coupling approach. His arguments can be summarized as follows. In the weak-coupling picture, at the QCP,

$$\chi(\mathbf{q}, \omega)^{-1} = \frac{[\delta + a^2(\mathbf{q} - \mathbf{Q})^2 - i\omega/\Gamma]}{\chi_0}, \tag{4.42}$$

where χ_0 is the uniform magnetic susceptibility, a is of the order lattice spacing, δ a quantity characterizing the distance from the QCP, and Γ characterizes the characteristic energy scale of spin fluctuations. Thus, the characteristic spin correlation length and time are given by

$$\xi \sim \frac{a}{\delta^{1/2}}, \quad \tau \sim \frac{1}{\Gamma\delta}, \tag{4.43}$$

so that $z = 2$. According to the Hertz–Millis theory [64], in dimension $d > 4 - z = 2$, the magnetic fluctuations can be regarded as a sort of overdamped exchange boson which propagates a long-range interaction given by

$$V(q) = J(q)^2 \chi(q, \omega). \tag{4.44}$$

This type of theory is inadequate for QCP in heavy fermion metals because of the following reasons:

(i) The theory predicts that the critical fluctuations will give rise to an anomalous specific heat $\Delta C_V/T \propto T^{(d-z)z} \propto T^{1/2}$ in three dimensions. Coleman has pointed out that in both, $CeNi_2Ge_2$ [66] and $CeCu_{6-x}Au_x$ [67], C_v/T has a logarithmic temperature dependence.
(ii) The theory does not lead to a breakdown of Fermi-liquid behavior since the quasiparticles are well defined on most of the Fermi surface, except in the vicinity of the "hot lines" that are linked to the magnetic **Q** vector on the Fermi surface (Figure 4.10). On the "hot lines", the quasiparticle scattering rate is $\Gamma_k \propto T$, but away from the "hot lines", the quasiparticle scattering rate has the classic Fermi-liquid form $\Gamma_k \propto T^2$ [68, 69].

Hlubina and Rice [69] have also pointed out that the resistivity ρ in a system with antiferromagnetic spin fluctuations, predicted to be $\rho \propto T^{3/2}$, will be "shorted out" below a certain temperature by "cold" portions of the Fermi surface which do not have strong spin fluctuations. They predict that below a certain crossover temperature, $\rho \propto T^{3/2}$ will

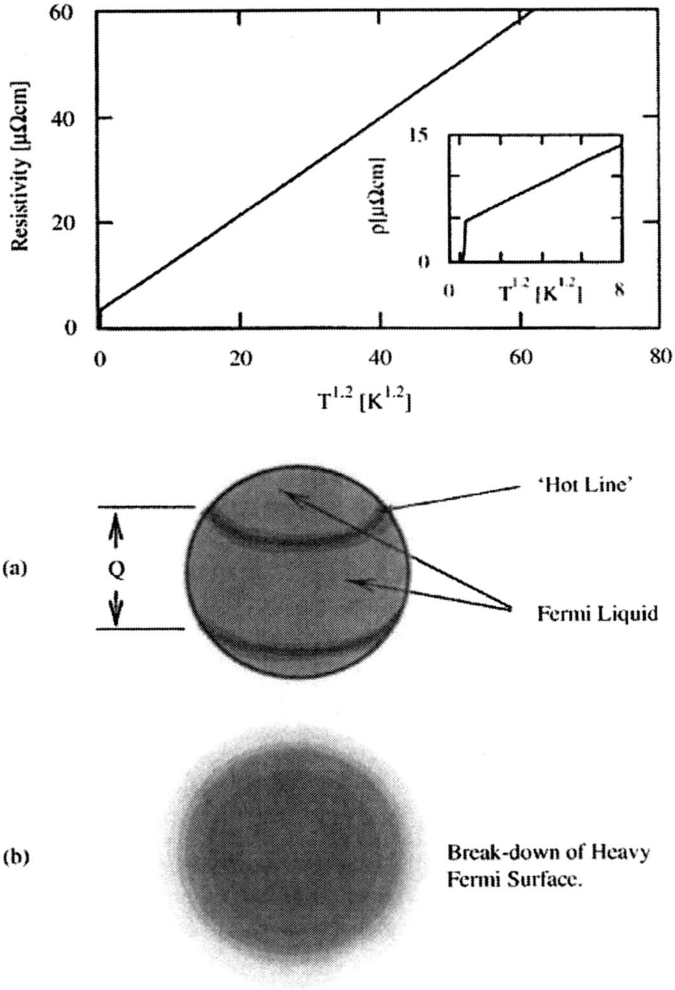

Figure 4.12. The anomalous resistivity of $CePd_2Si_2$ at the critical pressure [70]. (a) Magnetic scattering leads to "hot lines" around the Fermi surface, separated by regions of normal Formal liquid which would short-circuit the conductivity at low temperatures. (b) The absence of any T^2 regime in the resistivity suggests that sharply defined quasiparticles are eradicated from the entire Fermi surface. Reproduced with the permission of Elsevier from Ref. [61].

change to $\rho \propto T^2$, which is characteristic of Fermi-liquid behavior. In Figure 4.12, these details have been incorporated succintly by Coleman [61].

4.8. Strong-coupling theories

We shall discuss three elegant but diverse strong-coupling theories developed in recent years by three different groups.

4.8.1. Fractionalized Fermi liquids

Senthil et al. [71] have recently shown the existence of non-magnetic translation-invariant small-Fermi-surface states with a focus on two-dimensional Kondo lattices. When the local-moment system settles into a fractionalized spin liquid (FL^*) due to intermoment interactions, these states are obtained. A weak Kondo coupling to conduction electrons leaves a sharp (but small) Fermi surface of quasiparticles (FL) whose volume counts the conduction electron density but the structure of the spin liquid is undisturbed. These states have fractionalized excitations which coexist with conventional Fermi-liquid-like quasiparticles.

The effective-field theory of the spin-liquid states is a gauge theory in its deconfined state. The emergent gauge group is either Z_2 (dimension $d \geq 2$) or $U(1)$ ($d \geq 3$). The Z_2 states were discussed in detail in Ref. [71]. In a subsequent paper, Senthil et al. [72] have considered the three-dimensional lattice by using $U(1)$ states. Senthil et al. have focused on a three-dimensional $U(1)$ spin-liquid state with fermionic spinons which form a Fermi surface. The excitations in the $d = 3$ $U(1)$ spin-liquid phase are neutral spin-1/2 spinons, a gapless (emergent) gauge photon, and a gapped point defect (the "monopole"). Since the spinons minimally coupled to the photons, they interact through emergent long-ranged interactions. The gauge structure of the $U(1)$ spin-liquid state is stable to a weak Kondo coupling to conduction electrons. The $U(1)$ FL^* state consists of a spinon Fermi surface coexisting with a separate Fermi surface of conduction electrons. Senthil et al. [72] have used a mean-field theory to describe a $U(1)$ FL^* state and its transition to a heavy FL.

They have considered a three-dimensional Kondo–Heisenberg model on a cubic lattice:

$$H = \sum_k \epsilon_{\mathbf{k}} C^{\dagger}_{\mathbf{k}\alpha} C_{\mathbf{k}\alpha} + \frac{J_K}{2} \sum_r \vec{S}_r \cdot C^{\dagger}_{r\alpha} \vec{\sigma}_{\alpha\alpha'} C_{r\alpha'} + J_H \sum_{\langle rr' \rangle} \vec{S}_r \cdot \vec{S}_{r'}. \tag{4.45}$$

Here $C_{\mathbf{k}\alpha}$ is the conduction electron destruction operator, $\vec{S}_r$ are the spin-1/2 local moments, and summation over repeated spin indices α is implied. In a fermionic "slave-particle" representation of the local moments

$$\vec{S}_r = \frac{1}{2} f^{\dagger}_{r\alpha} \vec{\sigma}_{\alpha\alpha'} f_{r\alpha'}, \tag{4.46}$$

where $f_{r\alpha}$ is a spinful fermion destruction operator at site r. The decoupling of the Kondo and the Heisenberg exchange is made using two auxiliary fields by a saddle-point approximation, and the mean-field Hamiltonian is

$$\begin{aligned} H_{\mathrm{mf}} = &\sum_{\mathbf{k}} \epsilon_{\mathbf{k}} C^{\dagger}_{\mathbf{k}\alpha} C_{\mathbf{k}\alpha} - \chi_0 \sum_{\langle rr' \rangle} (f^{\dagger}_{r\alpha} f_{r'\alpha} + \mathrm{H.C.}) + \mu_{\mathrm{f}} \sum_r f^{\dagger}_{r\alpha} f_{r\alpha} \\ &- b_0 \sum_{\mathbf{k}} (C^{\dagger}_{\mathbf{k}\alpha} f_{\mathbf{k}\alpha} + \mathrm{H.C.}). \end{aligned} \tag{4.47}$$

Here b_0 and χ_0 have been assumed to be real and additional constants to H have been dropped. The mean-field parameters b_0, χ_0, and μ_{f} are obtained from

$$1 = \langle f^{\dagger}_{r\alpha} f_{r\alpha} \rangle, \tag{4.48}$$

$$b_0 = \frac{J_K}{2}\langle C^{\dagger}_{r\alpha} f_{r\alpha}\rangle, \tag{4.49}$$

$$\chi_0 = \frac{J_H}{2}\langle f^{\dagger}_{r\alpha} f_{r'\alpha}\rangle, \tag{4.50}$$

where r, r' are nearest neighbors.

At zero temperature, in the Fermi-liquid (FL) phase, χ_0, b_0, and μ_0 are non-zero. In the FL* phase, $b_0 = \mu_0 = 0$ but $\chi_0 \neq 0$. In this state, the conduction electrons are decoupled from the local moments and form a small Fermi surface. The local-moment system is described as a spin fluid with a Fermi surface of neutral spinons.

The mean field has been diagonalized by the transformation [72]

$$C_{\mathbf{k}\alpha} = u_{\mathbf{k}}\gamma_{\mathbf{k}\alpha+} + v_{\mathbf{k}}\gamma_{\mathbf{k}\alpha-}$$

and

$$f_{\mathbf{k}\alpha} = v_{\mathbf{k}}\gamma_{\mathbf{k}\alpha+} - u_{\mathbf{k}}\gamma_{\mathbf{k}\alpha-}. \tag{4.51}$$

The Hamiltonian can be written in terms of the new fermionic operators $\gamma_{\mathbf{k}\alpha\pm}$,

$$H_{\mathrm{mf}} = \sum_{\mathbf{k}\alpha} E_{\mathbf{k}+}\gamma^{\dagger}_{\mathbf{k}\alpha+}\gamma_{\mathbf{k}\alpha+} + E_{\mathbf{k}-}\gamma^{\dagger}_{\mathbf{k}\alpha-}\gamma_{\mathbf{k}\alpha-}, \tag{4.52}$$

where

$$E_{\mathbf{k}\pm} = \frac{\epsilon_{\mathbf{k}} + \epsilon_{\mathbf{kf}}}{2} \pm \sqrt{\left(\frac{\epsilon_{\mathbf{k}} - \epsilon_{\mathbf{kf}}}{2}\right)^2 + b_0^2}. \tag{4.53}$$

Here $\epsilon_{\mathbf{kf}} = \mu_{\mathrm{f}} - \chi_0\sum_{a=1,2,3}\cos(k_a)$. $u_{\mathbf{k}}$,$v_{\mathbf{k}}$ are determined by

$$u_{\mathbf{k}} = -\frac{b_0 v_{\mathbf{k}}}{E_{\mathbf{k}+} - \epsilon_{\mathbf{k}}}, \quad u_{\mathbf{k}}^2 + v_{\mathbf{k}}^2 = 1. \tag{4.54}$$

For the FL* phase, $b_0 = \mu_{\mathrm{f}} = 0$ but $\chi_0 \neq 0$. The conduction electron dispersion ϵ_k determines the electron Fermi surface and is small. The spinon Fermi surface encloses one spinon per site and has volume half that of the Brillouin zone. Senthil et al. [72] have assumed that the conduction electron filling is less than half and the electron Fermi surface does not intersect the spinon Fermi surface. In the FL phase near the transition (small b_0), there are two bands corresponding to $E_{\mathrm{k}\pm}$: one derives from the c electrons with f character (c band) while the other derives from the f particles with weak c character (f band).

As shown in Figure 4.13, for small b_0, the Fermi surface consists of two sheets since both bands intersect the Fermi energy. The total volume is large since it includes both local moments and conduction electrons. When b_0 decreases to zero, the transition moves

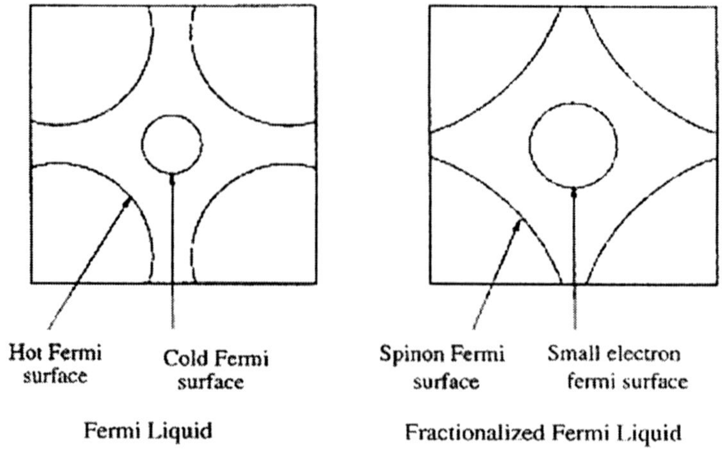

Figure 4.13. Fermi-surface evolution from FL to FL*: the FL phase has two Fermi-surface sheets (the cold c and the hot f sheet) close to the transition. The quasiparticle residue Z on the hot sheet vanishes on approaching the transition. The f-sheet becomes the spinon Fermi surface while the c-sheet is the small conduction-electron Fermi surface on the FL* side. Reprinted with the permission of the American Physical Society from Ref. [72].

to FL*, the c-Fermi surface expands in size to match onto the small Fermi surface of FL*, while the f-Fermi surface shrinks to match onto the spinon Fermi surface of FL*.

The quasiparticle weight Z close to the FL–FL* transition has been calculated in the mean-field theory. The electron Green's function is

$$G(k, i\omega_\nu) = \frac{u_{\mathbf{k}}^2}{i\omega_\nu - E_{\mathbf{k}+}} + \frac{v_{\mathbf{k}}^2}{i\omega_\nu - E_{\mathbf{k}-}}. \tag{4.55}$$

At the Fermi surface of the c band (dispersion $E_{\mathbf{k}\pm}$, quasiparticle residue Z $= u_{\mathbf{k}}^2$), $E_{k+} \approx \epsilon_k \approx 0$,

$$E_{k+} \approx \epsilon_{\mathbf{k}} + \frac{b_0^2}{\epsilon_{\mathbf{k}} - \epsilon_{\mathbf{k}f}} \Rightarrow u_{\mathbf{k}} \approx -\frac{J_{\mathrm{H}}}{b_0} v_{\mathbf{k}}. \tag{4.56}$$

From equations (4.50) and (4.52), one obtains $Z \approx 1$ on the c-Fermi surface. For the f-Fermi surface, $Z = v_{\mathbf{k}}^2$ and $|\epsilon_{\mathbf{k}} - \epsilon_{\mathbf{kf}}| \approx t \gg J_{\mathrm{H}}$, where t is the conduction electron bandwidth,

$$E_{k+} \approx \epsilon_{\mathbf{k}} + \frac{b_0^2}{\epsilon_{\mathbf{k}} - \epsilon_{\mathbf{kf}}} \Rightarrow u_{\mathbf{k}} \approx -\frac{t}{b_0} v_{\mathbf{k}}. \tag{4.57}$$

The expression for Z is

$$Z = v_{\mathbf{k}}^2 \approx \left(\frac{b_0}{t}\right)^2. \tag{4.58}$$

The quasiparticle residue remains non-zero on the c-Fermi surface (cold) but decreases continuously to zero on the f-Fermi surface (hot) on moving from FL to FL*. This is illustrated in Figure 4.13.

The picture obtained in the mean-field theory is modified in several ways by fluctuation effects. While the heavy Fermi liquid is stable to fluctuations, they are interesting in the FL* state. These are described by a $U(1)$ gauge theory minimally coupled to a spinon Fermi surface. Senthil et al. [72] have shown that the action for the f particles describes a Fermi surface of spinons coupled to a compact gauge field. They have assumed that the system has a deconfined phase which has a Fermi surface of spinons coupled minimally to a gapless "photon" ($U(1)$ gauge field). There is also a gapped monopole excitation. Two static spinons interact with each other through an emergent long-range $1/r$ Coulomb interaction. This is the $U(1)$ FL* phase and putting back a small coupling between the c and f particles will not change the deconfined nature of this phase and the monopole gap will be preserved. The gauge fluctuations are overdamped in the small-q limit. This form of the gauge-field action leads to a $T \ln(1/T)$ singularity in the low temperature in the $U(1)$ gauge FL* phase.

The effects of fluctuations beyond the mean-field theory at the phase transition between FL and $U(1)$ FL* phases was considered by Senthil et al. [72] by arguing that since the important changes near the transition occur at the hot Fermi surface, the c fields can be integrated out from the action (equation (16) in Ref. [72]). They also partially integrate out the f excitations well away from the hot Fermi surface which also changes the b effective action in their equation (16). Their phase diagram of the action is shown in Figure 4.14. The horizontal axis, represented by J_K, is accessed by varying μ_b, which is interpreted as the chemical potential of the boson [72]. The quantum-critical point between the FL and $U(1)$ FL* phases occurs at $\mu_b = 0$, $T = 0$.

Senthil et al. [45] have recently presented an elegant brief review of their work, focusing on the nature of the FL* state, with $\langle b \rangle = 0$ obtained by increasing J_H. First, it is a Fermi liquid in the sense that the c electrons form a conventional Fermi surface with sharp

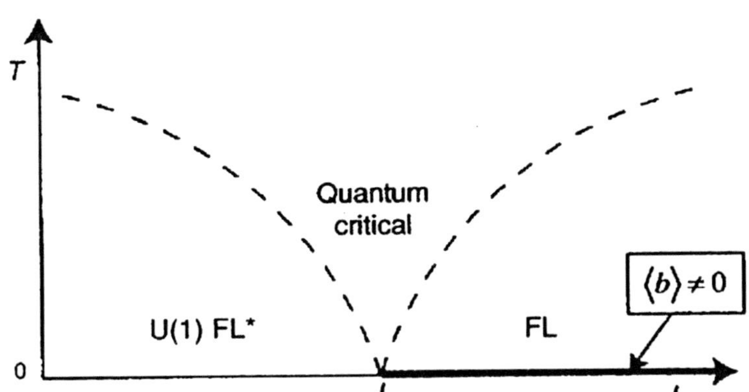

Figure 4.14. Crossover phase diagram for the vicinity of the d-3 quantum transition involving breakdown of Kondo screening. Reprinted with the permission of the American Physical Society from Ref. [72].

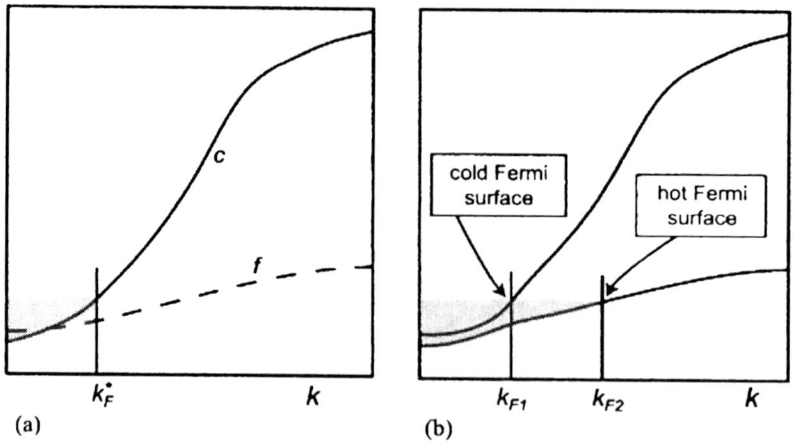

Figure 4.15. (a) The FL* state with a cold Fermi surface at $\mathbf{k}_F^*$. The f fermions form a spin liquid and are schematically represented by the dashed dispersion. (b) The FL state which obeys the conventional Luttinger theorem, but has two Fermi surfaces. Reprinted with the permission of Elsevier from Ref. [45].

electron-like quasiparticles at $\mathbf{k}_F^*$, as shown in Figure 4.15(a). This violates the Luttinger theorem since the volume enclosed by this Fermi surface contains only the electrons. However, closely linked with this non-Luttinger Fermi volume is the fate of the spin moments on the f electrons. These form a "spin-liquid" state, which does not break lattice translational symmetries and fully preserve spin rotation invariance which implies that there is no magnetic moment.

They have also presented in Figure 4.15, a description of the QCP between FL and FL* phases. In the FL phase, as one moves toward the FL* phase, the Fermi surfaces at k_{F1} and k_{F2} behave differently. The Fermi surface at k_{F1} evolves into the Fermi surface at k_{F*} in the FL* state, and remains "cold" at the QCP. There is no strong scattering of these quasiparticles and the electron quasiparticle residue remains finite across the transition. However, the Fermi surface at k_{F2} remains "hot". The lifetime of the quasiparticles becomes short and the quasiparticle residue ultimately vanishes at the QCP. The dynamics of the quasiparticles exhibits NFL behavior at the QCP and this behavior extends across the entire "hot" Fermi surface.

4.8.2. Local quantum criticality in heavy fermion metals

Si and co-workers [73–77] as well as Sun and Kotliar [78, 79] have described the local quantum criticality in heavy fermion metals by using EDMFT approaches described in Chapter 3. Si et al. have considered the Kondo-lattice model:

$$H = \sum_{ij,\sigma} t_{ij} C_{i\sigma}^{\dagger} C_{j\sigma} + \sum_{i} J_K \mathbf{S}_i \cdot \mathbf{S}_{c,i} + \sum_{ij} \frac{I_{ij}}{2} \mathbf{S}_i \cdot \mathbf{S}_j. \tag{4.59}$$

The tight-binding parameters t_{ij} determine the dispersion $\in_{\mathbf{k}}$ so that the bare conduction-band DOS are

$$\rho_0(\in) = \sum_{\mathbf{k}} \delta(\in - \in_{\mathbf{k}}). \tag{4.60}$$

$\mathbf{S}_i$ is a spin-1/2 local moment at each lattice site i which interacts with the spin $\mathbf{s}_{c,i}$ of the conduction (c) electrons. They have considered only positive J_K (antiferromagnetic) values, which can lead to Kondo quenching of the local moments. I_{ij} is the RKKY interaction. The Kondo lattice model is mapped in the EDMFT method, described in Chapter 3, to an effective impurity Kondo model. The effective impurity Hamiltonian is given by

$$H_{\text{imp}} = J_{\text{K}}\mathbf{S}\cdot\mathbf{s}_{\text{c}} + \sum_{p,\sigma} E_p C^{\dagger}_{p\sigma} C_{p\sigma} + g\sum_{p} \mathbf{S}\cdot(\vec{\phi}_p + \vec{\phi}^{\dagger}_{-p}) + \sum_{p} w_p \vec{\phi}^{\dagger}_p \cdot \vec{\phi}_p. \tag{4.61}$$

This is the Bose–Fermi Kondo Hamiltonian which describes a local moment coupled to the spin of a fermionic bath ($C_{p\sigma}$) and to a dissipative vector-bosonic bath ($\vec{\phi}_p$). The vector bosons describe a fluctuating magnetic field generated by the RKKY interaction by the local moments at all other lattice sites. The coupling constant g and the dispersions E_{p} and w_{p}, which can be obtained [74] from the relationships defined in equation (3.46) are such that integrating out the two baths in the Hamiltonian representation yield the effective action.

The essence of the work of Si et al. [73] has been very briefly but elegantly described in a few diagrams. In Figure 4.16(a), they have described the Kondo-lattice model.

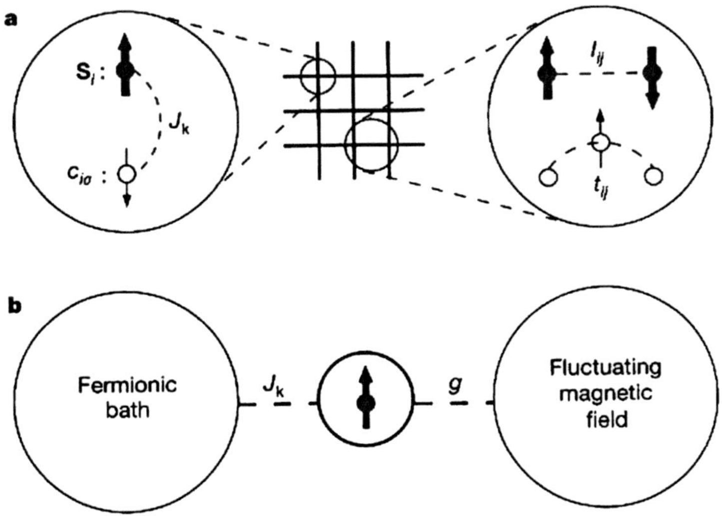

Figure 4.16. (a) The Kondo-lattice model. (b) The effective-impurity Kondo model to which the Kondo-lattice model is mapped in EDMFT (described in detail in Chapter 3). Reprinted with the permission of MacMillan Publishers Ltd., from Nature, Ref. [73], Copyright (2007).

At each lattice site, a spin-1/2 local moment $\mathbf{S}_i$ interacts with a local conduction electron orbital C_{ij} through an antiferromagnetic coupling constant J_K. The hopping amplitude between the conduction electrons at site i and j is t_{ij}. Figure 4.16(b) shows the effective impurity Kondo model to which the Kondo-lattice model is mapped in EDMFT (see Chapter 3). The bosonic bath describes a fluctuating magnetic field. The couplings to the two baths are J_K and g, respectively.

In Figure 4.17, Si et al. [73] have shown a conventional QPT in Kondo lattices. δ, the tuning parameter, is the ratio of the RKKY interactions to the Kondo energy scale T_K. When δ is small, the local moments are completely screened by the spins of the conduction electrons, and the system is a paramagnetic metal. When δ is increased, the magnetic correlations become more pronounced, and the dynamical spin susceptibility $\chi(\mathbf{Q}, \omega)$ diverges at the QCP $\delta = \delta_c$ which separates a paramagnetic metal from an antiferromagnetic metal. An increase in δ causes an increase in the coupling g between the local moment and the fluctuating magnetic field. At $\delta = \delta^c_{loc}$, g/T_K would reach a threshold value where the local Kondo problem becomes critical. The fluctuating magnetic field just succeeds in preventing the fermionic bath from quenching the local moment thereby yielding a singular local susceptibility. At this second critical point, a local energy scale E^*_{loc} (explained in Chapter 3 and shown in the figure as a dashed line), which serves as an effective Fermi energy scale, vanishes. When $\delta > \delta_c$, there exists a finite temperature T_N (solid line) below which the system is an antiferromagnetic metal. In this case, $\delta_c < \delta^c_{loc}$, i.e., E^*_{loc} is finite at the QCP. In Figure 4.17, the quantum-critical regime, which is described by the traditional theory, is denoted by a wavy line.

Si et al. [73] have also shown that when the spin fluctuations are two-dimensional, there is a second type of QCP in $\delta_c = \delta^c_{loc}$. This is shown in Figure 4.18 in which it is shown that E^*_{loc} also vanishes at the QCP. The long-wavelength fluctuations of the order parameter and the local fluctuations originating from the local moments coexist. The lattice system reaches its ordering transition at the point where the local problem also becomes critical. This transition, which is locally critical, is referred to as locally critical QPT.

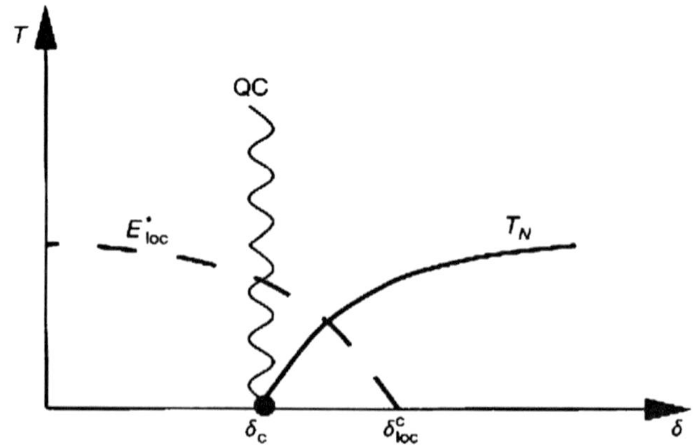

Figure 4.17. Diagram of a conventional quantum phase transition in a Kondo lattice. Reprinted with the permission of MacMillan Publishers Ltd., from Nature, Ref. [73], Copyright (2007).

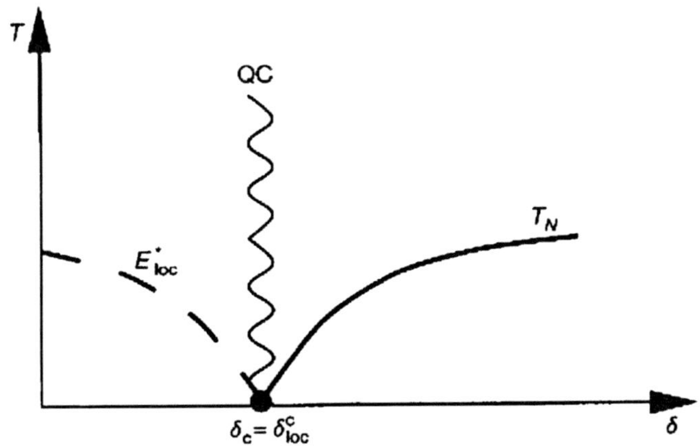

Figure 4.18. Diagram of a locally critical quantum phase transition in Kondo lattice. The local and spatially extended critical degrees of freedom coexist in the quantum-critical (QC) regime. Reprinted with the permission of MacMillan Publishers Ltd., from Nature Ref. [73]. Copyright (2007).

In a recent paper, Si [80] has considered the global magnetic phase diagram of Kondo-lattice systems. He has shown that the Kondo lattices can have two types of magnetic metal phases, AF_S and AF_L, in addition to paramagnetic metal phase PM_L. There is a second-order QPT (when quasiparticle residues vanish) which goes directly from AF_S to PM_L. At this magnetic QCP, the excitations include the fluctuations associated with a critical Kondo screening. One form of such type of QCP is the local quantum criticality.

The global magnetic phase diagram of the Kondo lattice [80] is obtained by considering the Kondo lattice

$$H = H_{\mathrm{f}} + H_{\mathrm{c}} + H_{\mathrm{K}}, \tag{4.62}$$

where H_{f} is the Hamiltonian for the local moment f, H_{c} is the conduction electron Hamiltonian, and H_{K} is the interaction (Kondo) Hamiltonian:

$$H_{\mathrm{f}} = \frac{1}{2}\sum_{ij} I^a_{ij} S^a_i S^a_j. \tag{4.63}$$

Here I^a_{ij} is the RKKY interaction (antiferromagnetic) between the spin-1/2 moments (one per site) and $a = x, y, z$ are spin projections. Si [80] has also introduced a term G, which describes the degree of frustration ($G = I_{\mathrm{nnn}}/I_{\mathrm{nn}}$, the ratio of the antiferromagnetic next-nearest-neighbor interaction over the nearest neighbor one). Increasing G corresponds to a decrease in the strength of the Neel order:

$$H_{\mathrm{c}} = \sum_{\mathbf{k}\sigma} \epsilon_{\mathbf{k}} C^{\dagger}_{\mathbf{k}\sigma} C_{\mathbf{k}\sigma}, \tag{4.64}$$

where there are x conduction electrons per site ($0 < x < 1$) and the bandwidth is W.

$$H_K = \sum_i J_K \mathbf{S}_i \cdot \mathbf{s}_{c,i}, \tag{4.65}$$

where the Kondo interaction J_K is antiferromagnetic and the spin terms have been defined earlier.

The magnetic phase diagram is shown in Figure 4.19. The horizontal axis labels $j_K \equiv J_K/W$, while the vertical axis describes the local-moment magnetism which is decoupled from the conduction electrons. There are three phases, PM_L, AF_L, and AF_S. In the paramagnetic phase, by using Luttinger's theorem, it can be shown that there are $(1 + x)$ electrons per unit cell within the paramagnetic Brillouin zone [81]. This is the PM_L phase which describes a heavy Fermi liquid. When $J_K/W(\ll I/W) \ll 1$, the local moments stay charge neutral and do not contribute to the electronic excitations. This phase is termed as AF_S since the Fermi surface is small. When the system moves from the AF_S phase to the PM_L phase, the Kondo screening is realized and the Neel order eventually goes away. However, the Kondo screening develops before the Neel order disappears. The Fermi

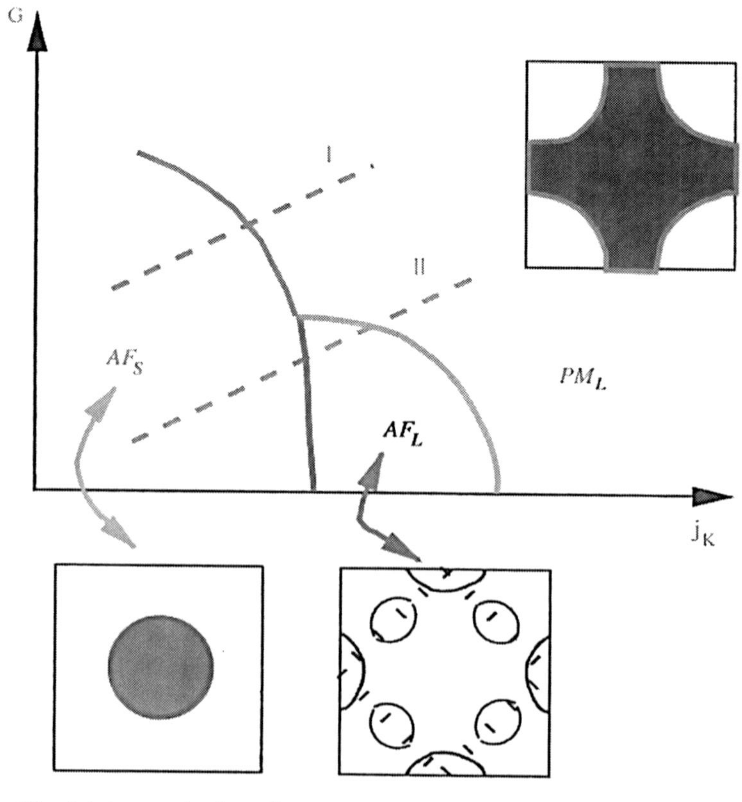

Figure 4.19. The global magnetic phase diagram of the Kondo lattice at zero temperature. $j_K = J_K/W$, $G = I_{nnn}/I_{nn}$, and I/W is kept fixed at a small value. The three phases AF_L, AF_S, and PM_L have distinct Fermi surfaces. "I" and "II" label two types of transition. Reproduced with the permission of Elsevier from Ref. [80].

surface of this intermediate magnetically ordered state is large since the local moments become a part of the electron fluid. This phase is denoted as the AF_L phase. The AF_L and the AF_S phases are separated by a Lifshitz transition. The direct transition between AF_S and PM_L phases is labeled as the type I transition and the magnetic quantum transition between the AF_L and PM_L phases is labeled as the type II transition in Figure 4.19.

The effective Kondo-screening scale of the lattice (coherence temperature) is finite at the type II magnetic transition and the QCP is of the Hertz–Moriya–Millis type [62, 63, 65]. However, the type I magnetic transition goes directly from the AF_L to the PM_L phase. The transition is of second order if z_L and z_S, the quasiparticle residues of the large Fermi surface in the PM_L and the small Fermi surface in the AF_S phases, go to zero as the transition is reached from respective sides. The Kondo singlets disintegrate and the coherence temperature vanishes as the QCP is approached from the PM_L side. The local quantum-critical theory is one of the mechanisms which explains how the quasiparticles are destroyed. The Fermi surface undergoes a sudden reconstruction at the QCP and as the continuous vanishing of z_L and z_S implies, the effective mass diverges as the QCP is approached from both sides.

4.8.3. *The underscreened Kondo model*

Coleman and Pepin [82] have used the Schwinger boson representation of the local moments to study the underscreened Kondo-impurity model (UKM). The UKM describes the screening of a local moment from spin S to spin $S^* = S - 1/2$ [83]. This residual moment decouples from the surrounding Fermi sea at low temperatures. The UKM has been studied by using strong-coupling expansion [2, 3] and numerical renormalization group [84] and diagonalized by using Bethe ansatz [6, 85–87] but none has addressed the possibility of a breakdown of the Landau Fermi-liquid behavior.

The UKM model can be written as

$$H = \sum_{\mathbf{k}\alpha} \epsilon_{\mathbf{k}} C^{\dagger}_{\mathbf{k}\alpha} C_{\mathbf{k}\alpha} + J\vec{S} \cdot \psi^{\dagger}_{\alpha} \vec{\sigma}_{\alpha\beta} \psi_{\beta}. \tag{4.66}$$

Here S denotes a spin $S > 1/2$, $C^{\dagger}_{\mathbf{k}\alpha}$ creates a conduction electron with wave vector $\mathbf{k}$, spin component α, and $\psi^{\dagger}_{\alpha} = \sum_{\mathbf{k}} C^{\dagger}_{\mathbf{k}\alpha}$ creates a conduction electron at the impurity site. Coleman and Pepin [82] have reformulated the UKM as an $SU(N)$-invariant Coqblin–Schrieffer model which enabled them to make a large-N expansion. The resulting Hamiltonian is

$$H = \sum_{\mathbf{k}\alpha} \epsilon_{\mathbf{k}} C^{\dagger}_{\mathbf{k}\alpha} C_{\mathbf{k}\alpha} + \frac{J}{N} \sum_{\alpha\beta} (\psi^{\dagger}_{\beta} b_{\beta})(b^{\dagger}_{\alpha} \psi_{\alpha}) - MB \tag{4.67}$$

Here the spin indices α and β have N values and the constraint $n_b = 2S$ is imposed to represent spin S. The local-moment magnetization $M = g_N[b^{\dagger}_{\uparrow} b_{\uparrow} - \tilde{S}]$, where the first spin component $\sigma = 1 \equiv \uparrow$ and $\tilde{S} = 2S/N$. The prefactor $g_N = N/2(N - 1)$ is chosen so that at maximum polarization, when $n_{\uparrow} = 2S$, $M = S$. Parcollet and Georges [88] have treated a

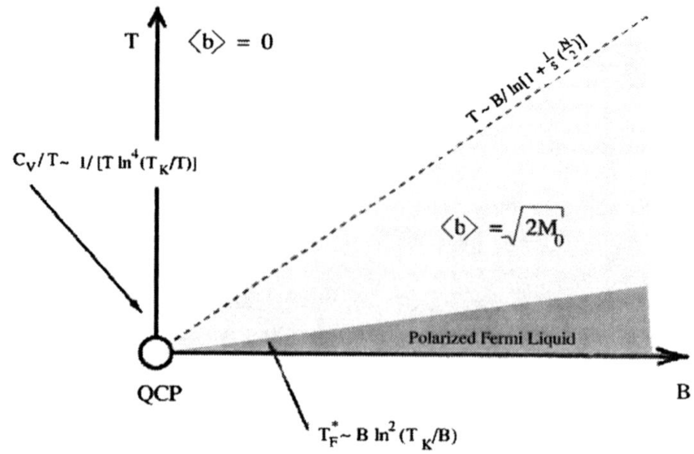

Figure 4.20. Schematic phase diagram of the large-N limit of the underscreened Kondo model. Reprinted with the permission of the American Physical Society from Ref. [82].

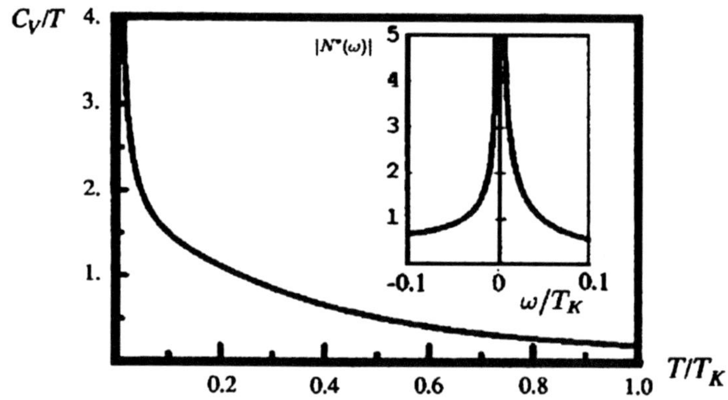

Figure 4.21. Zero-field specific heat capacity of the underscreened Kondo model for the case $\tilde{S} = 1/2$, showing the $1/T\ln^4(T_K/T)$ divergence at low temperature. The singular DOS $N^*(\omega)$ is shown in the inset. Reproduced with the permission of the American Physical Society from Ref. [82].

multichannel formulation of this model within an integral equation formalism. The summary of the results obtained by Coleman and Pepin [82] who have used a path-integral method is shown in Figure 4.20. The residual ferromagnetic interaction between a partially screened spin and the electro sea destabilizes the Landau Fermi liquid, forming a singular Fermi liquid with a $1/[T\ln^4(T_K/T)]$ divergence in the low-temperature specific heat coefficient C_V/T. A magnetic field B tunes this system back into Landau Fermi liquid with a Fermi temperature proportional to $B\ln^2(T_K/B)$.

In Figure 4.21, the specific heat coefficient is plotted as a function of T/T_K. The singular specific heat coefficient indicates that the fluid excitations in zero field has a singular DOS which makes it obvious that this state is not a Landau Fermi liquid. These singularities are a consequence of the singular energy and temperature dependence of the coupling constant.

Coleman and Pepin [82] have also shown that in a finite magnetic field, the DOS $N^*(\tilde{B})$, where $\tilde{B} = g_N B$, is non-singular at zero energy. They have derived,

$$T_F^* \sim N^*(B)^{-1} = B\ln^2\left(\frac{T_K}{B}\right). \tag{4.68}$$

Thus, T_F^* can be identified as the Fermi temperature which means that in a finite field the characteristic temperature of the Fermi liquid is the field itself. At $T \leq \tilde{B}$, $C_V = \gamma(\tilde{B})T$, where $\gamma = (\pi^2 k_B^2/3)N^*(B)$, and will be linear with temperature as shown in Figure 4.22. However, as the field is reduced to zero, $N^*(\omega)$ becomes singular, and the fluid becomes a singular Fermi liquid.

Thus, Coleman and Pepin's theory [82] shows how a field-tuned Fermi liquid with a characteristic scale grows linearly with the applied magnetic field. The low-temperature upturn in the specific heat and the appearance of B as the only scale of the problem are features observed in quantum-critical heavy electron systems [89–91].

Recently, Pepin [92] has introduced a new effective theory for the QCP of $YbRh_2(Si_{0.95}Ge_{0.05})_2$ starting from the Kondo–Heisenberg model at high energies. This effective theory shows the emergence of a new excitation at low energies, fermionic and spinless in nature (the χ fermions) characterizing the Kondo-bound state. The χ fermion forms a band when high-energy degrees of freedom are integrated out. The total Fermi-surface volume evolves from a small Fermi surface to a big Fermi surface as the χ-fermion band fills up. Using this model, Pepin has been able to reproduce many striking experimental results of $YbRh_2(Si_{0.95}Ge_{0.05})_2$.

Recently, Stewart [93] has updated his original review on NFL behavior of d- and f-electron metals. He has compiled a very large number of recent experimental data of resistivity, susceptibility, and specific heat. It indicates the intense activity in this field, but as we have shown, there is no unanimity on the theory of NFL models of heavy fermion systems. Hopefully, a complete theoretical understanding will emerge from the elegant theoretical models discussed above.

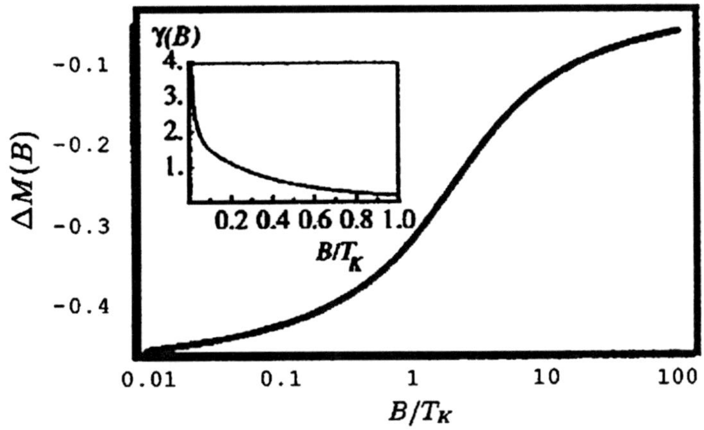

Figure 4.22. Ground-state magnetization, $\Delta M = S - M$, of the underscreened Kondo model for $\tilde{S} = 0.5$. The field-dependent linear specific heat capacity $\gamma(B)$ is shown in the inset. Reproduced with the permission of the American Physical Society from Ref. [82].

References

[1] L.D. Landau, Sov. Phys. JETP **3**, 920 (1957); **5**, 920 (1957); **8**, 70 (1959).
[2] P. Nozieres, J. Low Temp. Phys. **17**, 31 (1974); J. Physique **39**, 1117 (1978).
[3] P. Nozieres, A. Blandin, J. Phsique **41**, 193 (1980).
[4] K.G. Wilson, Rev. Mod. Phys. **47**, 773 (1975).
[5] N. Andrei, Phys. Rev. Lett. **45**, 379 (1980).
[6] N. Andrei, K. Furuya, and J.H. Lowenstein, Rev. Mod. Phys. **55**, 331 (1983).
[7] A.M. Tsvelick and P.B. Wiegmann, Adv. Phys. **32**, (1983).
[8] F.D.M. Haldane, Phys. Rev. Lett. **40**, 416 (1978).
[9] T.V. Ramakrishnan, in *Valence Instabilities*, eds. P. Wachter and H. Boppart, p. 351 (North-Holland, Amsterdam, 1982).
[10] Y. Kuramoto, Z. Phys. B **53**, 271 (1983).
[11] P. Coleman, Phys. Rev. B **28**, 5255 (1983).
[12] H. Razahfimandimby, P. Fulde, and J. Keller, Z. Phys. **54**, 111 (1984).
[13] N. d'Ambrumenil and P. Fulde, J. Magn. Magn. Mater. **47–48**, 1 (1985).
[14] J.M. Ziman, Proc. Phys. Soc. **86**, 337 (1965).
[15] D.G. Pettifor, J. Phys. C: Solid State Phys. **5**, 97 (1972).
[16] V. Heine, Phys. Rev. **153**, 673 (1967).
[17] P.K. Misra and J. Callaway, J. Phys. Condens. Matter **1**, 4485 (1989).
[18] Z. Zou and P.W. Anderson, Phys. Rev. Lett. **57**, 2073 (1986).
[19] S.K. Misra, P.K. Misra, and S.D. Mahanti, Phys. Rev. B **26**, 1902 (1982).
[20] G.S. Tripathi, L.K. Das, P.K. Misra, and S.D. Mahanti, Phys. Rev. B **25**, 309 (1982).
[21] P.A. Lee, T.M. Rice, J.W. Serene, L.J. Sham, and J.W. Wilkins, Comments Solid State Phys. B **12**, 99 (1986).
[22] G.R. Stewart, Rev. Mod. Phys. **56**, 755 (1984).
[23] K. Kadowaki and S.B. Woods, Solid State Commun. **58**, 507 (1986).
[24] M.T. Beal Monod, Physica B **109** & **110**, 1837 (1982).
[25] G.E. Brodale, R.A. Fischer, Norman E. Phillips, G.R. Stewart, and A.L. Giorgi, Phys. Rev. Lett. **57**, 234 (1986).
[26] P. Nozieres, *The Theory of Interacting Fermi Systems* (Benjamin, New York, 1964).
[27] (a) C.M. Varma, Phys. Rev. Lett. **55**, 2723 (1985); (b) C.M. Varma, Z. Nussinov, and W. Saarloos, Phys. Rep. **361**, 267 (2002).
[28] C.M. Varma and Y. Yafet, Phys. Rev. B **13**, 2950 (1976).
[29] C.M. Varma, W. Weber, and L.J. Randall, Phys. Rev. B **33**, 215 (1986).
[30] A. Auerbach and K. Levin, Phys. Rev. Lett. **57**, 877 (1986).
[31] A.J. Millis and P.A. Lee, Phys. Rev. B **35**, 3394 (1987).
[32] N. Read and D.M. Newns, J. Phys. C **16**, 3273 (1983).
[33] N. Read, J. Phys. C **18**, 2651 (1985).
[34] P. Coleman, Phys. Rev. B **29**, 3035 (1985).
[35] P. Coleman, in *Theory of Heavy-Fermions and Valence Fluctuations*, eds. T. Kasuya and T. Saso (Springer Series in Solid State Sciences, Vol. 62, Springer-Verlag, New York, 1985).
[36] A.A. Abrikosov, L.P. Gorkov, and I.E. Dzyaloshinski, *Methods of Quantum Field Theory in Statistical Physics* (Dover, New York, 1975).
[37] K. Levin and O.T. Valls, Phys. Rev. B **20**, 105 (1979).

[38] S. Doniach and S. Engelsberg, Phys. Rev. Lett. **17**, (1966).
[39] W.F. Brinkman and S. Engelsberg, Phys. Rev. **169**, (1968).
[40] E. Riedel, Z. Phys. **210**, 417 (19680).
[41] N. d'Ambrumenil and P. Fulde, in *Theory of Heavy-Fermion and Valence Fluctuations*, eds. T. Kasuya and T. Saso (Springer Series in Solid State Sciences, Vol. 62, Springer-Verlag, New York, 1985).
[42] P. Nozieres, Ann. Phys. (Paris) **10**, 19 (1985).
[43] P. Nozieres, Eur. Phys. B **6**, 447 (1998).
[44] S. Burdin, A. Georges, and D.R. Grempel, Phys. Rev. Lett. **85**, 1048 (2000).
[45] T. Senthil, S. Sachdeb, and M. Vojta, Physica B **359–361**, 9 (2005).
[46] J.M. Luttinger, Phys. Rev. **119**, 1153 (1960).
[47] C.L. Seaman, M.B. Maple, B.W. Lee, S. Ghamaty, M.S. Torikachvili, J.S. Kang, L.Z. Liu, J.W. Allen, and D.L. Cox, Phys. Rev. Lett. **67**, 2882 (1991).
[48] G. R. Stewart, Rev. Mod. Phys. **73**, 797 (2001).
[49] D.L. Cox, Phys. Rev. Lett. **59**, (1987) 1240.
[50] P.D. Sacramento and P. Schlottmann, Phys. Lett. A **142**, 245 (1989).
[51] P. Nozieres and A. Blandin, J. Phys. (Paris) **41**, 193 (1980).
[52] N. Andrei and C. Destri, Phys. Rev. Lett. **52**, 364 (1984).
[53] A. Tsvelik and P.B. Wiegmann, Z. Phys. B: Condens. Matter. **54**, 201 (1984).
[54] I. Affleck and A.W.W. Ludwig, Nucl. Phys. B **352**, 849 (1991).
[55] G.R. Stewart, Z. Fisk, and J. Willis, Phys. Rev. B **28**, 172 (1983).
[56] S.M. Shapiro, A.I. Goldman, G. Shirane, D.E. Cox, Z. Fisk, and S.L. Smith, J. Magn. Magn. Mater. **52**, 418 (1985).
[57] S. Horn, E. Holland-Moritz, M. Loewenhaupt, F. Steglich, H. Scheuer, A. Benoit, and J. Flouquet, Phys. Rev. B **23**, 3171 (1981).
[58] D.L. Cox, Physica B **186–188**, 312 (1993).
[59] A.P. Ramirez, P. Chandra, P. Coleman, Z. Fisk, J.L. Smith, and H.R. Ott, Phys. Rev. Lett. **73**, 3018 (1994).
[60] S. Sachdev, *Quantum Phase Transitions* (Cambridge University Press, Cambridge, 1999).
[61] P. Coleman, Physica B **259–261**, 353 (1999).
[62] J.A. Hertz, Phys. Rev. B **14**, 1165 (1976).
[63] A.J. Millis, Phys. Rev. B **48**, 7183 (1993).
[64] U. Zulicke and A.J. Millis, Phys. Rev. B **51**, 8996 (1995).
[65] T. Moriya and T. Takimoto, J. Phys. Soc. Jpn. **64**, 960 (1995).
[66] F. Steglich, B. Buschinger, P. Gegenwart, M. Lohmann, R. Helfrich, C. Langhammer, P. Hellmann, L. Donnevert, S. Thomas, A. Link, C. Geibel, M. Lang, G. Sparn, and W. Assmus, J. Phys. Condens. Matter **8**, 9909 (1996).
[67] H. von Lohhneysen, J. Phys. Condens. Matter. **8**, 9689 (1996).
[68] O.O. Bernal, D.E. MacLaughlin, H.G. Lukefahr, and B. Andraka, Phys. Rev. Lett. **75**, 2023 (1995).
[69] R. Hlubina and T.M. Rice, Phys. Rev. B **51**, 9253 (1995).
[70] S.R. Julian, C. Pfleiderer, F.M. Grosche, N.D. Mathur, G.J. McMullan, A.J. Diver, I.R. Walker, and G.G. Lonzarich, J. Phys. Condens. Matter **8**, 9675 (1996).
[71] T. Senthil, S. Sachdev, and M. Vojta, Phys. Rev. Lett. **90**, 216403 (2003).
[72] T. Senthil, M. Vojta, and S. Sachdev, Phys. Rev. B **69**, 035111 (2004).

[73] Q. Si, S. Rabello, K. Ingersent, and J.L. Smith, Nature **413**, 804 (2001).
[74] Q. Si, S. Rabello, K. Ingersent, and J.L. Smith, Phys. Rev. B **68**, 115103 (2003).
[75] D.R. Grempel, Q. Si, Phys. Rev. Lett. **91**, 026401 (2003).
[76] Z-X Zhu, D.R. Grempel, and Q. Si, Phys. Rev. Lett. **91**, 156404 (2003).
[77] Q. Si et al., J. Phys. Condens. Matter. **17**, R1025 (2005).
[78] P. Sun and G. Kotliar, Phys. Rev. Lett. **91**, 037209 (2003).
[79] P. Sun and G. Kotliar, Phys. Rev. B **71**, 245104 (2005).
[80] Q. Si, Physica B **378–380**, 23 (2006).
[81] N.E. Bickers, Rev. Mod. Phys. **59**, 845 (1987).
[82] P. Coleman and C. Pepin, Phys. Rev. B **68**, 220405 (2003).
[83] D.C. Mattis, Phys. Rev. Lett. **19**, 1478 (1967).
[84] D.M. Cragg and P. Lloyd, J. Phys. C **12**, L215 (1979).
[85] V.A. Fateev and O.B. Wiegmann, Phys. Lett. **81A**, 179 (1981).
[86] M. Tsvelik and P.B. Wiegmann, Adv. Phys. **32**, 453 (1983).
[87] V.T. Rajan, J.H. Lowenstein, and N. Andrei, Phys. Rev. Lett. **49**, 497 (1982).
[88] O. Parcollet and A. Georges, Phys. Rev. Lett. **79**, 4665 (1997).
[89] M. Grosche et al., J. Phys. Condens. Matter **12**, 533 (2000).
[90] A. Schroeder et al., Nature (London) **407**, 351 (2000).
[91] J. Custers, P. Gegenwart, H. Wilhelm, K. Neumaier, Y. Tokiwa, O. Tovarelli, C. Geibel, F. Steglich, C. Pepin, and P. Coleman, Nature (London) **424**, 524 (2003).
[92] C. Pepin, Phys. Rev. Lett. **94**, 066402 (2005).
[93] G.R. Stewart, Rev. Mod. Phys. **78**, 743 (2006).

Chapter 5

Metamagnetism in Heavy Fermions (Experimental Review)

5.1. Introduction

The name metamagnetism was originally introduced for antiferromagnetic (AF) materials where at low temperatures, for a critical value of the magnetic field (H), the spin flips [1] which gives rise to a first-order phase transition. This was extended to paramagnetic (Pa) systems where field reentrant ferromagnetism (F) would appear in itinerant magnetism [2, 3]. Eventually, it was used to describe a crossover inside a persistent paramagnetic state between low-field Pa phase and an enhanced paramagnetic polarized (PP) phase.

It is always possible to reproduce in itinerant systems a strong non-linearity of M in H by an effective density of states $\rho(E)$ picture with a pseudogap near the Fermi-level E_F. For a given value of H_M, the magnetic field will drive $\rho(E_F)$ to a maximum by polarizing the system. In the symmetric case, the relative shift of field shift of spin-up and spin-down density of states will reach the optimal decoupling condition.

The interest in metamagnetism grew considerably after experimental discovery of itinerant metamagnetism in the tetragonal heavy fermion paramagnetic compound $CeRu_2Si_2$. In heavy fermion systems, the f electrons are located near the border between itinerant and local-moment behavior as shown in Doniach's phase diagram [4] (Figure 1). In his Kondo necklace model, Doniach had considered a one-dimensional analog of a system of conduction electrons exchange coupled to a localized spin in each cell of a lattice. He had suggested that a second-order transition from an AF to a Kondo spin-compensated ground state would occur as the exchange coupling constant J would be increased to a critical value J_c. For J near to, and slightly smaller than J_c, there would exist antiferromagnets with very weak, "nearly quenched" moments, even though the f electrons are in a state with a well-defined local non-zero spin state. The existence of this transition can be understood by comparing the binding energy of the Kondo singlet

$$W_K \sim N(0)^{-1} e^{-1/N(0)J}, \tag{5.1}$$

with that of a Rudderman–Kittel–Kasuya–Yosida (RKKY) AF state

$$W_{AF} \sim CJ^2 N(0). \tag{5.2}$$

Here, $N(0)$ is the density of conduction electron state and C is a dimensionless constant which depends on the band structure. As shown in Figure 5.1, for $JN(0)$ less than a

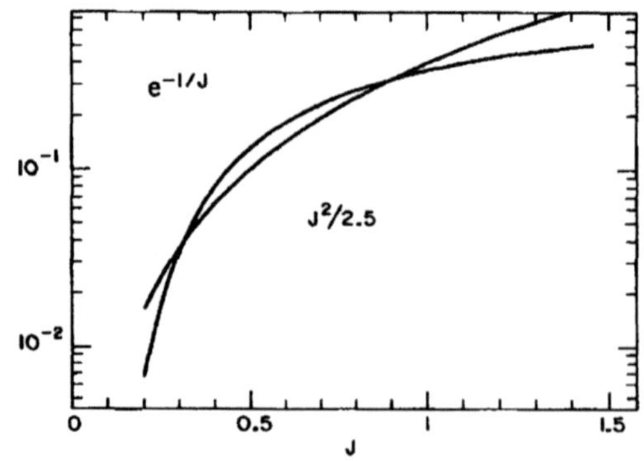

Figure 5.1. Comparison of AF with Kondo energies. Reproduced with the permission of Elsevier from Ref. [4].

critical value, the RKKY state dominates, while above this, the Kondo singlet binding dominates [4]. In Figure 5.1, the RKKY binding again takes over at large J, but the weak-coupling formula (5.1) breaks down in this regime.

While many compounds clearly fall on either side of this divide, $CeRu_2Si_2$ takes on a special role in this context. It is believed that its metamagnetic transition (MMT) is accompanied by an abrupt localization of the f electron on the high-field side of the transition, a notion that is largely based on the interpretation of the de Haas–van Alphen (dHvA) data [5]. We will discuss this in more detail later.

5.2. $CeRu_2Si_2$

$CeRu_2Si_2$ is characterized by a rather large value of the coefficient γ of its low-temperature specific heat ($\gamma = 350$ mJ/mol K^2) [6] and by a strong non-linearity of its magnetization (M) with the magnetic field H [7, 8]. A sharp magnetic inflection appears in $M(H)$ at low temperatures for the field applied along the c axis of the tetragonal structure. At $T \to 0$, this MMT occurs for a field value $H_M = 7.7$ T. This transition has generated a lot of interest since the change in magnetization is directly related to the variation of the effective mass m^* and compressibility κ. In addition, on both sides of H_M, a Pauli paramagnetic ground state seems to exist.

In the axial lattice of $CeRu_2Si_2$, inside a single-impurity Kondo frame, the bare Ce ion has an anisotropic doublet $S = 1/2$ crystal-field ground state mainly formed by the pure $|\pm 5/2\rangle$ component of its $J = 5/2$ angular momentum. The Kondo temperature T_K is near 20 K which is less than the crystal-field splitting $\Delta \sim 200$ K but comparable to the intersite exchange coupling [8]. For $T < T_K$, the single-site Kondo approach has to be replaced by a Kondo lattice since the magnetic and electronic coherence would govern all behaviors.

When the lattice is polarized with a magnetic field (H), a strong non-linear effect appears in the magnetization, $M(H)$. As shown in Figure 5.2, a marked inflection point occurs for $H_M \sim 7.8$ T. The location of H_M is well defined [8, 9]. There is an enhancement of the average effective mass $\langle m^* \rangle$ as derived from specific heat measurements.

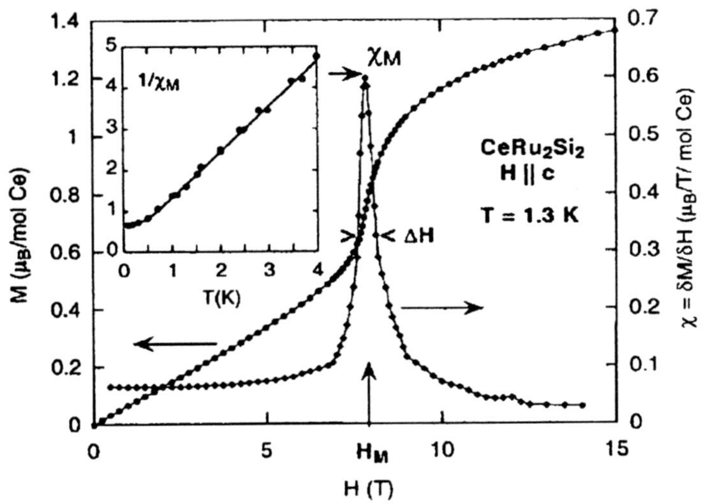

Figure 5.2. Low-temperature magnetization $M(H)$ of $CeRu_2Si_2$. The inset shows the temperature dependence of the differential susceptibility χ_M. Reproduced with the permission of Elsevier from Ref. [10].

It is also associated with huge magnetostriction and softening of the lattice. The differential susceptibility $\chi_M = (\partial M/\partial H)$ at H_M on cooling has a Curie law dependence down to 1 K and its flattening occurs only below 300 mK. The inset of Figure 5.2 shows the temperature dependence of χ_M.

Figure 5.3 represents the field variation of the coefficient $\gamma \sim \langle m^* \rangle$ of the linear γT term of the specific heat (extrapolated at $T = 0$ K) and of the field derivative of the magnetorestriction $V(H)$. Paulson et al. [11] have made an attempt to scale in the vicinity of H_M the field dependence of γ by a $\sqrt{H - H_M}$ law with the idea that the magnetic field plays the role of a control parameter δ as concentration x or pressure to tune the system through AF quantum critical point (QCP) [12]. However, deviations appear for $\delta = (|H - H_M|/H_M)$ lower than 0.3.

It may be noted that $CeRu_2Si_2$ is close to transition from a Pauli paramagnetic to a long-range ordered magnetic ground state. This is clearly demonstrated by the occurrence of antiferromagnetism in lanthanum-doped alloys, $Ce_{1-x}La_xRu_2Si_2$, of concentration of x higher than a critical value x_c close to 0.07. Figure 5.4 shows the moment amplitude m_0 and the ordering temperature of the series $Ce_{1-x}La_xRu_2Si_2$ [13–15].

The (H,T) phase diagram [1] is shown in Figure 5.5. H_c and H_a are MMT lines. H_a is a crossover field between two magnetic structures, while H_c corresponds to the destruction of long-range AF order. H_M is the metamagnetic-like crossover field observed in the paramagnetic state and subsists even below T_N in ordered alloys. The H_M and H_c lines seem to join each other at $T = 0$ (for $x = 0.1$).

The proximity of $CeRu_2Si_2$ to a magnetic instability leads to a high value of the electronic Gruneisen parameter, $\Gamma = 180$. Since $\Gamma = -\partial \log T^*/\partial \log V$, where T^* is the characteristic temperature of the heavy fermion ground state, any small volume change will be magnified by a factor equal to the large value of Γ [16].

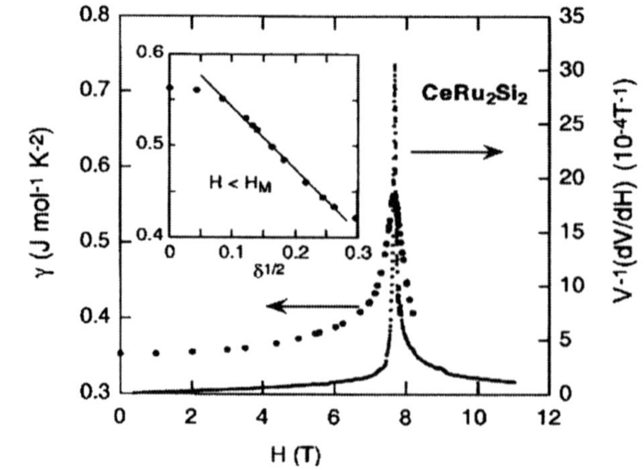

Figure 5.3. Field variation of $\gamma = (C/T)_{T\to 0}$ and of the H derivative of the magnetorestriction in $CeRu_2Si_2$. The inset shows γ as a function of the tuning parameter δ. Reprinted with the permission of Elsevier from Ref. [10].

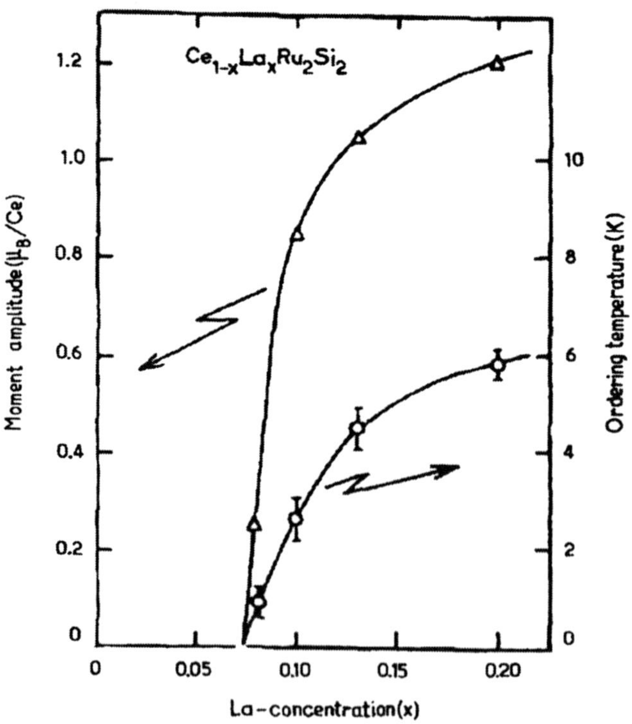

Figure 5.4. Variation of the ordered moment amplitude and of the Neel temperature in the series $Ce_{1-x}La_xRu_2Si_2$. Reprinted with the permission of Elsevier from Ref. [9].

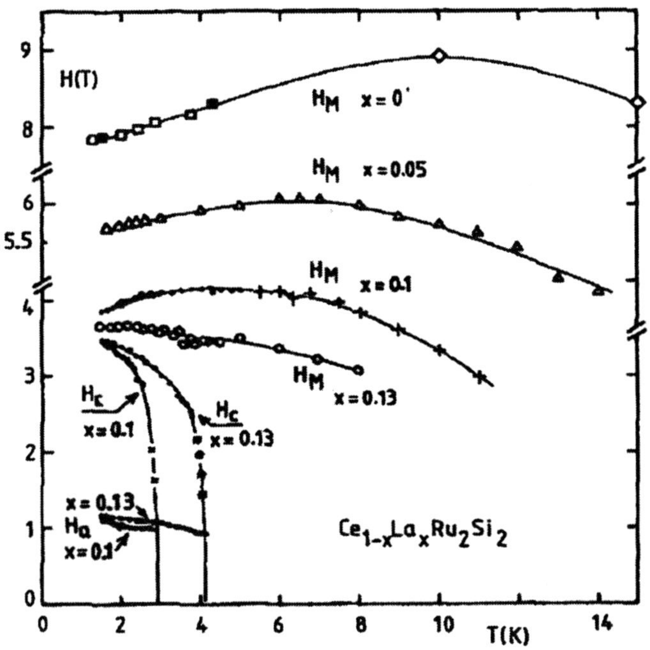

Figure 5.5. (H,T) phase diagram of $Ce_{1-x}La_xRu_2Si_2$. Reprinted with the permission of Elsevier from Ref. [9].

The thermal expansion (α) experiments at constant magnetic field [11] show that three different domains occur in the (H,T) phase diagram shown in Figure 5.6 which is drawn by the contour line $T_\alpha(H)$ where α reaches its optimum. The dashed areas visualize the low-temperature regime with the weak (Pa) and the highly polarized phase (PP) below and above H_M. These features have been confirmed by magnetization, specific heat, and transport measurements [17–21].

The viewpoint that the MMT in $CeRu_2Si_2$ is accompanied by an abrupt localization of the f electron on the high-field side was based on the dHvA data [5]. The problems with this viewpoint were recently discussed by Daou et al. [22]. (a) The MMT seems to be a crossover rather than a real transition [10, 17], while Luttinger's theorem applied to f localization requires a discrete change in Fermi surface volume. (b) Both the magnetization (0.7 μ_B per Ce [8] and the specific heat 500 mJ/mol K^2 [23]) are significantly different from the local-moment [24] analog $CeRu_2Ge_2$ (2.15 μ_B and 20 mJ/mol K^2 [25]) on the high-field side of the MMT. These large discrepancies persist up to the highest fields in which $CeRu_2Si_2$ has been studied. (c) The 4f localization scenario by analyzing the dHvA data accounts for observation of only 20% of the observed specific heat [26] thereby implying that there is "missing mass". (d) The concept of a "large" Fermi surface, which includes the f electrons, and a "small" Fermi surface which does not, is blurred for spin-polarized systems.

The experimental results of magnetoresistance (MR) and Hall effect for $CeRu_2Si_2$, obtained by Daou et al. [22], are shown in Figure 5.7.

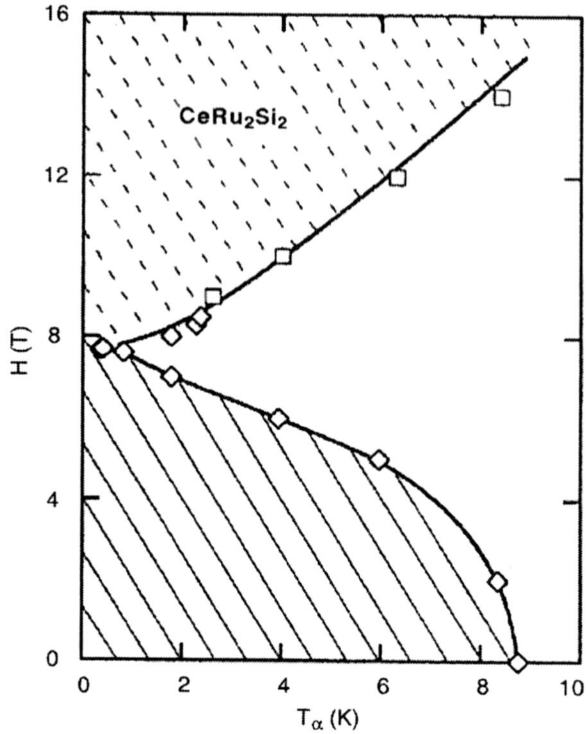

Figure 5.6. The crossover phase pseudodiagram $T_\chi(H)$ derived from the thermal expansion measurements [◇]. The high-field data [□] are the temperature of the C/T maxima observed in Ref. [11]. Reprinted with the permission of Elsevier from Ref. [10].

In Figure 5.7, one can see that the peak in ρ_{xx} at the MMT, associated with magnetic scattering, initially becomes sharp but turns into a kink at the lowest temperature. The peak in ρ_{xy} disappears completely and turns into a downward kink at the lowest temperatures. As shown in the insets of Figure 5.7, there is no discontinuity or hysteresis in either the magnetoresistivity or the Hall resistivity as the MMT is crossed. If there would have been 4f localization transition at the MMT, there would have been a discontinuity as well as huge changes in MR.

We shall now discuss the experimental results for some more heavy fermion systems, a few of which have similar MMT as $CeRu_2Si_2$ but different properties beyond MMT.

5.3. $Sr_3Ru_2O_7$

The Ce spin has a strong Ising character, and the Fermi surface has no low-dimensional character in $CeRu_2Si_2$. However, $Sr_3Ru_2O_7$ is the $n = 2$ member of the Ruddlesden–Popper series of layered perovskites $Sr_{n+1}Ru_nO_{3n+1}$ whose properties are dominated by layers of

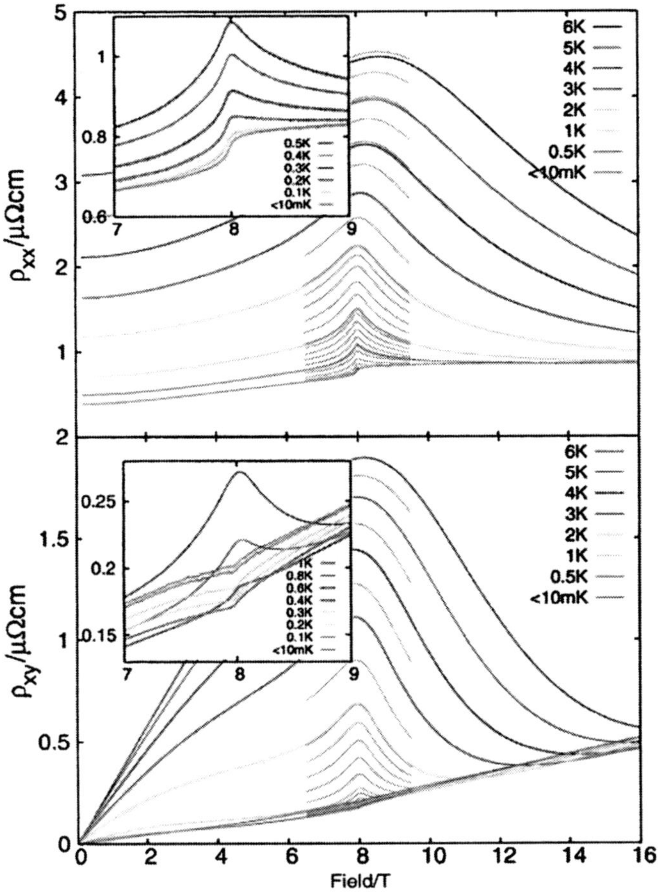

Figure 5.7. In-plane ($j \| ab$), $B \| c$) transverse magnetoresistivity (ρ_{xx}, upper panel) and Hall effect (ρ_{xy}, lower panel) for $CeRu_2Si_2$. The base T is around 6.5 mK. Reprinted with the permission of the American Physical Society from Ref. [22].

octahedral ruthenium oxide . The four electrons in the ruthenium 4d t_{2g} orbitals hybridize with the oxygen 2p to create metallic bands that are narrow but still sufficiently broad to avoid insulating transitions. Thus, $Sr_3Ru_2O_7$ has a Ru–O bilayer per formula unit and the Ru effective spin has mainly a Heisenberg character. The (H,T) "crossover" phase diagram is rather similar to that derived for $CeRu_2Si_2$ [27, 28].

The magnetization results of Perry et al. [27] for $Sr_3Ru_2O_7$ are summarized in Figure 5.8. There is a rapid superlinear rise in magnetization in a field of 5.5 T for magnetic fields applied in the *ab*-plane. The metamagnetism is broadened as the temperature is raised. The inset of Figure 5.8 shows the data for magnetic fields along the *c* axis and there is some anisotropy.

Perry et al. [27] have studied the effect of metamagnetism on the MR at a series of temperatures below 10 K. Their results are shown in Figure 5.9. For $B \| c$, the very broad

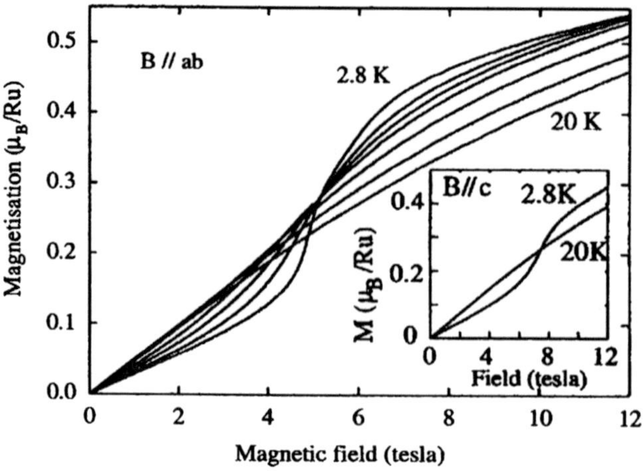

Figure 5.8. The magnetization of single-crystal $Sr_3Ru_2O_7$ for magnetic fields applied in the *ab*-plane. Metamagnetism is seen for all temperatures below 10 K, centered on a field of approximately 5.5 T. Inset: data for magnetic fields along the *c* axis where the metamagnetic field is around 7.7 T. Reproduced with the permission of the American Physical Society from Ref. [27].

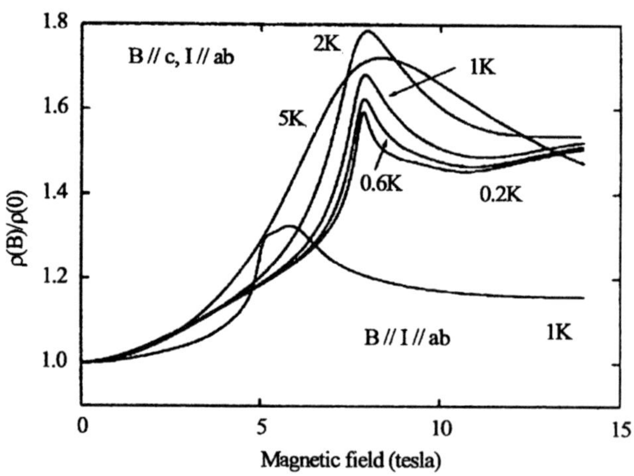

Figure 5.9. The magnetoresistance (MR) of single-crystal $Sr_3Ru_2O_7$ at a series of low temperatures. Reproduced with the permission of the American Physical Society from Ref. [27].

peak at 5 K sharpens considerably at lower temperatures. For in-plane fields, the low-temperature MR gives evidence of for some peak splitting, as shown by data at 1 K for this orientation. The transport measurements show that the MMT field is essentially independent for any direction of the applied field.

In metallic systems such as $Sr_3Ru_2O_7$, the change in magnetization is due to a rapid change from a paramagnetic state at low fields to a more highly polarized state at high

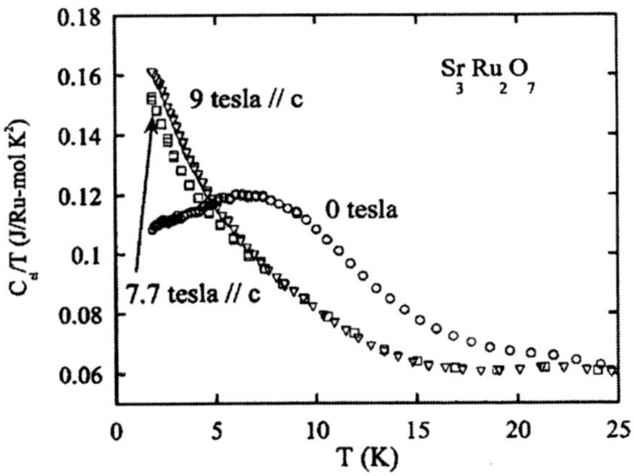

Figure 5.10. The electronic specific heat divided by temperature for $Sr_3Ru_2O_7$ at low temperatures for magnetic fields applied parallel to *c*. Divergent behavior occurs when the field is tuned to the metamagnetic field, identified from Figure 5.9 to be 7.7 T. Reproduced with the permission of the American Physical Society from Ref. [27].

fields via a crossover or a phase transition. However, some of the basic physics is same in both cases. The observation of itinerant metamagnetism demonstrates the existence of strong ferromagnetic coupling in the system. The low-temperature critical points are likely in both scenarios. A crossover is usually linked to close proximity in phase space by a QCP which can be reached by the application of another control parameter. The transition would be expected to be of first order if the metamagnetism is due to a $T = 0$ phase transition along the magnetic field axis. This first-order transition line might terminate in a critical point at a very low temperature. Thus, it is important to know as to whether there is any evidence of critical fluctuations associated with the metamagnetism of $Sr_3Ru_2O_7$.

In Figure 5.10, Perry et al. [27] have plotted electronic specific heat divided by temperature (C_{el}/T) of $Sr_3Ru_2O_7$ as a function of temperature for magnetic fields applied parallel to *c*. In zero field, C_{el}/T rises as the temperature falls below 20 K, but it crosses over at lower temperatures. The application of fields parallel to the *c* axis suppresses this crossover and C_{el}/T rises steeply at lower temperatures.

The above data provide sufficient evidence for close association between metamagnetism and critical fluctuations in $Sr_3Ru_2O_7$. The moment is generated by a spontaneous polarization of the spin-up and spin-down Fermi surfaces in itinerant ferromagnets. In an itinerant metamagnet, the extra moment appears at the metamagnetic field due to the same basic mechanism. So the observed fluctuations can be thought of as fluctuations of the Fermi surface itself. Although itinerant electron metamagnetism has been observed in several systems such as MnSi [29] and $CeRu_2Si_2$ [16], the experimental results of $Sr_3Ru_2O_7$ provide clear evidence of magnetic field driving a system closer to criticality.

In recent experiments, Perry et al. [30] have used ultra clean single crystals of $Sr_2Ru_2O_7$. They have observed multiple first-order MMTs as well as unusual behavior near the metamagnetic field. There is a strong enhancement of the resistivity in a narrow field window, with a minimum in the resistivity as a function of temperature below 1 K that becomes more pronounced as the disorder level decreases. The region of anomalous

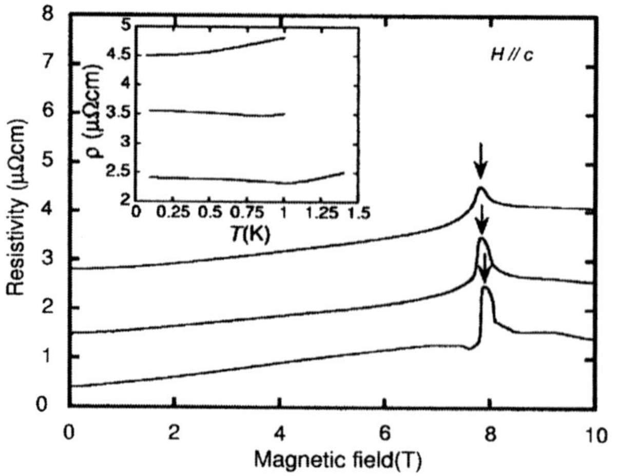

Figure 5.11. The resistivity (ρ) versus magnetic field in single crystals of $Sr_3Ru_2O_7$ for different values of ρ_{res}. Inset: temperature sweeps at the fields indicated by arrows in each case. Reproduced with the permission of the American Physical Society from Ref. [30].

behavior is bounded at low temperatures by two first-order phase transitions. In Figure 5.11, the resistivity (ρ) versus magnetic field of $Sr_3Ru_2O_7$ for three samples (of increasing purity) is shown. The temperature of each curve is between 0.06 and 0.1 K and the magnetic field is perpendicular to the RuO_2 planes. As the sample purity increases, the same basic peak remains as the dominant feature, but more structure develops. One observes a pronounced dip at 7.5 T, and the central peak is seen to have very steep sides at 7.8 and 8.1 T.

The inset of Figure 5.11 shows the results of sweeping the temperature at fixed fields. The top trace shows the unusual T^{-3} dependence previously reported for $\rho_{res} = 2.8\ \Omega$ cm [31]. The temperature dependence of ρ at the MMT changes to a negative gradient at low temperatures, with a minimum that increases systematically with temperature, as the sample gets purer.

In Figure 5.12, the variation of ρ and the ac magnetic susceptibility, $\chi(\omega)$, of samples with $\rho_{res} < 0.5\ \mu\Omega$ cm are shown. Here χ' and χ'' are the real and imaginary parts of χ. The two steep features in ρ coincide with peaks in both χ' and χ''. The peaks in χ'' indicate the presence of dissipation which can be interpreted in terms of hysteresis at a first-order phase boundary.

The summary of results of Figures 5.11 and 5.12 is that decreasing the levels of disorder leads to a negative gradient in $\rho(T)$ near the metamagnetic field and to the resolution of several pronounced features in ρ instead of the single peak seen in samples with $\rho_{res} = 2.8\ \mu\Omega$ cm. In χ, three peaks are observed at low T, two of which signal the crossing of first-order phase boundaries. The key experimental findings are that the enhancement of ρ in the metamagnetic region grows as the disorder scattering decreases and that the narrow field range over which the residual resistivity is enhanced is bounded at either side by first-order phase transitions.

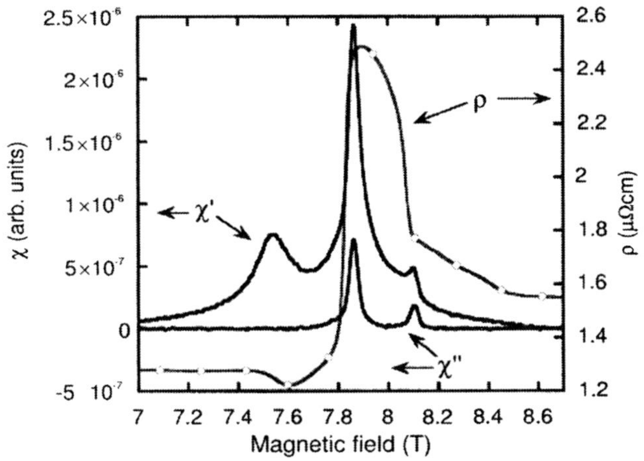

Figure 5.12. The susceptibility (χ) and resistivity (ρ) versus magnetic field across the metamagnetic field. The measurements were made at $T = 20$ mK on different crystals, both having $\rho_{res} < 0.5$ μΩ cm. The ac excitation field was 0.32 G at a frequency of 83.4 Hz. Reproduced with the permission of the American Physical Society from Ref. [30].

In a subsequent paper, Borzi et al. [32] have reported the dHvA effect on the itinerant metamagnet $Sr_3Ru_2O_7$. Using extremely highly pure sample, they have shown that the oscillations can be resolved in the field ranges 5 T $< B <$ 7.5 T and 8.3 T $< B <$ 15 T with sufficient precision to enable studies of the field dependence of the quasiparticle mass and Fermi surface topography. The picture that emerges is one of the metamagnetism being consistent with a Stoner-like transition of the itinerant system, accompanied by a strong enhancement of the quasiparticle mass as the transition is approached from either above or below. The system's properties appear to be governed by the presence of a QCP.

The dHvA effect was observed in four samples, with the field parallel to c but the results shown in Figure 5.13 are from samples showing the largest oscillatory amplitude and hence the best noise-to-signal ratio. The real part of the magnetic susceptibility is shown from 5 to 15 T. On the logarithmic scale, the tiny oscillations are visible below and above the transition. However, they are almost indistinguishable on a linear ordinate scale (inset of Figure 5.13).

The first-order magnetic transitions obtained in Ref. [30] correspond to the main peak and the high-field shoulder in Figure 5.13. Borzi et al. [32] did not find any evidence of first-order behavior associated with the peak at 7.55 T. They have also performed a field-dependent mass analysis in $Sr_3Ru_2O_7$ and resolved a mass change for the two sheets for which the signal could be tracked over an adequate range of temperature and field. Their results are presented in the sample dataset in Figure 5.14.

The pronounced increase in the effective mass of the Landau quasipartcles, as shown in Figure 5.14, can be understood in terms of the fluctuations close to a QCP. It may be noted that the dHvA experiment of Borzi et al. [32] does not resolve the question as to how the quantum oscillations evolve and behave in the region between 7.5 and 8.3 T.

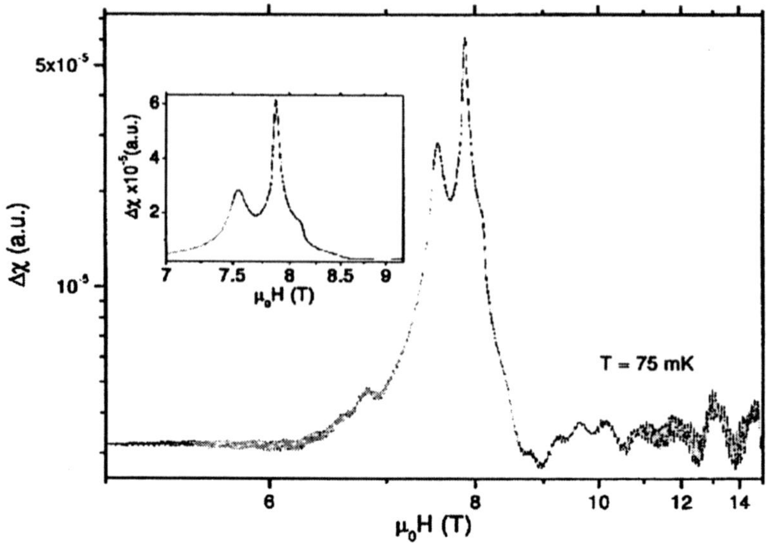

Figure 5.13. The real part of the differential magnetic susceptibility through the MMT in $Sr_3Ru_2O_7$. The ordinates are on a logarithmic scale, while the abscissas are on a reciprocal scale since quantum oscillations are periodic in a reciprocal field. Reproduced with the permission of the American Physical Society from Ref. [32].

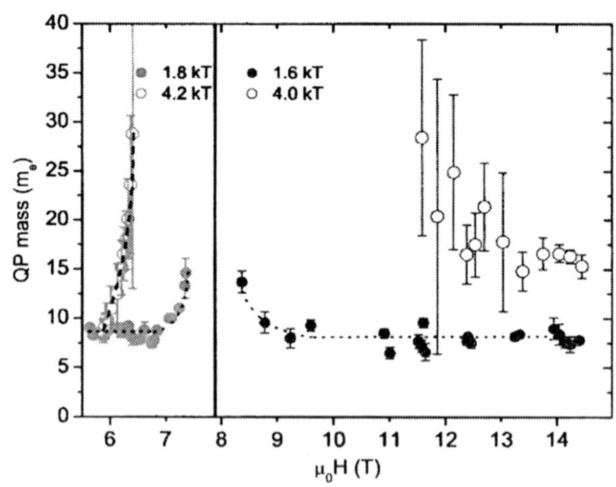

Figure 5.14. Effective mass as a function of field for the peaks at 1.8(6) and 4.2(0) kT for the low-field (high-field) states. The dotted lines are guide to the eye. Reproduced with the permission of the American Physical Society from Ref. [32].

5.4. $CeCu_{6-x}Au_x$

Pure $CeCu_6$ is a heavy fermion (γ = 1670 mJ/mol K^2), with neither long-range magnetic order nor superconductivity. It has well-behaved but highly renormalized Fermi-liquid properties. It exhibits magnetic correlations as observed by inelastic neutron

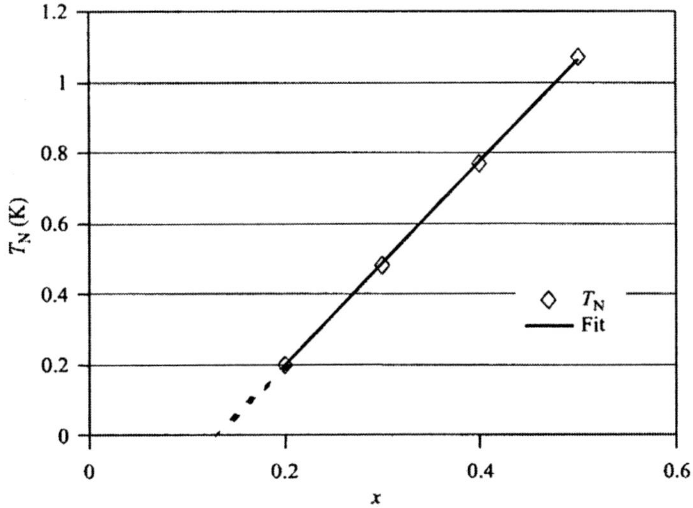

Figure 5.15. Variation of Neel temperature as a function of Au concentration x, from single-crystal diffraction data. A linear fit gives $x_c^{mag} = 0.13 \pm 0.01$. Reproduced with the permission of Elsevier from Ref. [36].

scattering [33, 34]. This proximity to magnetic order of "non-magnetic" $CuCu_6$ leads to an instability of the Kondo ground state with respect to AF order when replacing some Cu by Au, which is one of the most intensely studied systems [35] since it is considered to be classic case of a QCP. In Figure 5.15, the results of variation of Neel temperature as a function of Au concentration x are shown [36].

The intersite magnetic correlations in $CeCu_6$ developing before 1 K are destroyed by a magnetic field of $B \approx 2$ T [37] as evidenced by a distinct maximum in the derivative of the magnetization M, dM/dB versus B ($B \| c$). This maximum vanishes for $T > 0.6$ K. Figure 5.16 shows the detailed temperature dependence of this transition which occurs at $B_M = 1.7$ T. B_M, which corresponds to the field at which correlations are no longer observable in inelastic neutron scattering [34], remains independent of T even though the maximum of dM/dB at B_M becomes weaker and gradually disappears.

It is evident from Figure 5.16 that even at $T = 0.15$ K, the dM/dB maximum is very weak which corresponds to a small intensity of correlations with respect to on-site fluctuations. This distinguishes $CeCu_6$ from other systems such as $CeRu_2Si_2$ and UPt_3.

5.5. UPt_3

In UPt_3, superconductivity coexists with AF ordering [38] which occurs below $T_N = 5$ K. Another interesting phenomenon is an MMT which was observed for the first time in the high-field magnetization experiment at a critical field H_c ($\approx$20 T) for the field perpendicular to the [0001] (c axis) direction [39, 40]. The high-field magnetization results of Sugiyama et al. [41] are shown in Figure 5.17.

The MMT is broad even below $T_N \approx 5$ K and persists up to about 30 K, which is close to a characteristic temperature $T_{\chi max}$ ($\approx$20 K) where the magnetic susceptibility has a

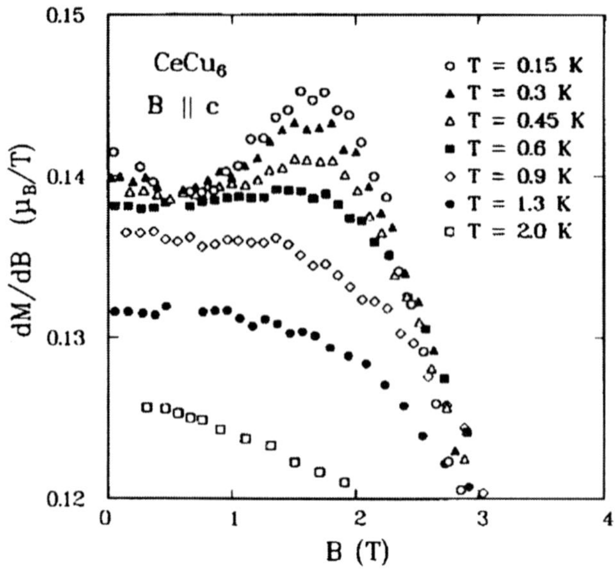

Figure 5.16. Magnetic susceptibility dM/dB of $CeCu_6$ versus applied magnetic field B ($B \| c$) at various temperatures. Reproduced with the permission of Elsevier from Ref. [37].

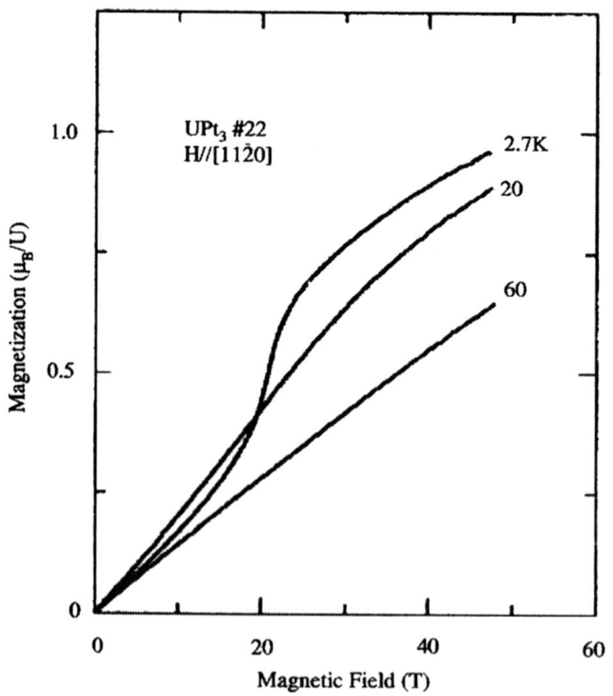

Figure 5.17. Magnetization curves in UPt_3. Reproduced with the permission of the American Physical Society from Ref. [41].

maximum in the temperature dependence [40]. This is in contrast with the URu_2Si_2 [42] and UPd_2Al_3 [43] behavior, where the MMT is sharp below T_N and becomes broad above T_N as in UPt_3. Thus, the ordered state in UPt_3 is different from that of URu_2Si_2 and UPd_2Al_3. In some sense, the MMT in UPt_3 is similar to $CeRu_2Si_2$ and $CeCu_6$ without magnetic ordering.

5.6. UPd_2Al_3

UPd_2Al_3 is a heavy fermion superconductor which is AF with Neel temperature $T_N = 14$ K. The results for magnetization in UPd_2Al_3 at 1.3, 13, and 35 K in the field along [11$\bar{2}$0] direction with the hexagonal structure are shown in Figure 5.18 [44]. The sharp MMT occurs at a critical field of 18 T when observed below T_N but becomes broad above T_N.

The phase diagram of UPd_2Al_3 is shown in Figure 5.19. The dotted line indicates the field dependence of T_N [45]. The MMT is found above T_N, up to a $T_{\chi\,max}$ of 35 K. This MMT presumably occurs due to a change of the 5f electron character, i.e., from an f-itinerant heavy fermion state to an approximately localized one. Hence, the MMT also occurs below $T_{\chi\,max}$ and the high-field state above H_c can be connected to the high-temperature state in the zero magnetic field.

The temperature dependence of the magnetization at 10, 20, 30, 40, and 50 T is shown in the lower part of Figure 5.19 [44]. The magnetization above H_c increases monotonically with decrease in temperature. However, the magnetization below H_c decreases below $T_{\chi\,max}$. A saturated magnetic moment above 30 T below 20 K and a steep increase of the magnetization below 10–20 K at 20 T might indicate a ferromagnetic character [44].

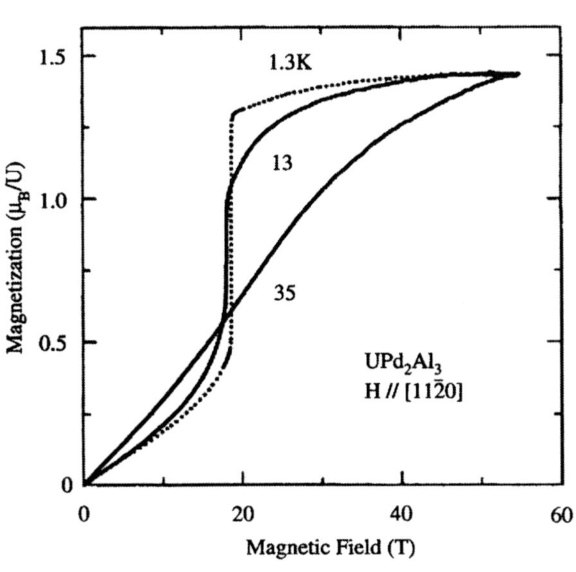

Figure 5.18. Typical magnetization curves of UPd_2Al_3. Reprinted with the permission of Elsevier from Ref. [44].

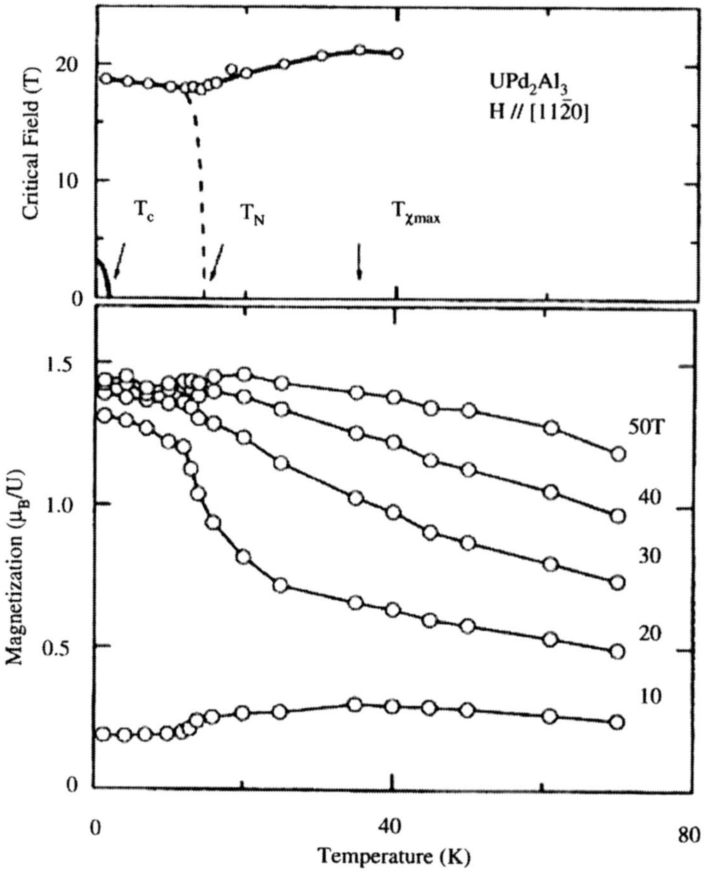

Figure 5.19. Phase diagram (top) and the temperature dependence of the magnetization (bottom) of UPd_2Al_3. Reprinted with the permission of Elsevier from Ref. [44].

5.7. URu_2Si_2

The ground states of URu_2Si_2 are particularly complex. URu_2Si_2 belongs to a class of strongly correlated metals in which f electrons develop a distinctly itinerant character [46]. Coulomb interactions cause the quasiparticle effective masses to be heavily renormalized, making the energetic rewards for forming ordered ground states much higher than normal metals [47, 48]. Upon cooling, prior to becoming superconducting at $T_c \approx 1.2$ K [46], this material undergoes a second-order phase transition at $T_o = 17.5$ K [10], the nature of which remains controversial [49, 50]. Due the absence of a microscopic description or origin, the term "hidden order" (HO) has been coined. The presence of an itinerant f-electron Fermi surface [51, 52] also provides URu_2Si_2 with the essential conditions for itinerant electron magnetism (IEM) [53] since the f electrons can revert to a localized behavior upon their alignment in strong magnetic fields.

In a recent paper, Harrison et al. [54] have shown that the multiple transitions in URu_2Si_2 occur since the magnetic field first destroys the HO phase before creating a new field-induced reentrant phase [55] in the vicinity of the metamagnetic phase transition. Their phase diagram is shown in Figure 5.20. There is a pronounced asymmetry between the occupancy of itinerant spin-up and spin-down f-electron states due to IEM [53] since the population of the spin-up component suddenly sinks below the Fermi energy ε_F at a magnetic field B_M. The magnetization measurements reveal that the magnetic-field-induced phase is accompanied by the opening of a gap in the spin-up f-electron band at B_M. In Figure 5.20, a single metamagnetic crossover at high temperatures splits into two transitions below ~6 K, giving rise to the domed "reentrant" HO phase.

The magnetization M of URu_2Si_2 measured in pulsed magnetic fields of up to 44 T at several temperatures [54] is shown in Figure 5.21(a). The presence of a single broad maximum in χ above $T \approx 6$ K in Figure 5.21(b) indicates the existence of a IEM of a type similar to that observed in $CeRu_2Si_2$ [8, 9], $Sr_3Ru_2O_7$ [27], and UPt_3 [40, 44]. Figure 5.21(c)

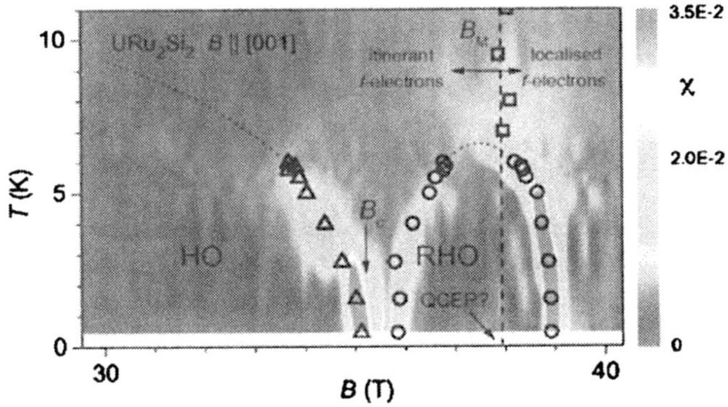

Figure 5.20. Intensity plot of the differential susceptibility in URu_2Si_2. The symbols square, triangle, and circle mark metamagnetic crossover field $B_M \approx 38$ T, and transition into and out of HO and reentrant HO (RHO) phases. Reproduced with the permission of Elsevier from Ref. [49].

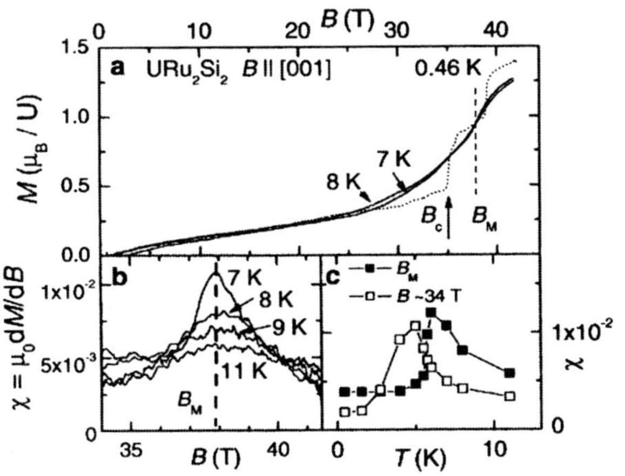

Figure 5.21. (a) M of URu_2Si_2 at different temperatures for B applied along the c axis. (b) χ in the vicinity of B_M at several temperatures above the reentrant ordering temperature. (c) χ at B_M and at $B \approx 34$ T as a function of temperature. Reproduced with the permission of the American Physical Society from Ref. [54].

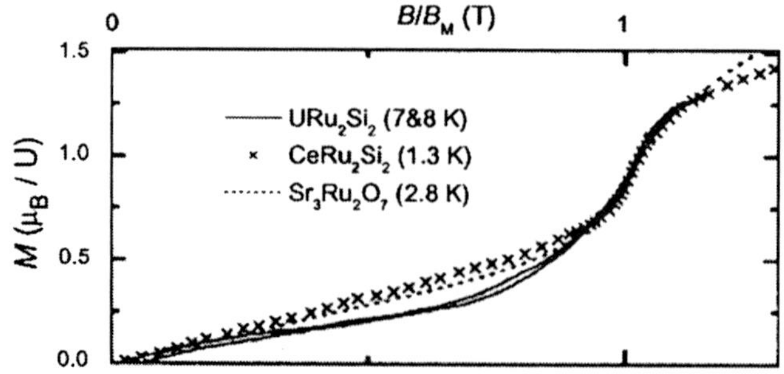

Figure 5.22. A comparison of itinerant electron metamagnetism in URu_2, $CeRu_2Si_2$, and $Sr_3Ru_2O_7$. Reproduced by the permission of Elsevier from Ref. [49].

shows the rapid increase in χ at B_M with decreasing temperature (filled squares). The maximum in χ vanishes below $T \approx 6$ K on entering the field-induced ordered phase recently identified in specific heat measurements [55]. The fact that B_M occurs within the field-induced ordered phase implies that fluctuations associated with the metamagnetic critical end point [28].

In Figure 5.22, a comparison of itinerant electron metamagnetism in URu_2Si_2 [54], $CeRu_2Si_2$ [8, 10, 12], and $Sr_3Ru_2O_7$ [27, 28] has been shown [49]. Here $B_M = \mu_0 H_M$ is the field at which the transition or crossover occurs. The value of the moment has been renormalized only in the case of $Sr_3Ru_2O_7$.

As shown in Figure 5.22, there are good reasons to believe that the underlying heavy fermions state of URu_2Si_2, from which the HO condenses, is more similar to other heavy fermion systems than was hitherto assumed.

In Figure 5.23, Harrison et al. [49] have sketched a generic-phase diagram for heavy fermion compounds, generalized to include the magnetic field.

In Figure 5.23, the symbols FM, PM, and AFM refer to ferromagnetic, paramagnetic, and AF, respectively. x and p refer to doping and/or pressure, respectively, which may be offset with respect to zero, depending on the material. The HO phase, which occurs only in URu_2Si_2, is not shown in the diagram. At ambient pressure, UPt_3 and $CeRu_2Si_2$ are close to exhibiting a QCP at $B = 0$, possibly occupying the gray circle region. Ambient pressure URu_2Si_2, because of its incipient antiferromagnetism, probably occurs at slightly lower x or p where it is overcome by superconductivity.

When a magnetic field is applied to URu_2Si_2, various types of quantum criticality can be produced either by the suppression of the parent AF phase, the HO phase or by the induction of itinerant electron metamagnetism. The existence of many competing interactions in URu_2Si_2 leads to a unique complex phase diagram as a function of the magnetic field as shown in Figure 5.24(a). This phase diagram was compiled from resistivity ρ measurements made as a function of temperature and field [56].

The phase transitions which exist in the magnetization [54], specific heat [55], or ultrasound velocity [57] lead to extrema in $d\rho/dT$ and $d\rho/dH$. This is apparent in Figure 5.24(b) upon eliminating the regions in H versus T space occupied by ordered spaces. At magnetic fields higher than 39 T, only below a characteristic temperature T^* (open circles), $\rho = \rho_0 + aT^n$ where $n \approx 2$, yielding a Fermi-liquid behavior. When $\mu_0 H_0 \sim 37$ T, the Fermi-liquid parameter a appears to diverge since a fit of $a^{-1/2}$ intercepts the

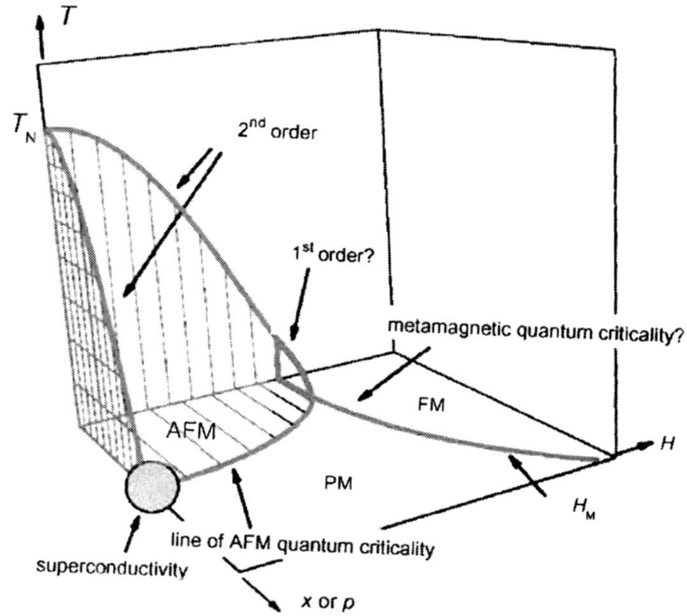

Figure 5.23. Generic-phase diagram of heavy fermion systems, generalized to include the magnetic field. Reproduced with the permission of Elsevier from Ref. [49].

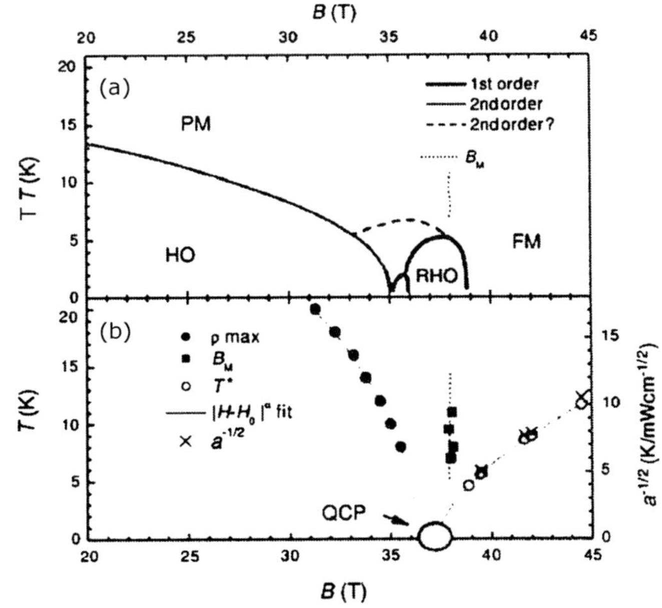

Figure 5.24. (a) The phase diagram of URu_2Si_2 as a function of magnetic field and temperature extracted from Ref. [56] in which the actual data points are omitted for clarity. (b) An assemblage of physical properties of URu_2Si_2 outside the ordered phases, together with the power law fits. All data points extrapolate to a metamagnetic quantum critical point (QCP) at 37 ± 1 T. Reprinted with the permission of Elsevier from Ref. [49].

$T = 0$ axis in a $|H - H_0|^\alpha$ fashion where $\alpha \sim 0.7$. At $T > T^*$, the resistivity crosses over to a non-Fermi-liquid behavior with $n \leq 1$. At $T > 7$ K and fields lower than 37 T, the MR has a broad maximum. Its T versus H loci (filled circles) intercepts the field axis at $\mu_0 H_0 \sim 37$ T with $\alpha \sim 0.5$.

In a recent paper, Kim et al. [58] have presented a brief summary of electrical transport, magnetization, and specific heat measurements on $U(Ru_{1-x}Rh_x)_2Si_2$ ($x = 0, 2, 2.5, 3$, and 4%) in magnetic fields up to 45 T. The results indicate the formation of multiple phases near their putative metamagnetic QCPs at low temperatures. They have reported a clear correlation between the field dependence of the phases and that of the QCP. Their results are summarized in Figure 5.25.

In Figure 5.25, the ordered phases are shaded for clarity. PFL and HFL refer to a polarized Fermi liquid and a heavy Fermi liquid state. The crossover temperature T^* delineate the low-temperature T^2 resistivity behavior to high-temperature sub-linear behavior. HO is the hidden order (which is destroyed at $x = 4\%$), ρ_{max} is the maxima in the magnetoresistivity, and B_M is the metamagnetic field. For $x = 0\%$, while the dominant phase is destroyed around 35.2 T, a new phase III was found, which was interpreted as a possible reentrant phase [54]. In addition, phases II and V were identified as weak features [56]. When x is increased, phase II expanded while the HO phase as well as phases III and V are suppressed. At $x = 4\%$, the HO phase vanishes and only phase II (shaded black) is stabilized between 26 and 38 T. The details of the methodology using which these results were obtained are explained in Ref. [58].

5.8. $CePd_2Si_2$

Sheikin et al. [59] have done magnetic torque measurements to study the dHvA effects in the heavy fermion system $CePd_2Si_2$. They observed a distinct anomaly in the field dependence of the torque at about 10 T for $B \| a$ manifests a transition between two magnetic states. The transition is also seen in MR measurements, where a sudden drop of the resistivity is observed at the same field, $B_M \sim 10$ T at 100 mK. The transition moves to a slightly lower field at $T = 4.2$ K, which indicates that B_M, the field at which the MMT occurs, has a weak temperature dependence at low temperatures. Based on the above data and those of preliminary study of the AC specific heat, they have sketched a magnetic phase diagram of this compound which is shown in Figure 5.26. The phase diagram contains both the AF and MMT lines. The AF transition temperature T_N decreases with magnetic field, in agreement with previously reported results [60]. This suggests that T_N would eventually be suppressed to zero at higher field as indicated in the gray line in Figure 5.26. An MMT between two AF phases (denoted as AF1 and AF2 in Figure 5.26) is found at $B_M \sim 10$ T at low temperature which shifts to lower field when the temperature increases.

B_M is found to have a weak angular dependence when the magnetic field is tilted in the basal plane. In addition, the field dependence of T_N does not match that of Ref. [60]. Above the MMT, in the intermediate AF state, six different dHvA frequencies were detected when the magnetic field was applied in the basal plane. The corresponding effective masses range from 6 to 23 m_e. This was compared with the band structure calculations for the paramagnetic state. Sheikin et al. [59] have taken into account the

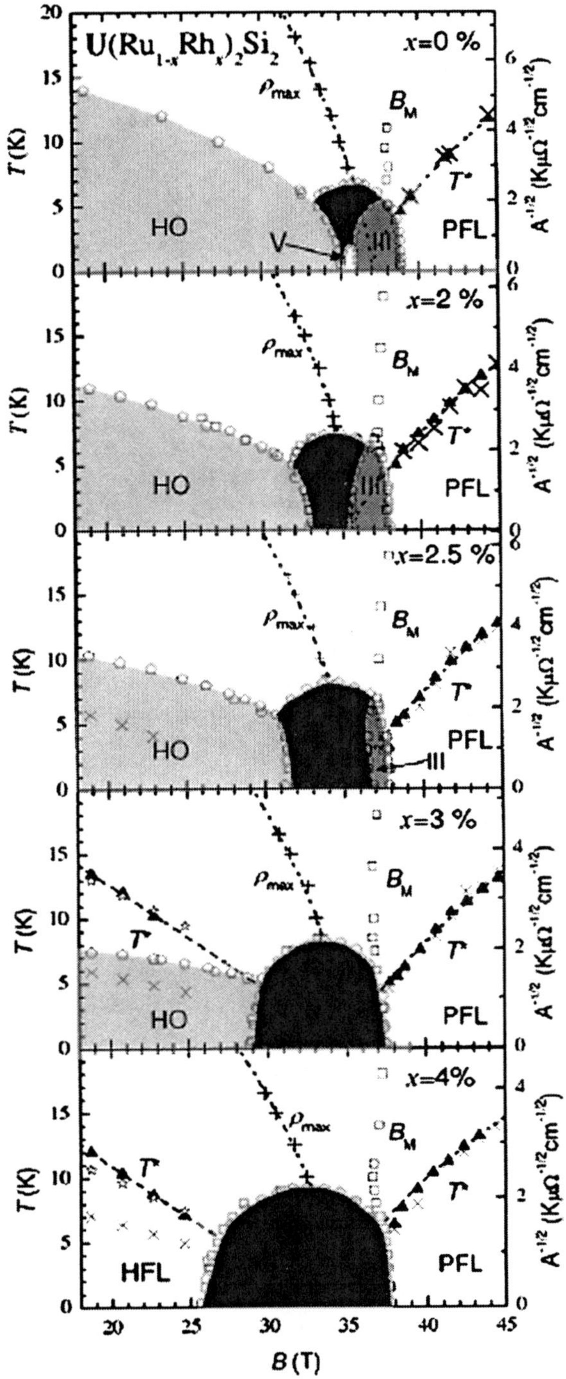

Figure 5.25. High-magnetic-field region of the phase diagram of $U(Ru_{1-x}Rh_x)_2Si_2$ for $x = 0, 2, 2.5, 3$, and 4% in which the ordered phases are shaded for clarity. Reproduced with the permission of Elsevier from Ref. [58].

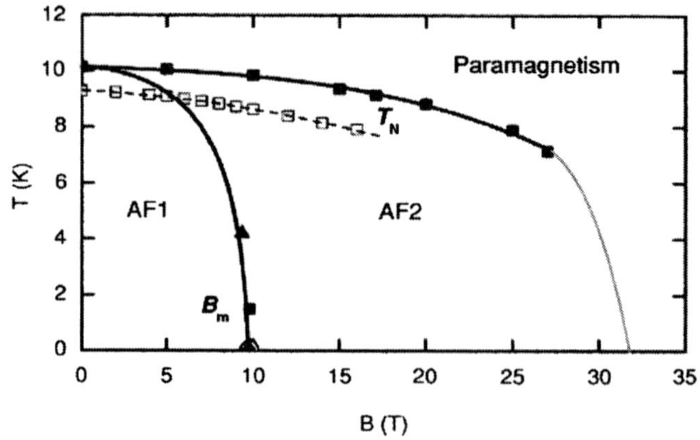

Figure 5.26. Magnetic phase diagram of $CePd_2Si_2$ is sketched for $B \| a$ based on the magnetic torque (open circle), MR (filled triangles), and AC specific heat (filled squares) measurements. The AF transition temperature T_N as a function of magnetic field from the previous study (Ref. [60] is also shown as open squares). Reproduced with the permission of the American Physical Society from Ref. [59].

modification of the Brillouin zone in the AF phase and the corresponding changes in the Fermi surfaces, and concluded that the f electrons seem to remain itinerant even above B_M, which is very unusual for Ce-based heavy compounds.

5.9. $YbRh_2Si_2$

$YbRh_2Si_2$ has a Kondo temperature of about 25 K located very close to a QCP related to a very weak AF order below $T_N = 70$ mK [61]. A tiny critical magnetic field of $B_c = 0.06$ T, applied in the easy magnetic plane perpendicular to the tetragonal c axis, is sufficient to suppress the AF order [62]. Tokiwa et al. [63] have performed dc-magnetization measurements at temperatures down to 40 mK and fields up to 11.5 T at both ambient pressure as well as hydrostatic pressure up to 1.3 GPa. At ambient pressure a kink at $B^* = 9.9$ T indicates a new type of field-induced transition from itinerant to a localized 4f state. This transition is different from the MMT observed in other heavy fermion compounds. Here, ferromagnetic, rather than AF, correlations dominate below B^*.

In Figure 5.27, the results of hydrostatic pressure experiments reveal a clear correspondence of B^* to the characteristic spin fluctuation temperature T_0, estimated by fitting the zero-field low-temperature specific heat with $C(T)/T = -D\ln(T/T_0)$. Since at ambient pressure T_0 matches with the single-ion Kondo temperature T_K [61], the pressure dependence of T_0 is assumed to represent that of the Kondo temperature T_K. Tokiwa et al. [63] have used the specific-heat data under hydrostatic pressure reported by Mederle et al. [64]. As shown in Figure 5.27, there is a correlation between T_0 and B^*. The exponential decrease with increasing pressure is compatible with the Kondo temperature $T_K \propto \exp[-1/J_{cf}D_c(\varepsilon_F)]$, where J_{cf} is the conduction electron exchange integral and $D_c(\varepsilon_F)$ is the conduction electron density of states at the Fermi energy.

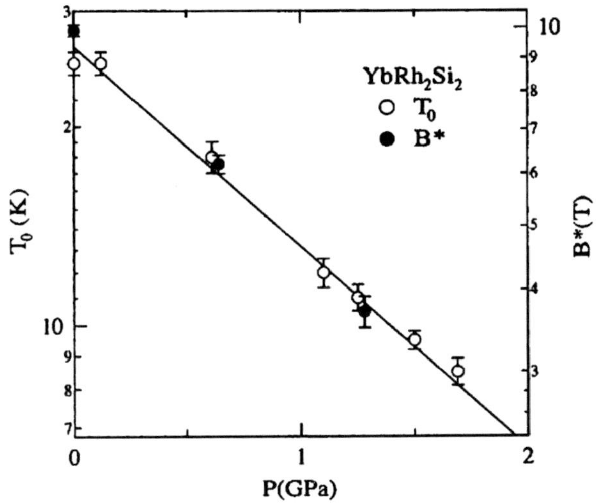

Figure 5.27. Pressure dependence of the characteristic spin fluctuation temperature T_0 (left axis) and B^* (right axis) for $YbRh_2Si_2$. Solid line represents $\exp(-0.7\ GPa^{-1} \times P)$ dependence. Reproduced with the permission of the American Physical Society from Ref. [63].

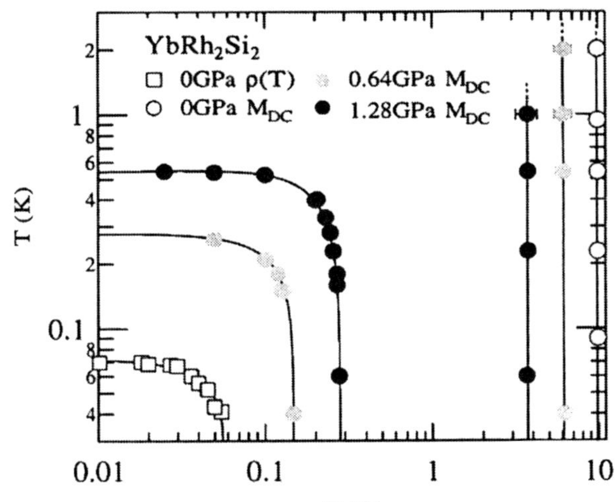

Figure 5.28. Temperature–magnetic field phase diagram of $YbRh_2Si_2$ for $B \perp c$ as log T versus log B. White, gray, and black symbols indicate points at 0, 0.64, and 1.28 GPa, respectively. Circles and squares are data points from magnetization and resistivity measurements. Lines are guides to the eye. Reproduced with the permission of the American Physical Society from Ref. [63].

In Figure 5.28, the results of Tokiwa et al. [63] has been summarized in the temperature–magnetic field phase diagram. The phase boundary of the AF state was determined from the kinks in both constant-temperature and constant-field scan, respectively.

In all $M(B)$ measurements under pressure, the field B^* at which the kink occurs is independent of temperature. The anomaly broadens rapidly with increasing temperature, and the kink disappears around 2 K at ambient pressure and 1 K at the highest pressure. Thus, the transition occurs only in the coherent regime at $T \ll T_K(P)$.

Because of the very weak RKKY intersite interaction in $YbRh_2Si_2$, evidenced by the very low-ordering temperature T_N, the AF correlations cannot persist at fields needed to destroy the Kondo interaction in this system [63]. Thus, in contrast to the metamagnetic-like behavior in the other HF systems resulting from the crossover from an AF correlated itinerant to a ferromagnetically polarized localized 4f moment state, the B^* anomaly in $YbRh_2Si_2$ may result from an itinerant to localized transition with FM polarization at both sides [63].

5.10. $CeIr_3Si_2$

Muro et al. [65] have recently measured the electrical resistivity, specific heat, and magnetization of $CeIr_3Si_2$. Their results indicate that $CeIr_3Si_2$ is a new Kondo-lattice compound displaying successive magnetic transitions at $T_{N1} = 4.1$ K and $T_{N2} = 3.3$ K. The transition at T_{N2} is ferrimagnetic because of a sudden increase of $M(T)/B$ and a hysteresis in $M(B)$ at 1.9 K. This is plotted in Figure 5.29.

The isothermal magnetization $M(B)$ is plotted in Figure 5.30. $M(B)$ shows four MMTs at 0.6, 0.9, 1.2, and 1.43 T at 1.9 K. There is only one MMT at $T = 3.7$ and 5 K.

In order to study the magnetic structure at each step of the multistep metamagnetism, Muro et al. [65] are now doing neutron diffraction measurements on single-crystal samples of $CeIr_3Si_2$.

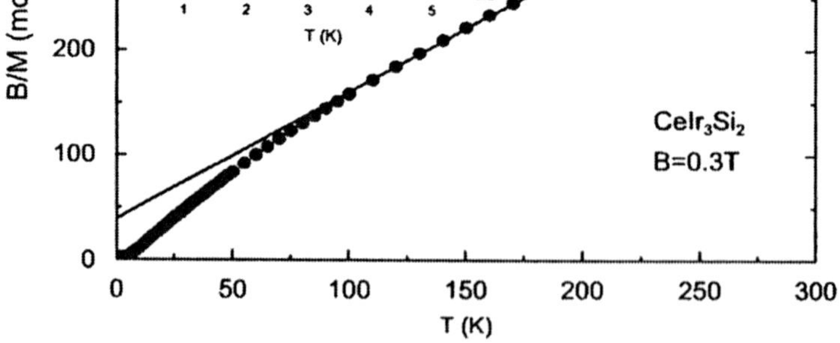

Figure 5.29. The temperature dependence of inverse susceptibility B/M of $CeIr_3Si_2$ measured under the magnetic field of 0.3 T. The solid line represents the Curie–Weiss law. Inset shows $M(T)/B$ below 5.5 K under 0.5 and 0.3 T. Dashed lines indicate T_{N1} and T_{N2}. Reproduced with the permission of Elsevier from Ref. [65].

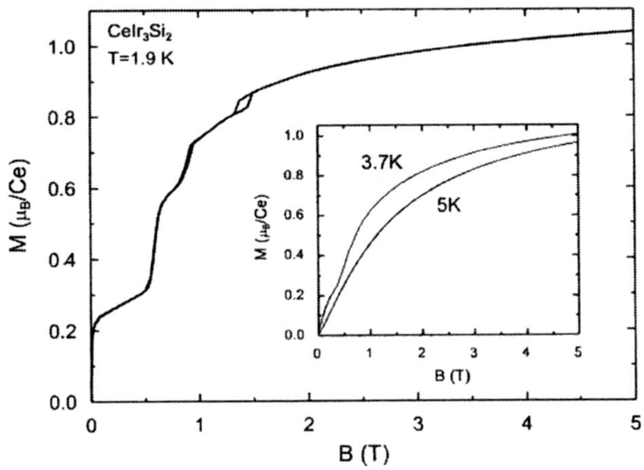

Figure 5.30. Isothermal magnetization *M*(*B*) of $CeIr_3Si_2$ at 1.9 K. The inset shows *M*(*B*) at 3.7 and 5 K. Reproduced with the permission of Elsevier from Ref. [65].

References

[1] E. Stryjewski and N. Giordano, Adv. Phys. **26**, 487 (1977).
[2] E.P. Wohflarth and P. Rhodes, Philos. Mag. **7**, 1817 (1962).
[3] H. Yamada, Phys. Rev. B **47**, 11211 (1993).
[4] S. Doniach, Physica B **91**, 231 (1977).
[5] H. Aoki, S. Ujii, A.K. Albsessard, and Y. Onuki, Phys. Rev. Lett. **71**, 2110 (1993).
[6] R.A. Fisher, C. Marcenat, N.E. Phillips, P. Haen, F. Lapierre, P. Lejay, J. Flouquet, and J. Voiron, J. Low Temp. Phys. **84**, 49 (1991).
[7] M.J. Besnus, J.P. Kappler, P. Lehmann, and A. Mayer, Solid State Commun. **55**, 779 (1985).
[8] P. Haen, J. Flouquet, F. Lapierre, P. Lejay, and G. Remenyi, J. Low Temp. Phys. **67**, 331 (1987).
[9] J. Flouquet, S. Kambe, L.P. Regnault, P. Haen, J.P. Brison, F. Lapierre, and P. Lejay, Physica B **215**, 77 (1995).
[10] J. Flouquet, P. Haen, S. Raymond, D. Aoki, and G. Knebel, Physica B **319**, 251 (2002).
[11] C. Pauelson, A. Lacerda, L. Puech, P. Haen, P. Lejay, J. L. Tholence, J. Flouquet, and A. de Visser, J. Low Temp. Phys. **81**, 317 (1990).
[12] A. Lacerda, A. de Visser, P. Haen, P. Lejay, and J. Flouquet, Phys. Rev. B **40**, 8759 (1989).
[13] S. Quezel, P. Burlet, J.L. Jacoud, L.P. Regnault, J. Rossat-Mignod, C. Vettier, P. Lejay, and J. Flouquet, J. Magn. Magn. Mater. **76–77**, 403 (1988).
[14] J.M. Mignot, J.L. Jacoud, L.P. Regnault, J. Rossat-Mignod, P. Haen, P. Lejay, Ph. Boutrouille, B. Hennion, and D. Petrigrand, Physica B **163**, 611 (1990).
[15] J.M. Mignot, L.P. Regnault, J.L. Jacoud, J. Tossat-Mignod, P. Haen, and P. Lejay, Physica B **171**, 357 (1991).

[16] J.M. Mignot, J. Flouquet, P. Haen, F. Lapierre, L. Puech, and J. Voiron, J. Magn. Magn. Mater. **76–77**, 97 (1988).
[17] T. Sakakibara, T. Tayama, K. Matssuhira, H. Mitamura, H. Amitsuka, K. Maezawa, and O. Onuki, Phys. Rev. B **51**, 12030 (1995).
[18] A. Aoki, T.D. Matsuda, H. Sugawara, H. Sato, H. Ohkani, R. Settai, Y. Onuki, E. Yanamoto, Y. Haga, and A.V. Andreev, J. Magn. Magn. Mater. **177–181**, 271 (1998).
[19] K. Heuser, E.W. Scheidt, T. Schreiner, Z. Fisk, and G.R. Stewart, J. Low Temp. Phys. **118**, 235 (2000).
[20] S. Kambe, H. Suedrow, J. Flouquet, P. Lejay, and P. Haen, Solid State Commun. **96**, 175 (1995).
[21] S. Kambe, S. Raymond, L.-P. Regnault, J. Flouquet, P. Lejay, and P. Haen, J. Low Temp. Phys. **102**, 477 (1996).
[22] R. Daou, C. Bergemann, and S.R. Julian, Phys. Rev. Lett. **96**, 026401 (2006).
[23] H.P. van der Meulen, A. de Visser, J.J.M. Franse, T.T.J.M. Beandscot, J.A.A.J. Perenbrom, H. van Kempen, A. Lacerda, P. Lejay, and J. Flouquet, Phys. Rev. B **44**, 814 (1991).
[24] C.A. King and G.G. Lonzarich, Physica B **206**, 29 (1995).
[25] A. Bohm, R. Caspary, U. Gotturick, A. Grauel, U. Habel, M. Lang, M. Nowak, R. Schefzyk, W. Schiebeling, and H. Spille, J. Magn. Magn. Mater. **76–77**, 150 (1988).
[26] F.S. Tauz, S.R. Julian, G.J. McMullan, and G.G. Lonzarich, Physica B **206**, 29 (1995).
[27] R.S. Perry, L.M. Galvin, S.A. Grigera, L. Capogna, A.J. Schofild, A.P. Mackenzie, M. Chiao, S.R. Julian, S.I. Ikeda, S. Nakatsuji, Y. Maeno, and C. Pfleiderer, Phys. Rev. Lett. **86**, 2661 (2001).
[28] S.A. Grigera, R.S. Perry, A.J. Schofield, M. Chiao, S.R. Julian, G.G. Lonzarich, S.I. Ikeda, and A.P. Mackenzie, Science **294**, 329 (2001).
[29] C. Thessieu, C. Pfleiderer, A.N. Stepanov, and J. Flouquet, J. Phys. Condens. Matter **9**, 6677 (1997).
[30] R.S. Perry, K. Kitagawa, S.A. Grigera, R.A. Borzi, A.P. Mackenzie, K. Ishida, and Y. Maeno, Phys. Rev. Lett. **92**, 166602 (2004).
[31] S. Raymond, P. Haen, R. Calmeczuk, S. Kambe, B. Fak, P. Lejay, T. Fukuhara, and J. Flouquet, J. Phys. Condens. Matter **11**, 5547 (1999).
[32] R.A. Borzi, S.A. Grigera, R.S. Perry, N. Kitagawa, Y. Maeno, and A.P. Mackenzie, Phys. Rev. Lett. **92**, 216403 (2004).
[33] G. Aeppli, H. Yoshizawa, Y. Endoh, E. Bucher, J. Hufnagl, Y. Onuki, and T. Komatsubara, Phys. Rev. Lett. **57**, 122 (1986).
[34] J. Rossat-Mignod, L.P. Regnault, J.L. Jacoud, C. Vetter, P. Lejay, J. Flouquet, E. Walker, D. Jaccard, and A. Amato, J. Magn. Magn. Mater. **76–77**, 376 (1988).
[35] A. Schroder, G. Aeppli, R. Coldea, M. Adams, O. Stockert, H.V. Lohneysen, E. Bucher, R. Ramzashvili, and P. Coleman, Nature **407**, 351 (2000).
[36] R.A. Robinson, D.J. Gossens, M.S. Torikachvili, K. Kakurai, and H. Okumara, Physica B **385–386**, 38 (2006).
[37] H. V. Lohneysen, H.G. Schlager, and A. Schroder, Physica B **186–188**, 590 (1993).

[38] G. Aeppli, A. Goldman, G. Shirane, E. Bucher, and M-ch Lux-Steiner, Phys. Rev. Lett. **58**, 808 (1987).

[39] A. de Visser, H. Nakotte, L.T. Tai, A.A. Menovsky, S.A.M. Mentive, G.J. Nieuwenhuys, and J.A. Mydosh, Physica B **179**, 84 (1990).

[40] P.H. Frings and J.J.M. Franse, Phys. Rev. B **31**, 4355 (1985).

[41] K. Sugiyama, M. Nakashima, D. Aoki, K. Kindo, N. Kimura, H. Aoki, T. Komatsubara, S. Uji, Y. Haga, E. Yamamoto, H. Harima, and Y. Onuki, Phys. Rev. B **60**, 9248 (1999).

[42] K. Sugiyama, H. Fuke, K. Kindo, S. Shimohata, A.A. Menovsky, J.A. Mydosh, and M. Date, J. Phys. Soc. Jpn. **59**, 3331 (1990).

[43] K. Oda, T. Kumada, K. Sugiyama, N. Sato, T. Komatsubara, and M. Date, J. Phys. Soc. Jpn. **63**, 3115 (1994).

[44] K. Sugiyama, M. Nakashima, M. Futoh, H. Ohkuni, T. Inoue, K. Kindo, N. Kimura, E. Yamamoto, Y. Haga, T. Honma, R. Settai, and Y. Onuki, Physica B **281–282**, 244 (2000).

[45] A. de Visser, H.P. van der Muller, L.T. Tai, and A.A. Menovsky, Physica B **199–200**, 100 (1994).

[46] T.T.M. Palstra, A.A. Menovsky, J. van den Berg, A.J. Dirkmaat, G.J. Niewenhuys, and J.A. Mydosh, Phys. Rev. Lett. **55**, 2727 (1985).

[47] A.C. Hewson, *The Kondo Problem in Heavy Fermions* (Cambridge University Press, Cambridge, UK, 1993).

[48] G.R. Stewart, Rev. Mod. Phys. **56**, 755 (1984).

[49] N. Harrison, K.H. Kim, M. Jaime, and J.A. Mydosh, Physica B **346–347**, 92 (2004).

[50] P. Chandra, P. Coleman, J.A. Mydosh, and V. Tripathi, Nature (London) **417**, 831 (2002).

[51] H. Ohkuni, T. Ishida, Y. Inada, Y. Haga, E. Yamamoto, Y. Onuki, and S. Takahasi, J. Phys. Soc. Jpn. **66**, 945 (1997).

[52] N. Keller, S.A.J. Wiegers, J.A.A.J. Perenboom, A. de Visser, A.A. Menovosky, and J.J.M. Franse, J. Magn. Magn. Mater. **176–177**, 298 (1998).

[53] D.M. Edwards and A.C.M. Green, Z. Phys. B **103**, 243 (1997).

[54] N. Harrison, M. Jaime, and J.A. Mydosh, Phys. Rev. Lett. **90**, 096402 (2003).

[55] M. Jaime, K.H. Kim, G. Jorge, S. McCall, and J.A. Mydosh, Phys. Rev. Lett. **89**, 287201 (2002).

[56] K.H. Kim, N. Harrison, M. Jaime, G.S. Boebinger, and J.A. Mydosh, Phys. Rev. Lett. **91**, 256401 (2003).

[57] K.H. Kim, N. Harrison, H. Amitsuka, M. Jaime, and J.A. Mydosh, Physica B **378–380**, 31 (2006).

[58] A. Suslov et al., cond-mat/0212158.

[59] J. Sheikin, A. Groger, S. Raymond, D. Jaccard, D. Aoki, H. Harima, and J. Flouquet, Phys. Rev. B **67**, 094420 (2003).

[60] I. Sheikin, Y. Wang, F. Bouquet, P. Lejay, and A. Junod, J. Phys. Condens. Matter **14**, L543 (2002).

[61] O. Trovarelli, C. Geibel, S. Mederle, C. Langhammer, F.M. Grosche, P. Gegenwart, M. Lang, G. Spam, and F. Steglich, Phys. Rev. Lett. **85**, 626 (2000).

[62] P. Gegenwart, J. Custers, C. Geibel, K. Neumaier, T. Tayama, K. Tenya, O. Tovarelli, and F. Steglich, Phys. Rev. Lett. **89**, 056402 (2002).
[63] Y. Tokiwa, P. Gegenwart, T. Rado, J. Ferstl, G. Sparn, C. Geibel, and F. Steglich, Phys. Rev. Lett. **94**, 226402 (2005).
[64] S. Mederle, R. Borth, C. Geibel, F.M. Grosche, G, Sparn, G. Trovarelli, and F. Steglich, J. Phys. Condens. Matter **14**, 10731 (2002).
[65] Y. Muro, Y. Ohno, T. Okada, and K. Motoya, J. Magn. Magn. Mater. **310**, 389 (2007).

Chapter 6

Theory of Metamagnetism in Heavy Fermions

6.1. Review of theoretical models

In 1936, Neel [1] had predicted the magnetic field conditions under which an antiferromagnet would show an abrupt decoupling between the direction of antifrromagnetism and the easy axis, known as "spin-flopping" or "magnetization turnover". Later, he extended [2] this analysis to behavior termed "metamagnetic transition" [3] in which an antiferromagnetic array abruptly transforms to a fully aligned ferromagnetic array at a critical field. In this review, we will concentrate on theoretical models applied to metamagnetism in heavy fermions.

Evans [4] used quasiparticle band descriptions to explain the metamagnetic transition in $CeCu_2Si_2$ [5–7]. Suvasini et al. [8] have used a fully relativistic spin-polarized linear muffin-tin orbital (SPRLMTO) method [9, 10] to make first-principle local density approximation (LDA) band structure calculations with the magnetic field H above the metamagnetic critical field H_c. They have tried to explain the metamagnetic transition in UPt_3 [11, 12] using this method.

Norman [13] has solved the Eliashberg equations for magnetic fluctuations for the case of UPt_3. He has used the neutron data on UPt_3 of Aeppli et al. [14] which had indicated antiferromagnetic correlations in UPt_3. The spin-fluctuation theory concentrates almost exclusively on spins, and heavy quasiparticles feature only implicitly via a large linear term in the electronic specific heat calculated from the spin-fluctuation spectrum. Preliminary proposals for unified theories have been made by Coleman and Lonzarich [15] as well as by Edwards [16].

6.2. Strong-coupling spin-fluctuation theory in the high-field state

Later, Edwards [17] had proposed a theory, based on the Anderson lattice model, for the onset of metamagnetic transition, by using a strong-coupling spin-fluctuation theory via magnons, in the high-field state. An outline of Edward's theory is given below.

In the Anderson lattice model,

$$H = \sum_{\mathbf{k}\sigma} \varepsilon_{\mathbf{k}} n_{\mathbf{k}\sigma} + E_f \sum_i n_{i\sigma} + U \sum_i n_{i\uparrow} n_{i\downarrow} + \sum_{\mathbf{k}\sigma} V_{\mathbf{k}} (C^{\dagger}_{\mathbf{k}\sigma} f_{\mathbf{k}\sigma} + f^{\dagger}_{\mathbf{k}\sigma} C_{\mathbf{k}\sigma}), \tag{6.1}$$

where the symbols have been explained in earlier chapters. In a Ce system, the f-level E_f is well below the Fermi level with a doublet ground state separated from the first excited ionic state by considerably more than the impurity Kondo temperature. The spin is a pseudospin 1/2 labeling the doublet states but Edwards [17] has considered it as a real spin 1/2.

Edwards considered the Anderson lattice in the high-field state, due to an applied field H_a, and assumed that the number $n_\downarrow$ of $\downarrow$ spin f electrons per atom was negligible. The $\uparrow$ spin self-energy $\Sigma_\uparrow = 0$ and thus the $\uparrow$ spin quasiparticles have no mass enhancement. In order to calculate the $\downarrow$ spin self-energy $\Sigma_\downarrow$, and the one-particle properties, the Green function can be written as

$$\begin{bmatrix} G^\sigma_{\rm ff} & G^\sigma_{\rm fc} \\ G^\sigma_{\rm cf} & G^\sigma_{\rm cc} \end{bmatrix} = \begin{bmatrix} E - E_{\rm f} - \Sigma_\sigma & -V_{\mathbf{k}} \\ -V_{\mathbf{k}} & E - \varepsilon_{\mathbf{k}} \end{bmatrix}^{-1} \tag{6.2}$$

Using techniques similar to that for Hubbard model [18], the perturbation in U can be obtained from

$$\Sigma_\downarrow(k) = Un_\uparrow + U^2 \int \frac{d^4k'}{(2\pi)^4 i} \gamma(k\downarrow; k'\uparrow) G^\uparrow_{\rm ff}(k') \chi_{\rm f}(k - k'), \tag{6.3}$$

where $k \equiv (\mathbf{k}, E)$, $\chi_{\rm f}$ is the f-transverse susceptibility and γ is an irreducible vertex function. The vertex function was approximated by Edwards, using a Ward identity [18, 19],

$$\gamma(k\downarrow k'\uparrow) \simeq \gamma(k\downarrow; k\uparrow) \approx \frac{\Sigma_\downarrow(k)}{(Un_\uparrow)}. \tag{6.4}$$

Here, $U\gamma$ is the renormalized electron-magnon vertex function, $\chi_{\rm f}(q, \omega)$ is approximated by a single pole at the magnon energy $2\mu_B H_a + \omega_q$, and ω_q is neglected as a first approximation. This form of $\Sigma_\downarrow$ is shown in Figure 6.1, the second order of which is the high-field analog of paramagnon (spin-fluctuation) theory.

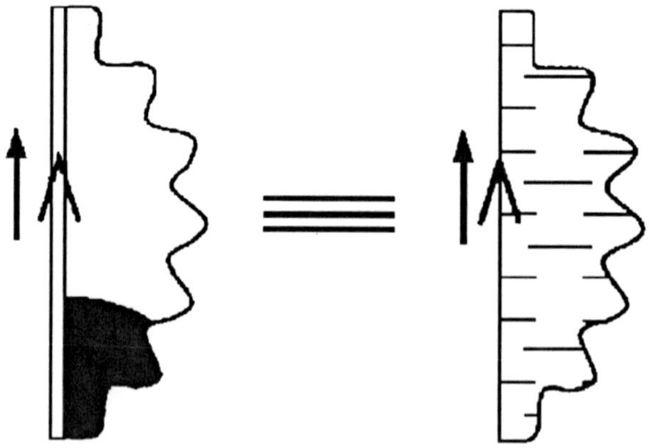

Figure 6.1. Diagrams (excluding Hartree–Fock one) for $\Sigma_\downarrow$. The wavy line is a magnon. Reproduced with the permission of Elsevier from Ref. [17].

Edward's calculations [17] on the high-field state of the Anderson lattice can be summarized as follows: (i) most of the ↓ spin spectral weight is at $E_f + U$, but some are above the ↓ Fermi surface in the form of flattened quasiparticle bands below an electron-magnon scattering continuum for $E > 2\mu_B H_a$; (ii) the ↓ spin quasiparticle mass increases with decreasing field; (iii) the onset of a metamagnetic transition from the high-field side is marked by a steep upturn in the rate of increase of the ↓ spin Fermi radius with decreasing field, which occurs when $2\mu_B H_a$ is of the order of $k_B T_K$; (iv) in the small V limit, the conduction-band self-energy reduces to the second-order perturbation result obtained with a Schrieffer–Wolff coupling $2V^2/|E_f|)\mathbf{S}\cdot\boldsymbol{\sigma}$; (v) the mass enhancement obtained by this localized spin model is also due to f hybridization at the Fermi surface.

6.3. Metamagnetic transition in a small cluster *t–J* model

Freericks and Falicov [20] have used the t–J model and calculated the many-body eigenstates of a small cluster in a magnetic field. They have shown that antiferromagnetic superexchange favors low-spin arrangements for the ground state while a magnetic field favors high-spin arrangements. The transition from the low-spin ground state to a high-spin ground state, as a function of magnetic field, passes through a range where there is a peak in the many-body density of states which qualitatively describes the metamagnetic transition.

The Freericks–Falicov theory [20] can be summarized as follows: the physics of the heavy fermion system is described by the lattice Anderson impurity model [21]

$$H_A = \sum_{\mathbf{k},\sigma} \varepsilon_{\mathbf{k}} a^{\dagger}_{\mathbf{k}\sigma} a_{\mathbf{k}\sigma} + \epsilon_f \sum_{i,\sigma} f^{\dagger}_{i\sigma} f_{i\sigma} + U \sum_i f^{\dagger}_{i\uparrow} f_{i\uparrow} f^{\dagger}_{i\downarrow} f_{i\downarrow} + \sum_{i,\mathbf{k},\sigma} [V_{i\mathbf{k}} f^{\dagger}_{i\sigma} a_{\mathbf{k}\sigma} + V^{*}_{i\mathbf{k}} a^{\dagger}_{\mathbf{k}\sigma} f_{i\sigma}] \tag{6.5}$$

in the large-U ($U \to \infty$) limit. All these parameters and operators have been described in Chapter 2. The hybridization matrix elements are assumed of the form

$$V_{i\mathbf{k}} = \frac{\exp(i\mathbf{R}_i \cdot \mathbf{k}) V g(k)}{\sqrt{N}}, \tag{6.6}$$

where $g(k)$ is the form factor, which is a dimensionless function of order 1, and N is the number of lattice sites. The Fermi-level E_F is defined as the maximum energy of the filled conduction-band states, in the limit $V \to 0$. The origin of the energy scale is chosen at $E_F = 0$ and the conduction-band density of states per site at the Fermi level is defined as ρ.

The Anderson Hamiltonian (6.5) can be mapped onto the large-U limit of the Hubbard Hamiltonian [22] which, in turn, may be mapped onto a t–J model [23–25]:

$$H_{t-J} = -\sum_{i,j,\sigma} t_{ij} (1 - f^{\dagger}_{i-\sigma} f_{i-\sigma}) f^{\dagger}_{i\sigma} f_{j\sigma} (1 - f^{\dagger}_{j-\sigma} f_{j-\sigma}) + \sum_{i,j} J_{ij} \mathbf{S}_i \cdot \mathbf{S}_j. \tag{6.7}$$

Here, the unrenormalized (bare) hopping matrix t_{ij} satisfies

$$t_{ij} = \sum_k \frac{V_{ik}^* V_{jk}}{\varepsilon_k - \varepsilon} = \frac{V^2}{N} \sum_k \frac{g^2(k)}{(\varepsilon_k - \varepsilon)} e^{-i\mathbf{k}\cdot(\mathbf{R}_i - \mathbf{R}_j)}, \tag{6.8}$$

and the antiferromagnetic superexchange is defined to be $J_{ij} \equiv 4|t_{ij}|^2/U$.

The canonical transformation which maps the Anderson model onto the t–J model is valid only within a narrow region of parameter space [20].

The lattice Anderson impurity model (equation (6.5)) has been studied for various small clusters with at most four sites and has been discussed extensively in Chapter 2. This approach to the many-body problem begins with the periodic crystal approximation with a small number of non-equivalent sites. The results for the tetrahedral cluster [26–32] (with one electron per site), extensively discussed in Chapter 2, indicate the formation of the heavy fermion state and its sensitivity to variations in parameters. When the band structure ε_k is such that the bottom of the band is at the Γ point of the fcc Brillouin zone, the ground state is a spin singlet (for a small range of ε) with nearly degenerate triplet and quintet excitations. The specific heat has a huge low-temperature peak and the magnetic susceptibility is large. A magnetically ordered heavy fermion state is sometimes observed when Γ is the top of the conduction band.

The mapping of the Anderson model onto the t–J model (equation (6.7)) reduces the Hilbert space by a factor of $(3/16)^N$, which allows larger clusters to be studied. A good example of treating the heavy fermion system in the t–J model lies in an eight-site fcc lattice cluster with seven electrons [33, 34]. The many-body states are degenerate at $J = 0$ ($J_{ij} = J$ when i, j are first nearest neighbors but $J = 0$ otherwise) but the degeneracy is lifted for finite J, with low-spin configuration favored (energetically) over high-spin configuration. A magnetic field in the z direction lifts the degeneracy even more since the many-body eigenstates have the energy

$$E(B) = E(0) - m_z g\mu_B B \equiv E(0) - m_z bJ, \tag{6.9}$$

where m_z is the z-component of spin, B the magnetic field, g the Lande g factor, μ_B the Bohr magneton, and b dimensionless magnetic field. The high-spin eigenstates are energetically favored in a strong magnetic field and level crossings occur as a function of b. The heavy fermion system is described by a ground state with nearly degenerate low-lying excitations of many different configurations. The antiferromagnetic superexchange pushes high-spin states up in energy with splittings on the order of J. The magnetic field pulls down these high-spin states and generates level crossings in the ground state. The magnetization and spin–spin correlation functions both change abruptly at the level crossings. This gives rise to the metamagnetic transition in heavy fermions.

We show the results of magnetization at temperature (the details of the method of numerical calculations are given in Ref. [20]) fixed at $T = J/k_B$ (high-temperature regime) in Figure 6.2. The magnetization smoothly changes from a value of 0 to 5/2 as a function of magnetic field, showing little structure.

The magnetization as a function of magnetic field at temperature $T = J/5k_B$ is shown in Figure 6.3. Since the temperature is smaller than the energy-level spacing, the magnetization shows steps at the various level crossings.

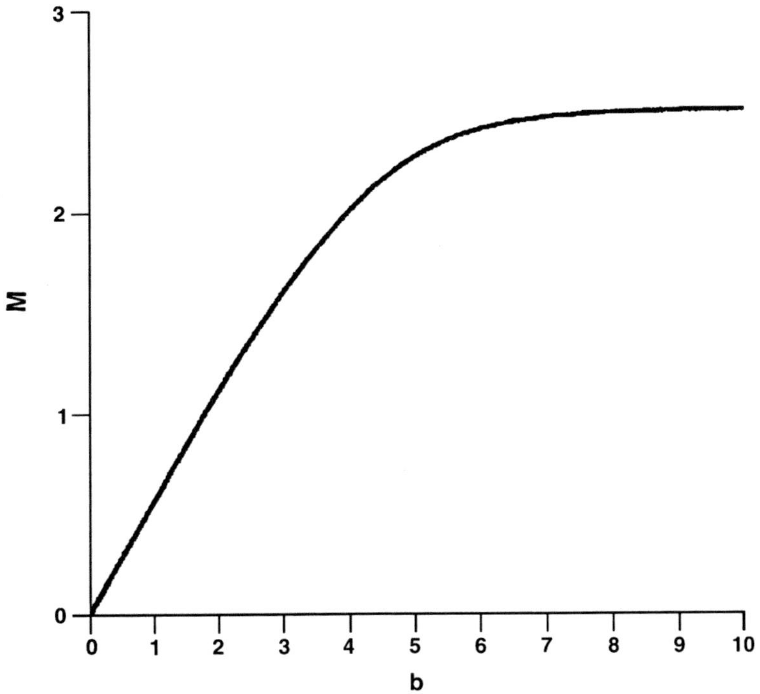

Figure 6.2. Magnetization as a function of magnetic field at $T = J/k_B$ for the heavy fermion model. Reproduced with the permission of the American Physical Society from Ref. [20].

These theoretical results obtained by calculating the magnetization from the many-body states qualitatively agree with the numerous experimental data (see Ref. [20] for details) on metamagnetism in heavy fermion systems.

Tripathi [35] has proposed, starting from a variant of the Hubbard Hamiltonian, a model for the magnetic-field-induced metamagnetic transition in an itinerant electron system. The coupling of the applied field with the spin magnetic moments of the electrons is described by two parts—a part which splits the spin degeneracy and the other which is responsible for spin fluctuations. The damping of the spin fluctuations at the high-field state coupled with the strong electron–electron interactions was shown to be responsible for the onset of metamagnetism. However, there was no attempt to compare the results with any experimental observation. Later, Tripathi et al. [36] extended this model to include electron–phonon interaction which was found to suppress metamagnetism in an itinerant electron system. They also considered the effect of electron–phonon interaction on the low-field spin susceptibility by considering electron self-energy as a function of frequency and magnetic field. They found that the modification brought about by the field dependence of the self-energy is cancelled by the mass enhancement arising out of the frequency dependence of the self-energy.

Ono [37] has investigated the metamagnetism of heavy fermions using the periodic Anderson model (PAM). His theory includes anisotropic hybridization and electron-lattice coupling. The latter coupling in the model appears to be responsible for an enhancement of metamagnetization and a large magnetostriction effect. Ohara et al. [38] have studied

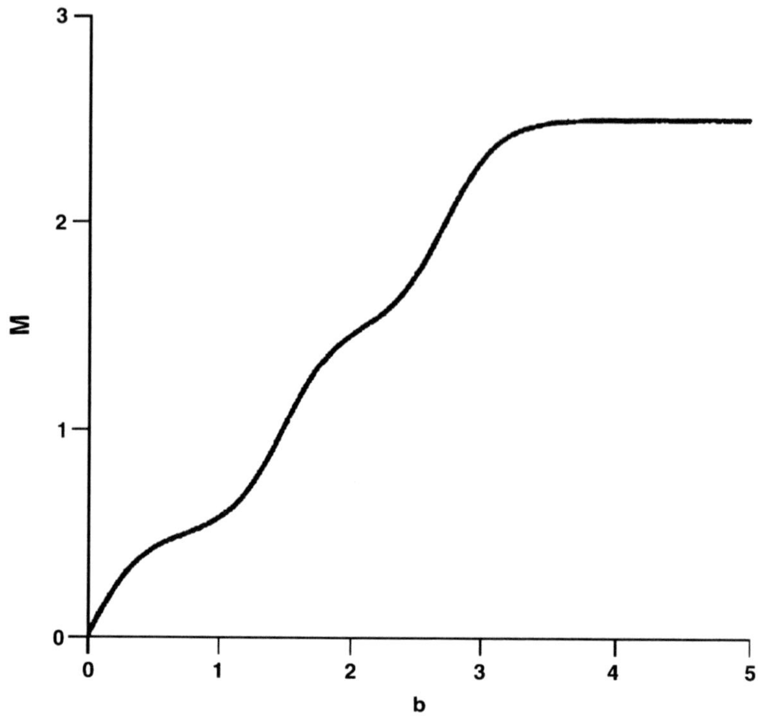

Figure 6.3. Calculated magnetization as a function of magnetic field at temperature $T = J/5k_B$ (low-temperature regime). Reproduced with the permission of the American Physical Society from Ref. [20].

the metamagnetic behavior of $CeRu_2Si_2$ by using the periodic Coqblin–Schrieffer [39] model with anisotropic c-f exchange interaction in the mean-field theory. The metamagnetic transition is identified with a singularity of the chemical potential at the field of metamagnetic transition.

6.4. Competition between local quantum spin fluctuations and magnetic-exchange interaction

Satoh and Ohkawa [40, 41] have shown, based on the PAM, that the competition between the quenching of the magnetic moments by local spin fluctuations and a magnetic-exchange interaction caused by a virtual exchange of pair excitations of quasiparticles in spin channels is responsible for the metamagnetic crossover in $CeRu_2Si_2$. They have also included the effect of the electron–lattice interaction. The strength of the exchange interaction is proportional to the bandwidth of quasiparticles and its sign changes with increasing magnetizations. They have found that it is antiferromagnetic in the absence of magnetizations but ferromagnetic in the metamagnetic crossover region. They have reproduced the experimental results of static quantities.

Their results can be summarized as follows. The PAM in a magnetic field H can be written as

$$H = \sum_{\lambda \mathbf{k}\sigma} E_\lambda(\mathbf{k}) a^\dagger_{\lambda\mathbf{k}\sigma} a_{\lambda\mathbf{k}\sigma} + \sum_{\mathbf{k}\sigma} (E_\mathrm{f} - \sigma H^*) f^\dagger_{\mathbf{k}\sigma} f_{\mathbf{k}\sigma}$$
$$+ \sum_{\lambda\mathbf{k}\sigma} [V_\lambda(\mathbf{k}) a^\dagger_{\lambda\mathbf{k}\sigma} f_{\mathbf{k}\sigma} + \mathrm{H.C.}] + \frac{1}{2} U \sum_{i\sigma} n_{i\sigma} n_{i-\sigma}, \tag{6.10}$$

with λ the band index of conduction electrons, $n_{i\sigma} = f^\dagger_{i\sigma} f_{i\sigma}$, and $H^* = m_0 H$, where m_0 is the saturation magnetization per f electron. The kinetic energy of conduction electrons $E_\lambda(\mathbf{k})$ and the f-electron level E_f are measured from the chemical potential.

In order to treat the local quantum spin fluctuations which are responsible for the quenching of the magnetic moments, Satoh and Ohkawa [41] have used the single-site approximation (SSA) [42] which is rigorous for paramagnetic states in infinite dimensions. Within the SSA, Green's function for f electrons and conduction electrons is given by

$$G_{\mathrm{ff}\sigma}(i\varepsilon_n, \mathbf{k}) = \frac{1}{i\varepsilon_n - E_\mathrm{f} - \tilde{\Sigma}_\sigma(i\varepsilon_n) - \sum_\lambda ((|V_\lambda(\mathbf{k})|^2)/(i\varepsilon_n - E_\lambda(\mathbf{k}))}, \tag{6.11}$$

where $\tilde{\Sigma}_\sigma$ is the single-site self-energy function and

$$G_{\lambda\lambda'\sigma}(i\varepsilon_n, \mathbf{k}) = \delta_{\lambda\lambda'} g_\lambda(i\varepsilon_n, \mathbf{k}) + g_\lambda(i\varepsilon_n, \mathbf{k}) V_\lambda(\mathbf{k}) G_{ff\sigma}(i\varepsilon_n, \mathbf{k}) V^*_{\lambda'}(\mathbf{k}) g_{\lambda'}(i\varepsilon_n, \mathbf{k}), \tag{6.12}$$

and $g_\lambda(i\varepsilon_n, \mathbf{k}) = [i\varepsilon_n - E_\lambda(\mathbf{k})]^{-1}$. Here n is an integer and $i\varepsilon_n$ an imaginary Fermi energy. The single-site self-energy function is obtained by solving a single-impurity Anderson model (SIAM) [42] which has the same Coulomb repulsion U and the same localized electron-level E_f as in equation (6.10). This is called a mapped Anderson model (MAM). The parameters of the MAM are determined through the mapping condition,

$$\tilde{G}_{\mathrm{ff}\sigma}(i\varepsilon_n) = \frac{1}{N} \sum_{\mathbf{k}} G_{\mathrm{ff}\sigma}(i\varepsilon_n, \mathbf{k}), \tag{6.13}$$

where N is the number of unit cells, and

$$\tilde{G}_{\mathrm{ff}\sigma}(i\varepsilon_n) = \frac{1}{i\varepsilon_n - E_\mathrm{f} - \tilde{\Sigma}_\sigma(i\varepsilon_n) - L(i\varepsilon_n)}, \tag{6.14}$$

with $L(i\varepsilon_n) = (1/\pi)\int_{-\infty}^{\infty} d\varepsilon \Delta(\varepsilon)/(i\varepsilon_n - \varepsilon)$. Here $\Delta(\varepsilon)$ is the hybridization energy of MAM. $\tilde{\Sigma}_\sigma(i\varepsilon_n)$ is obtained by solving the MAM numerically once a trial function for $\Delta(\varepsilon)$ is given.

Satoh and Ohkawa [41] have used the well-known results of the Kondo problem. They have used a Fermi-liquid description to expand the self-energy function

$$\tilde{\Sigma}_\sigma(i\varepsilon_n) = \tilde{\Sigma}_\sigma(+i0) + [1 - \tilde{\phi}_m] i\varepsilon_n + \cdots, \tag{6.15}$$

for small $|\varepsilon_n|$, where $\tilde{\phi}_m$ is a mass enhancement factor in the SSA. In the symmetrical case, the coherent part of equation (6.11) can be written as

$$G^{(c)}_{ff\sigma}(i\varepsilon_n, \mathbf{k}) = \frac{1}{\tilde{\phi}_m i\varepsilon_n - \Sigma_\lambda ((|V_\lambda(\mathbf{k})|^2)/(i\varepsilon_n - E_\lambda(\mathbf{k})))}. \tag{6.16}$$

The quasiparticles are defined as the poles of equation (6.16), i.e., the dispersion relation of quasiparticles are obtained by solving the following equation:

$$\tilde{\phi}_m z - \sum_\lambda \frac{|V_\lambda(\mathbf{k})|^2}{z - E_\lambda(\mathbf{k})} = 0. \tag{6.17}$$

The solutions are written as $z = \xi_\nu(\mathbf{k})$ with ν representing the branch of quasiparticles. It can be shown that in a magnetic field [41],

$$\tilde{\phi}_m z - \delta\tilde{\Sigma}_\sigma(m) - \sum_\lambda \frac{|V_\lambda(\mathbf{k})|^2}{z - E_\lambda(\mathbf{k})} = 0, \tag{6.18}$$

with $\delta\tilde{\Sigma}_\sigma(m)$ a magnetic part of the self-energy and m is the magnetization $m = \Sigma_\sigma \sigma\langle f^\dagger_{i\sigma} f_{i\sigma}\rangle$. Solutions of equation (6.18) are denoted by $\xi_{\nu\sigma}(\mathbf{k}, m)$. Thus, for the finite-field case, all arguments can be developed in parallel with the zero-field case. Comparing equations (6.17) and (6.18), they obtained [41]

$$\xi_{\nu\sigma}(\mathbf{k}, m) \approx \xi_\nu(\mathbf{k}) - \sigma\Delta E(m), \tag{6.19}$$

where

$$\sigma\Delta E(m) \equiv \frac{-\delta\tilde{\Sigma}_\sigma(m)}{\tilde{\phi}_m}. \tag{6.20}$$

It can be shown [41] that with the use of equation (6.20), the Luttinger's theorem gives

$$m = \sum_\sigma \sigma \int_{-\infty}^{0} d\varepsilon\, \rho^*_{\mathrm{f}}[\varepsilon + \sigma\Delta E(m)], \tag{6.21}$$

where the polarization of conduction electron has been ignored and ρ^* has been replaced with ρ^*_{f}, since the right hand side of equation (6.21) is a difference between the contributions of spin from up and down directions and hence only the low-energy part is relevant. ΔE_m, the magnetic part of the self-energy, can be determined as a function of m from equation (6.21).

Satoh and Ohkawa [41] have also studied the magnetic exchange interactions working between quasiparticles. They have shown that in the symmetrical case, the magnetic susceptibility can be written as

$$\chi_s(i\omega_l, \mathbf{q}, m) = \frac{\tilde{\chi}_s(i\omega_l, m)}{1 - (1/4)I_s(i\omega_l, \mathbf{q}, m)\tilde{\chi}_s(i\omega_l, m)}, \tag{6.22}$$

where $I_s(i\omega_l, \mathbf{q}, m)$ is the intersite interaction and $\tilde{\chi}_s(i\omega_l, m)$ the magnetic susceptibility of the MAM. Equation (6.22) is consistent with the physical picture of the Kondo lattice that local quantum spin fluctuations at each site are connected with one another by the intersite interaction.

It can also be shown that the main contribution to I_s is divided into two parts [43]

$$I_s(i\omega_l, \mathbf{q}, m) = J_s(\mathbf{q}) + J_Q(i\omega_l, \mathbf{q}, m). \tag{6.23}$$

Here J_s is an exchange interaction caused by the virtual exchange of high-energy spin excitations which do not depend on m. It consists of the conventional superexchange interaction, an extended superexchange interaction, and a Rudderman–Kittel–Kasuya–Yosida (RKKY) interaction. Since J_s depends on the whole band structure, Satoh and Ohkawa [41] have treated it as a phenomenological parameter. J_Q is due to the virtual exchange of low-energy spin excitations within quasiparticle bands. They have also studied the static and uniform component of J_Q abbreviated to $J_Q(m, x)$. They have also shown that the volume dependence of $J_Q(m, x)$ is the same as that of T_K, i.e., $J_Q(m, x) = e^{-x}J_Q(m, 0)$ and the magnitude of $J_Q(m, x)$ is of the order of k_BT_K. Since $J_Q(m, x)$ is scaled with T_K, the bandwidth of the quasiparticles, it is possible to do a single-parameter scaling [40]. In Figure 6.4, the theoretical results of magnetization and magnetostriction, calculated by Satoh and Ohkawa [41], for $CeRu_2Si_2$ by using a parameterized model, are compared with the experimental results [5.44].

Muoto [45] has studied the PAM away from the half-filling in the framework of the self-consistent second-order perturbation theory. He has studied the concentration dependence of the density of states spectrum, spin and charge susceptibilities, and magnetization processes. He has shown that the coefficient of the electronic specific heat and the spin susceptibility are both enhanced by strong correlations. Although the study does not directly pertain to metamagnetic transition, the magnetization curve and the above quantities have some bearing with the metamagnetic transition.

Meyer et al. [46] have investigated the SIAM, which has gained much interest in the context of the dynamical mean-field theory (DMFT) (discussed extensively in Chapter 3), by introducing a modified perturbation theory (MPT). Their approximation scheme yields reasonable results away from the symmetric case. Meyer and Nolting [47–49] have addressed the problem of field-induced metamagnetic transition and tried to compare the results with that observed in $CeRu_2Si_2$. They have used DMFT in combination with MPT to calculate the magnetization and the effective mass. However, their work is not free from ambiguities as regards the choice of the chemical potential. The results are qualitative and the calculated field at the metamagnetic transition is several hundreds of Tesla which is much too large compared with the experimental value.

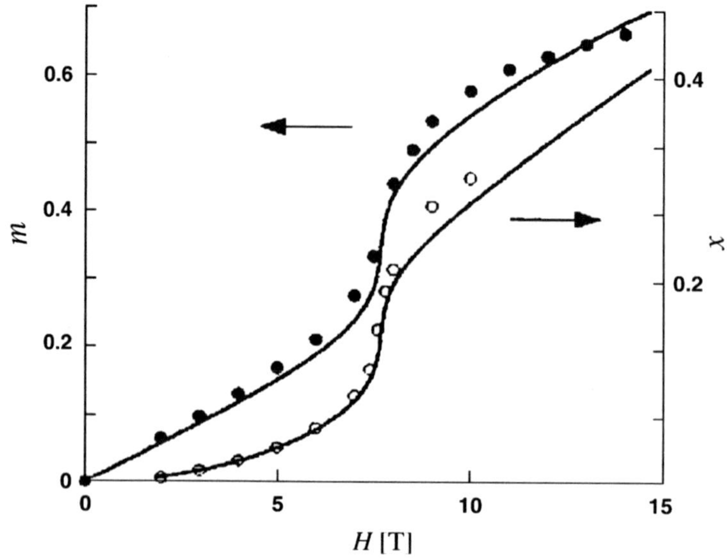

Figure 6.4. Magnetization and magnetostriction of $CeRu_2Si_2$ calculated in a parameterized model. Experimental data are shown in dots and circles, respectively, from Refs. [5] and [44]. Reproduced with the permission of the American Physical Society from Ref. [41].

6.5. Itinerant electrons and local moments in high and low magnetic fields

The foregoing remarks suggest that in spite of an extensive activity in metamagnetism, no consensus has been reached about mechanisms that contribute to this phenomenon. We shall present a theory of metamagnetism [50] starting from a variant of the periodic Anderson Hamiltonian in the presence of an applied magnetic field. In Section 6.5.1, we consider equations of motion for the magnetization and spin-fluctuation amplitudes in the high-field state where damping effects are considered phenomenologically. We solve these equations in Section 6.5.2 and obtain a non-linear equation for the magnetization as a function of the applied magnetic field. In Section 6.5.3, we analyze the itinerant electron paramagnetism in the presence of local moments and, in the process, derive an expression for the EPR (Knight) shift including the effect of electron–phonon interactions. In Section 6.5.4, we parameterize the model in order to compare with the experimental results for $CeRu_2Si_2$.

6.5.1. The model

We consider a variant of the periodic Anderson Hamiltonian in the momentum–space representation:

$$H_A = H_c + H_f + H_{hyb}, \tag{6.24}$$

where

$$H_{\mathrm{c}} = \sum_{\mathbf{k}\sigma} \varepsilon_{\mathrm{c}\mathbf{k}} C^{\dagger}_{\mathbf{k}\sigma} C_{\mathbf{k}\sigma} + U_{\mathrm{c}} \sum_{\mathbf{k}\mathbf{k}'\mathbf{q}} C^{\dagger}_{\mathbf{k}+\mathbf{q}\uparrow} C_{\mathbf{k}\uparrow} C^{\dagger}_{\mathbf{k}'-\mathbf{q}\downarrow} C_{\mathbf{k}'\downarrow}, \tag{6.25}$$

$$H_{\mathrm{f}} = \sum_{\mathbf{k}\sigma} \varepsilon_{\mathrm{f}\mathbf{k}} f^{\dagger}_{\mathbf{k}\sigma} f_{\mathbf{k}\sigma} + U_{\mathrm{f}} \sum_{\mathbf{k}\mathbf{k}'\mathbf{q}} f^{\dagger}_{\mathbf{k}+\mathbf{q}\uparrow} f_{\mathbf{k}\uparrow} f^{\dagger}_{\mathbf{k}'-\mathbf{q}\downarrow} f_{\mathbf{k}'\downarrow}, \tag{6.26}$$

and

$$H_{\mathrm{hyb}} = \sum_{\mathbf{k}\sigma} V_{\mathbf{k}\sigma} (C^{\dagger}_{\mathbf{k}\sigma} f_{\mathbf{k}\sigma} + f^{\dagger}_{\mathbf{k}\sigma} C_{\mathbf{k}\sigma}). \tag{6.27}$$

Here, H_{c} and H_{f} describe the conduction- and the f-electron Hamiltonians. $C^{\dagger}_{\mathbf{k}\sigma}(f^{\dagger}_{\mathbf{k}\sigma})$ and $C_{\mathbf{k}\sigma}(f_{\mathbf{k}\sigma})$ are the creation and annihilation operators for the conduction (f) electrons, respectively. $\varepsilon_{\mathrm{c}\mathbf{k}}$ and $\varepsilon_{\mathrm{f}\mathbf{k}}$ are the conduction- and f-electron energies while U_{c} and U_{f} are the respective correlation energies. The other terms have their usual meanings defined earlier and the f-electron spin is chosen same as the conduction-electron spin. When a magnetic field **H** is applied, the resulting interacting Hamiltonians for both the systems are

$$H_{\mathbf{H}\mathrm{c}} = \mu_{\mathrm{B}} H_z \sum_{\mathbf{k}} (C^{\dagger}_{\mathbf{k}\uparrow} C_{\mathbf{k}\uparrow} - C^{\dagger}_{\mathbf{k}\downarrow} C_{\mathbf{k}\downarrow}) + \mu_{\mathrm{B}} \sum_{\mathbf{k}} (C^{\dagger}_{\mathbf{k}\uparrow} C_{\mathbf{k}\downarrow} H_+ + C^{\dagger}_{\mathbf{k}\downarrow} C_{\mathbf{k}\uparrow} H_-) \tag{6.28}$$

and

$$H_{\mathbf{H}\mathrm{f}} = \mu_{\mathrm{B}} H_z \sum_{\mathbf{k}} (f^{\dagger}_{\mathbf{k}\uparrow} f_{\mathbf{k}\uparrow} - f^{\dagger}_{\mathbf{k}\downarrow} f_{\mathbf{k}\downarrow}) + \mu_{\mathrm{B}} \sum_{\mathbf{k}} (f^{\dagger}_{\mathbf{k}\uparrow} f_{\mathbf{k}\downarrow} H_+ + f^{\dagger}_{\mathbf{k}\downarrow} f_{\mathbf{k}\uparrow} H_-). \tag{6.29}$$

Here $H_{\pm} = H_x \pm iH_y$; which are responsible for spin fluctuations, and H_z is the z-component of the applied field which is responsible for the spin splitting of the energy levels. Assuming a mean-field approximation (MFA) for the electron correlation energies, the total Hamiltonian can be derived from equations (6.23)–(6.29) as follows:

$$\begin{aligned} H_{\mathrm{T}} = {} & \sum_{\mathbf{k}} \varepsilon_{\mathrm{c}\mathbf{k}\uparrow} C^{\dagger}_{\mathbf{k}\uparrow} C_{\mathbf{k}\uparrow} + \sum_{\mathbf{k}} \varepsilon_{\mathrm{c}\mathbf{k}\downarrow} C^{\dagger}_{\mathbf{k}\downarrow} C_{\mathbf{k}\downarrow} + \sum_{\mathbf{k}} \varepsilon_{\mathrm{f}\mathbf{k}\uparrow} f^{\dagger}_{\mathbf{k}\uparrow} f_{\mathbf{k}\uparrow} + \sum_{\mathbf{k}} \varepsilon_{\mathrm{f}\mathbf{k}\downarrow} f^{\dagger}_{\mathbf{k}\downarrow} f_{\mathbf{k}\downarrow} \\ & + \sum_{\mathbf{k}} [V_{\uparrow} (C^{\dagger}_{\mathbf{k}\uparrow} f_{\mathbf{k}\uparrow} + f^{\dagger}_{\mathbf{k}\uparrow} C_{\mathbf{k}\uparrow}) + V_{\downarrow} (f^{\dagger}_{\mathbf{k}\downarrow} C_{\mathbf{k}\downarrow} + C^{\dagger}_{\mathbf{k}\downarrow} f_{\mathbf{k}\downarrow})] \\ & + \mu_{\mathrm{B}} \sum_{\mathbf{k}} [(C^{\dagger}_{\mathbf{k}\uparrow} C_{\mathbf{k}\downarrow} + f^{\dagger}_{\mathbf{k}\uparrow} f_{\mathbf{k}\downarrow}) H_+ + (C^{\dagger}_{\mathbf{k}\downarrow} C_{\mathbf{k}\uparrow} + f^{\dagger}_{\mathbf{k}\downarrow} f_{\mathbf{k}\uparrow}) H_-], \end{aligned} \tag{6.30}$$

where

$$\varepsilon_{\mathrm{c}\mathbf{k}\uparrow} = \varepsilon_{\mathrm{c}\mathbf{k}} + U_{\mathrm{c}} n^{\mathrm{c}}_{\mathrm{c}\mathbf{k}\downarrow} + \mu_{\mathrm{B}} H_z, \tag{6.31}$$

$$\varepsilon_{\mathrm{ck}\downarrow} = \varepsilon_{\mathrm{ck}} + U_{\mathrm{c}} n^{\mathrm{c}}_{\mathrm{ck}\uparrow} - \mu_{\mathrm{B}} H_z, \tag{6.32}$$

$$\varepsilon_{\mathrm{fk}\uparrow} = \varepsilon_{\mathrm{fk}} + U_{\mathrm{f}} n^{\mathrm{c}}_{\mathrm{fk}\downarrow} + \mu_{\mathrm{B}} H_z, \tag{6.33}$$

$$\varepsilon_{\mathrm{fk}\downarrow} = \varepsilon_{\mathrm{fk}} + U_{\mathrm{f}} n^{\mathrm{c}}_{\mathrm{fk}\uparrow} - \mu_{\mathrm{B}} H_z. \tag{6.34}$$

Here $V_{\uparrow\downarrow}$ are assumed to be k-independent and have an implicit H_z dependence. To the first order in H_z,

$$V_{\uparrow\downarrow} = V \pm V' H_z. \tag{6.35}$$

In fact, the hybridization parameter, $V_{\mathbf{k}\sigma}$, depends on the magnetic field through both the spin (σ) and the vector operator (κ) which can be expressed [51, 52] as

$$\kappa = \mathbf{k} + i\left(\frac{e\mathbf{H}}{2\hbar c}\right) X \nabla_{\mathbf{k}}. \tag{6.36}$$

However, the magnetic field dependence through κ is an orbital effect and is generally small compared to the effect of spin. In equations (6.31)–(6.35),

$$n^{\mathrm{c}}_{\mathrm{ck}\uparrow\downarrow} = \langle C_{\mathbf{k}\uparrow\downarrow} C_{\mathbf{k}\uparrow\downarrow} \rangle, \tag{6.37}$$

$$n^{\mathrm{c}}_{\mathrm{fk}\uparrow\downarrow} = \langle f_{\mathbf{k}\uparrow\downarrow} f_{\mathbf{k}\uparrow\downarrow} \rangle, \tag{6.38}$$

and

$$V' = \left(\frac{\partial V}{\partial H_z}\right)_0. \tag{6.39}$$

We have followed a non-perturbative approach [50] within the MFA and assumed

$$\langle C^{\dagger}_{\mathbf{k}'+\mathbf{q}\uparrow\downarrow} C_{\mathbf{k}\uparrow\downarrow} \rangle = \langle C^{\dagger}_{\mathbf{k}\uparrow\downarrow} C_{\mathbf{k}\uparrow\downarrow} \rangle \delta_{\mathbf{k}',\mathbf{k}} \delta_{\mathbf{q},0}, \tag{6.40}$$

and

$$\langle C^{\dagger}_{\mathbf{k}+\mathbf{q}\uparrow\downarrow} C^{\dagger}_{-\mathbf{k}-\mathbf{q}\downarrow\uparrow} \rangle = \langle C^{\dagger}_{\mathbf{k}\uparrow\downarrow} C^{\dagger}_{-\mathbf{k}\downarrow\uparrow} \rangle \delta_{\mathbf{q},0}. \tag{6.41}$$

The validity of above approximations results from the fact that in a spatially homogeneous system, the conservation of momentum holds good [53]. Since magnetism is a non-perturbative phenomenon, a non-perturbative approach within the MFA seems to be reasonable. As has been done in our previous work [35, 36], the down-spin electron annihilation operators $C_{\mathbf{k}\downarrow}$ and $f_{\mathbf{k}\downarrow}$ are assumed to be equivalent to the down-spin hole creation operators $d^{\dagger}_{-\mathbf{k}\downarrow}$ and $b^{\dagger}_{-\mathbf{k}\downarrow}$, respectively, for the conduction (c) and f electrons. Equation (6.30) can be written in the electron–hole representation as

$$\begin{aligned} H_{\mathrm{T}} &= \sum_{\mathbf{k}} \varepsilon_{\mathrm{c}\mathbf{k}\uparrow} C^{\dagger}_{\mathbf{k}\uparrow} C_{\mathbf{k}\uparrow} - \sum_{\mathbf{k}} \varepsilon_{\mathrm{c}\mathbf{k}\downarrow} d^{\dagger}_{-\mathbf{k}\downarrow} d_{-\mathbf{k}\downarrow} + \sum_{\mathbf{k}} \varepsilon_{\mathrm{f}\mathbf{k}\uparrow} f^{\dagger}_{\mathbf{k}\uparrow} f_{\mathbf{k}\uparrow} - \sum_{\mathbf{k}} \varepsilon_{\mathrm{f}\mathbf{k}\downarrow} b^{\dagger}_{-\mathbf{k}\downarrow} b_{-\mathbf{k}\downarrow} \\ &+ \sum_{\mathbf{k}} [V_{\uparrow}(C^{\dagger}_{\mathbf{k}\uparrow} f_{\mathbf{k}\uparrow} + f^{\dagger}_{\mathbf{k}\uparrow} C_{\mathbf{k}\uparrow}) + V_{\downarrow}(d_{-\mathbf{k}\downarrow} b^{\dagger}_{-\mathbf{k}\downarrow} + b_{-\mathbf{k}\downarrow} d^{\dagger}_{-\mathbf{k}\downarrow})] \\ &+ \sum_{\mathbf{k}} \mu_{\mathrm{B}}[H_{+}(C^{\dagger}_{\mathbf{k}\uparrow} d^{\dagger}_{-\mathbf{k}\downarrow} + f^{\dagger}_{\mathbf{k}\uparrow} b^{\dagger}_{-\mathbf{k}\downarrow}) + H_{-}(d_{-\mathbf{k}\downarrow} C_{\mathbf{k}\uparrow} + b_{-\mathbf{k}\downarrow} f_{\mathbf{k}\uparrow})], \end{aligned} \tag{6.42}$$

and the necessary changes in equations (6.31)–(6.34) are

$$\varepsilon_{\mathrm{c}\mathbf{k}\uparrow} = \varepsilon_{\mathrm{c}\mathbf{k}} + U_{\mathrm{c}}[1 - n^{\mathrm{h}}_{\mathrm{c}\downarrow}(-\mathbf{k})] + \mu_{\mathrm{B}} H_{z}, \tag{6.43}$$

$$\varepsilon_{\mathrm{c}\mathbf{k}\downarrow} = \varepsilon_{\mathrm{c}\mathbf{k}} + U_{\mathrm{c}} n^{\mathrm{e}}_{\mathrm{c}\uparrow}(\mathbf{k}) - \mu_{\mathrm{B}} H_{z}, \tag{6.44}$$

$$\varepsilon_{\mathrm{f}\mathbf{k}\uparrow} = \varepsilon_{\mathrm{f}\mathbf{k}} + U_{\mathrm{f}}[1 - n^{\mathrm{h}}_{\mathrm{f}\downarrow}(-\mathbf{k})] + \mu_{\mathrm{B}} H_{z}, \tag{6.45}$$

and

$$\varepsilon_{\mathrm{f}\mathbf{k}\downarrow} = \varepsilon_{\mathrm{f}\mathbf{k}} + U_{\mathrm{f}} n^{\mathrm{e}}_{\mathrm{f}\uparrow}(\mathbf{k}) - \mu_{\mathrm{B}} H_{z}. \tag{6.46}$$

In equations (6.43)–(6.46),

$$n^{\mathrm{e}}_{\mathrm{c}\uparrow}(\mathbf{k}) = \langle C^{\dagger}_{\mathbf{k}\uparrow} C_{\mathbf{k}} \rangle, \tag{6.47}$$

$$n^{\mathrm{h}}_{\mathrm{c}\downarrow}(-\mathbf{k}) = \langle d^{\dagger}_{-\mathbf{k}\downarrow} d_{-\mathbf{k}\downarrow} \rangle, \tag{6.48}$$

$$n^{\mathrm{e}}_{\mathrm{f}\uparrow}(\mathbf{k}) = \langle f^{\dagger}_{\mathbf{k}\uparrow} f_{\mathbf{k}\downarrow} \rangle, \tag{6.49}$$

and

$$n^{\mathrm{h}}_{\mathrm{f}\downarrow}(-\mathbf{k}) = \langle b^{\dagger}_{-\mathbf{k}\downarrow} b_{-\mathbf{k}\downarrow} \rangle. \tag{6.50}$$

These are distribution functions for c- and f-type electrons and holes. The convenience of equation (6.42) is that the entire Hamiltonian is expressed in the k-space. Here $C^{\dagger}_{\mathbf{k}\uparrow}d^{\dagger}_{-\mathbf{k}}$ and $f^{\dagger}_{\mathbf{k}\downarrow}b^{\dagger}_{-\mathbf{k}\downarrow}$ are the spin-fluctuation operators for c bands and f bands. Particle–hole symmetry occurs when particles are mapped onto holes. However, we have mapped only the down-spin electrons on down-spin holes, keeping the up-spin electrons intact. Therefore, negative sign appears before the down-spin energies. The up-spin energies involve quantities like $(1 - n^{\mathrm{h}}_{\downarrow})$. The mapping only becomes a symmetry when both $n^{\mathrm{h}}_{\downarrow} = 1/2$ and the number of holes equals the number of particles.

6.5.2. *High-field ferromagnetic case*

The equations of motion for the number and spin-fluctuation operators can be obtained, using Heisenberg's equation [50]:

$$\frac{d}{dt}F = -\frac{i}{\hbar}[F, H_{\mathrm{T}}] \tag{6.51}$$

as

$$\begin{aligned}\frac{d}{dt}&\langle C^{\dagger}_{\mathbf{k}\uparrow}C_{\mathbf{k}\uparrow} + d^{\dagger}_{-\mathbf{k}\downarrow}d_{-\mathbf{k}\downarrow}\rangle \\ &= -\frac{i}{\hbar}\big[2\mu_{\mathrm{B}}(\langle C^{\dagger}_{\mathbf{k}\uparrow}d^{\dagger}_{-\mathbf{k}\downarrow}\rangle H_{1} - \langle d_{-\mathbf{k}\downarrow}C_{\mathbf{k}\uparrow}\rangle H_{-}) + \{V_{\uparrow}\langle C^{\dagger}_{\mathbf{k}\uparrow}f_{\mathbf{k}\uparrow} - f^{\dagger}_{\mathbf{k}\uparrow}C_{\mathbf{k}\uparrow}\rangle \\ &+ V_{\downarrow}\langle b^{\dagger}_{-\mathbf{k}\downarrow}d_{-\mathbf{k}\downarrow} - d^{\dagger}_{-\mathbf{k}\downarrow}b_{-\mathbf{k}\downarrow}\rangle\}\big] + \gamma_{cn}\langle C^{\dagger}_{\mathbf{k}\uparrow}C_{\mathbf{k}\uparrow} + d^{\dagger}_{-\mathbf{k}\downarrow}d_{\mathbf{k}\downarrow}\rangle,\end{aligned} \tag{6.52}$$

$$\begin{aligned}\frac{d}{dt}&\langle f^{\dagger}_{\mathbf{k}\uparrow}f_{\mathbf{k}\uparrow} + b^{\dagger}_{-\mathbf{k}\downarrow}b_{-\mathbf{k}\downarrow}\rangle \\ &= -\frac{i}{\hbar}\big[2\mu_{\mathrm{B}}\{\langle f^{\dagger}_{\mathbf{k}\uparrow}b^{\dagger}_{-\mathbf{k}\downarrow}\rangle H_{+} - \langle b_{-\mathbf{k}\downarrow}f_{\mathbf{k}\uparrow}\rangle H_{-} + \{V_{\uparrow}\langle f^{\dagger}_{\mathbf{k}\uparrow}C_{\mathbf{k}\uparrow} - C^{\dagger}_{\mathbf{k}\uparrow}f_{\mathbf{k}\uparrow}\rangle \\ &+ V_{\downarrow}\langle d^{\dagger}_{-\mathbf{k}\downarrow}b_{-\mathbf{k}\downarrow} - b^{\dagger}_{-\mathbf{k}\downarrow}d_{-\mathbf{k}\downarrow}\rangle\}\big] + \gamma_{\mathrm{f}n}\langle f^{\dagger}_{\mathbf{k}\uparrow}f_{\mathbf{k}\uparrow} + b^{\dagger}_{-\mathbf{k}\downarrow}b_{-\mathbf{k}\downarrow}\rangle,\end{aligned} \tag{6.53}$$

$$\begin{aligned}\frac{d}{dt}\langle C^{\dagger}_{\mathbf{k}\uparrow}d^{\dagger}_{-\mathbf{k}\downarrow}\rangle &= -\frac{i}{\hbar}\big[\{(\varepsilon_{\mathbf{k}\downarrow} + V_{\downarrow}) - (\varepsilon_{\mathbf{k}\uparrow} + V_{\uparrow})\}\langle C^{\dagger}_{\mathbf{k}\uparrow}d^{\dagger}_{-\mathbf{k}\downarrow}\rangle \\ &+ \mu_{\mathrm{B}}H_{-}\{\langle C^{\dagger}_{\mathbf{k}\uparrow}C_{\mathbf{k}\uparrow} + d^{\dagger}_{-\mathbf{k}\downarrow}d_{-\mathbf{k}\downarrow}\rangle - 1\}\big] + \gamma_{\mathrm{c,eh}}\langle C^{\dagger}_{\mathbf{k}\uparrow}d_{-\mathbf{k}\downarrow}\rangle,\end{aligned} \tag{6.54}$$

and

$$\begin{aligned}\frac{d}{dt}\langle f^{\dagger}_{\mathbf{k}\uparrow}b^{\dagger}_{-\mathbf{k}\downarrow}\rangle &= -\frac{i}{\hbar}\big[\{(\varepsilon_{\mathrm{f}\mathbf{k}\downarrow} + V_{\downarrow}) - (\varepsilon_{\mathrm{f}\mathbf{k}\uparrow} + V_{\uparrow})\}\langle f^{\dagger}_{\mathbf{k}\uparrow}b^{\dagger}_{-\mathbf{k}\downarrow}\rangle \\ &+ \mu_{\mathrm{B}}H_{-}\{\langle f^{\dagger}_{\mathbf{k}\uparrow}f_{\mathbf{k}\uparrow} + b^{\dagger}_{-\mathbf{k}\downarrow}b_{-\mathbf{k}\downarrow}\rangle - 1\}\big] + \gamma_{\mathrm{f,eh}}\langle f^{\dagger}_{\mathbf{k}\uparrow}b^{\dagger}_{-\mathbf{k}\downarrow}\rangle.\end{aligned} \tag{6.55}$$

Here, equations (6.52)–(6.55) have been obtained from equation (6.51) by making the assumption [50] that the average value of the four types of electron–hole operators is the same since we are interested in the variation of the magnetization as a function of the magnetic field:

$$\langle C^{\dagger}_{\mathbf{k}\uparrow} d^{\dagger}_{-\mathbf{k}\downarrow} \rangle = \langle f^{\dagger}_{\mathbf{k}\uparrow} b^{\dagger}_{-\mathbf{k}\downarrow} \rangle = \langle C^{\dagger}_{\mathbf{k}\uparrow} b^{\dagger}_{-\mathbf{k}\downarrow} \rangle = \langle f^{\dagger}_{\mathbf{k}\uparrow} d^{\dagger}_{-\mathbf{k}\downarrow} \rangle. \tag{6.56}$$

In addition, the constants γ_{cn}, γ_{fn} represent dampings of the c and f magnetizations and $\gamma_{\mathrm{c,eh}}$, $\gamma_{\mathrm{f,eh}}$ represent the dampings of the spin-fluctuation amplitudes. Metamagnetic transition is associated with a sharp rise in the magnetization and/or a hysteresis which is a non-equilibrium phenomenon. However, the system relaxes through the damping of the number and the spin-fluctuation amplitudes. These dampings occur partly due to the collision processes among the fermions (electrons and holes) and among the spin excitons (electron–hole pairs). Since the damping mechanisms are statistical in nature and are due to random processes, these have been included phenomenologically [50], which can also be justified from first principles by coupling the system to a heat bath [54].

Since equations (6.52)–(6.56) denote equations for physical quantities, for simplicity, we assume [50]

$$\langle f^{\dagger}_{\mathbf{k}\uparrow} f_{\mathbf{k}\uparrow} + b^{\dagger}_{-\mathbf{k}\downarrow} b_{-\mathbf{k}\downarrow} \rangle = n_{\mathrm{ef}\uparrow} + n_{\mathrm{hf}\downarrow}, \tag{6.57}$$

$$\langle C^{\dagger}_{\mathbf{k}\uparrow} C_{\mathbf{k}\uparrow} + d^{\dagger}_{-\mathbf{k}\downarrow} d_{-\mathbf{k}\downarrow} \rangle = n_{\mathrm{ec}\uparrow} + n_{\mathrm{hc}\downarrow}, \tag{6.58}$$

$$\langle f^{\dagger}_{\mathbf{k}\uparrow} b^{\dagger}_{-\mathbf{k}\downarrow} \rangle = A^{\dagger}_{\mathrm{fb}}, \tag{6.59}$$

and

$$\langle C^{\dagger}_{\mathbf{k}\uparrow} d^{\dagger}_{-\mathbf{k}\uparrow} \rangle = A^{\dagger}_{\mathrm{cd}}. \tag{6.60}$$

Here $n_{\mathrm{ec}\uparrow}$ and $n_{\mathrm{hc}\downarrow}$ describe the distribution functions of up-spin electrons and down-spin holes in the conduction band; and $n_{\mathrm{ef}\uparrow}$ and $n_{\mathrm{hf}\downarrow}$ represent the distribution functions for corresponding electrons and holes in the f band.

The explicit time dependence of these quantities is

$$\begin{aligned} A^{\dagger}_{\mathrm{fb}} &= a^{\dagger}_{\mathrm{fb}} e^{i\omega t}, \\ A^{\dagger}_{\mathrm{cd}} &= a^{\dagger}_{\mathrm{cd}} e^{i\omega t}, \\ H_{\mp} &= h_{\mp} e^{\pm i\omega t}. \end{aligned} \tag{6.61}$$

Using equation (6.61) in equations (6.54) and (6.55), we obtain

$$a^{\dagger}_{\mathrm{cd}} = \frac{\mu_{\mathrm{B}} h_{-}(1 - m_{\mathrm{c}\mathbf{k}})}{\hbar\omega - (1/N)U_{\mathrm{c}} \sum_{\mathbf{k}'} (1 - m_{\mathrm{c}\mathbf{k}'}) + V_{\mu} H_{z} + i\hbar\gamma_{\mathrm{c,eh}}}, \tag{6.62}$$

and

$$a_{\mathrm{fb}}^{\dagger} = \frac{\mu_{\mathrm{B}} h_{-}(1 - m_{\mathrm{f}\mathbf{k}})}{\hbar\omega - (1/N)U_{\mathrm{f}} \sum_{\mathbf{k}'} (1 - m_{\mathrm{f}\mathbf{k}'}) + V_{\mu} H_z + i\hbar\gamma_{\mathrm{f,eh}}}, \tag{6.63}$$

where

$$\begin{aligned} V_{\mu} &= -2(V' + \mu_{\mathrm{B}}), \\ m_{\mathrm{c}\mathbf{k}} &= n_{\mathrm{ec}\uparrow}(\mathbf{k}) + n_{\mathrm{hc}\downarrow}(-\mathbf{k}), \text{ and} \\ m_{\mathrm{f}\mathbf{k}} &= n_{\mathrm{ef}\uparrow}(\mathbf{k}) + n_{\mathrm{hf}\downarrow}(-\mathbf{k}). \end{aligned} \tag{6.64}$$

Here, $m_{\mathrm{c}\mathbf{k}}$ and $m_{\mathrm{f}\mathbf{k}}$ are the average magnetization functions in dimensionless forms for conduction and f bands for strong ferromagnets in which the up-spin electronic bands are occupied and down-spin bands not occupied, i.e., these are filled with down-spin holes. Using equation (6.61) in equations (6.52) and (6.53), one obtains

$$\gamma_{en} m_{\mathrm{c}\mathbf{k}} = -\frac{i}{\hbar} 2\mu_{\mathrm{B}} (a_{\mathrm{cd}}^{\dagger} h_{+} - a_{\mathrm{cd}} h_{-}) - \frac{i}{\hbar} (V_{\uparrow} A_{\mathrm{hy}}^{\mathrm{de}} + V_{\downarrow} A_{\mathrm{hy}}^{\mathrm{dh}}), \tag{6.65}$$

and

$$\gamma_{fn} m_{\mathrm{f}\mathbf{k}} = -\frac{i}{\hbar} 2\mu_{\mathrm{B}} (a_{\mathrm{fb}}^{\dagger} h_{+} - a_{\mathrm{fb}} h_{-}) + \frac{i}{\hbar} (V_{\uparrow} A_{\mathrm{hy}}^{\mathrm{de}} + V_{\downarrow} A_{\mathrm{hy}}^{\mathrm{dh}}). \tag{6.66}$$

Here,

$$A_{\mathrm{hy}}^{\mathrm{de}} = A_{\mathrm{hy}}^{\mathrm{e}} - A_{\mathrm{hy}}^{\mathrm{e}\dagger} \equiv \langle C_{\mathbf{k}\uparrow}^{\dagger} f_{\mathbf{k}\uparrow} - f_{\mathbf{k}\uparrow}^{\dagger} C_{\mathbf{k}\uparrow} \rangle \tag{6.67}$$

and

$$A_{\mathrm{hy}}^{\mathrm{dh}} = A_{\mathrm{hy}}^{\mathrm{h}} - A_{\mathrm{hy}}^{\mathrm{h}\dagger} \equiv \langle b_{-\mathbf{k}\downarrow}^{\dagger} d_{-\mathbf{k}\downarrow} - d_{\mathbf{k}\uparrow}^{\dagger} b_{\mathbf{k}\downarrow} \rangle. \tag{6.68}$$

We have [50] also made the following assumptions:

$$U_{\mathrm{c}}(1 - m_{\mathrm{c}\mathbf{k}}) \cong U_{\mathrm{f}}(1 - m_{\mathrm{f}\mathbf{k}}) \cong U_{\mathrm{av}}(1 - m_{\mathrm{av}}), \tag{6.69}$$

where

$$U_{\mathrm{av}} = \frac{1}{2}(U_{\mathrm{c}} + U_{\mathrm{f}}), \tag{6.70}$$

$$m_{\mathrm{av}} = \frac{1}{2}(m_{\mathrm{c}\mathbf{k}} + m_{\mathrm{f}\mathbf{k}}), \tag{6.71}$$

and

$$\gamma_{c,eh} = \gamma_{f,eh} = \gamma_{eh}. \tag{6.72}$$

From equations (6.62), (6.63), (6.65), and (6.66), we obtain

$$\gamma_m \sum_{\mathbf{k}} m_t(\mathbf{k}) = \frac{4\mu_B^2 \gamma_{eh} h_t^2 \Sigma_{\mathbf{k}}[2 - m_t(\mathbf{k})]}{[(V_\mu H_z/\hbar\gamma_{eh}) - ((U_c + U_f)/4\hbar\gamma_{eh})(1/N)\Sigma_{\mathbf{k}}\{2 - m_t(\mathbf{k})\}]^2 + 1}, \tag{6.73}$$

where

$$m_t(\mathbf{k}) = 2m_{av}(\mathbf{k}), \tag{6.74}$$

$$h_t^2 = H_x^2 + H_y^2 = h^2 - H_z^2, \tag{6.75}$$

and

$$\gamma_m = \gamma_{en} + \gamma_{fn}, \tag{6.76}$$

where γ_m is the damping of the magnetization.

From equations (6.72)–(6.75), we obtain

$$\frac{4\mu_B^2 h^2}{\hbar^2 \gamma_{eh}\gamma_m} = \frac{4\mu_B^2 H_z^2}{\hbar^2 \gamma_{eh}\gamma_m} + \frac{m}{(m_s - m)}\left[1 + \left(\frac{V_\mu H_z}{\hbar\gamma_{eh}} - \frac{\{U_c + U_f\}}{4\hbar\gamma_{eh}}\{m_s - m\}\right)^2\right], \tag{6.77}$$

where

$$m = \frac{1}{N\mu_B}\sum_{\mathbf{k}} m_t(\mathbf{k}) \tag{6.78}$$

and

$$m_s = \frac{1}{N\mu_B}\sum_{\mathbf{k}} 2. \tag{6.79}$$

Here, N is the number of unit cells, and m and m_s are the total and the saturation magnetic moments, respectively, of the atom in the units of μ_B. We note from equation (6.77) that $h \sim m$ is non-linear even in the absence of H_z, and is driven by the transverse part of the magnetic field. For low H_z, m is related to the Pauli spin susceptibility. In the next section, we analyze the low-field paramagnetic susceptibility of itinerant electrons in the presence of c-l hybridization.

6.5.3. Low-field paramagnetic susceptibility

We shall analyze the low-field paramagnetic susceptibility in the limit of $H_z \to 0$, in the presence of many-body and conduction-electron moment and local-moment interactions. The one-electron Green's function $\mathcal{G}(\mathbf{r}, \mathbf{r}', H_z, \mu, \zeta_l)$ in the presence of magnetic field H_z, localized magnetic moments μ_z satisfies the following equation:

$$(\zeta_l - \mathcal{H})\mathcal{G}(\mathbf{r},\mathbf{r}',H_z,\mu,\zeta_l) = \delta(\mathbf{r} - \mathbf{r}'), \tag{6.80}$$

where

$$\mathcal{H} = \mathcal{H}_0 + \mathcal{H}_{s-f} + \Sigma(\mathbf{r},\mathbf{r}',H_z,\mu,\zeta_l) + \mathcal{H}_{H_z}. \tag{6.81}$$

In equation (6.81), $\mathcal{H}_0$ is the unperturbed Hamiltonian, $\zeta_l = (2l + 1)i\pi/2 + \mu$; $l = 0, \pm 1, \pm 2, \ldots$, μ is the chemical potential; the mean-field hybridization Hamiltonian,

$$\mathcal{H}_{s-f} = \frac{1}{2}\sum_i \mathcal{J}(\mathbf{r} - \mathbf{R}_i)\sigma_z\langle S_{iz}\rangle, \tag{6.82}$$

$$\mathcal{H}_{H_z} = \mu_B\sigma_z H_z, \tag{6.83}$$

and $\Sigma_{\xi l}$ is the proper self-energy which, when expanded in powers of H_z and $\langle S_{iz}\rangle$, becomes

$$\Sigma_{\xi_l} = \Sigma_0 + H_z\Sigma_1 + \sum_i \langle S_{iz}\rangle\Sigma_{2i} + H_z^2\Sigma_3, \tag{6.84}$$

where

$$\Sigma_1 = \left(\frac{\partial\Sigma_{\xi_l}}{\partial H_z}\right)_{H_z\to 0}, \tag{6.85}$$

$$\Sigma_{2i} = \left.\frac{\partial\Sigma_{\xi_l}}{\partial\langle S_{iz}\rangle}\right|_{\langle S_{iz}\rangle\to 0}, \tag{6.86}$$

and

$$\Sigma_3 = \left(\frac{\partial^2\Sigma_{\xi_l}}{\partial H_z^2}\right)_{H_z\to 0}. \tag{6.87}$$

Assuming

$$\sum_i \langle S_{iz}\rangle = \frac{1}{2\mu_B}\chi_{cw}H_z, \tag{6.88}$$

equation (6.81) can be written with the help of equations (6.82)–(6.84) and (6.88) as

$$\mathcal{H} = \mathcal{H}_o + \mathcal{H}', \tag{6.89}$$

where

$$\mathcal{H}_o = \frac{p^2}{2m} + \Sigma_0, \tag{6.90}$$

and

$$\mathcal{H}' = \left[\mu_B \sigma_z + \Sigma_1 + \frac{1}{2\mu_B}\chi_{cw}\left\{\frac{1}{2}\mathcal{J}(\mathbf{r} - \mathbf{R})\sigma_z + \Sigma_2\right\}\right]H_z + H_z^2\Sigma_3. \tag{6.91}$$

Here χ_{cw} is the Curie–Weiss susceptibility, $\mathcal{J}$ the exchange integral, and σ_z the Pauli spin matrix. We now consider the thermodynamic potential [55–59]

$$\Omega = \frac{1}{\beta}[Tr\ln(-\mathcal{G}_{\xi_l}) - Tr\Sigma_{\xi_l}\mathcal{G}_{\xi_l} + \varphi(\mathcal{G}_{\xi_l})], \tag{6.92}$$

where $(\mathcal{G}_{\xi l})$ is the abbreviated notation for the one-particle Green's function. *Tr* is defined as $\Sigma_l Tr$, where trace refers to summation over a complete one-particle set, and the functional $\phi(\mathcal{G}_{\xi l})$ is defined as [55–59]

$$\varphi(\mathcal{G}_{\xi_l}) = \lim_{\lambda \to 1} Tr \sum_n \frac{\lambda^n}{2n} \Sigma^{(n)}(\mathcal{G}_{\xi_l})\mathcal{G}_{\xi_l}. \tag{6.93}$$

Here $\Sigma^{(n)}(\mathcal{G}_{\xi l})$ is the *n*th-order self-energy part, where only the interaction parameter λ occurring in equation (6.93) is used to determine the order. $\varphi(\mathcal{G}_{\xi l})$ is defined through the decomposition of $\Sigma^{(n)}(\mathcal{G}_{\xi l})$ into the skeleton diagrams. There are $2n\mathcal{G}_{\xi l}$ lines for the *n*th-order diagrams in $\varphi(\mathcal{G}_{\xi l})$. Differentiating $\varphi(\mathcal{G}_{\xi l})$ with respect to $\mathcal{G}_{\xi l}$ has the effect of "opening" any of the $2n$ lines of the *n*th-order diagram and each will give the same contribution when *Tr* is taken.

The spin susceptibility is given by

$$\chi_S = -\left.\frac{\partial^2\Omega}{\partial H_z^2}\right|_{H_z \to 0} = \chi_{qp} + \chi_{corr}, \tag{6.94}$$

where, from equations (6.92) and (6.94), one obtains

$$\chi_{qp} = -\frac{\partial^2}{\partial H_z^2}\left[\frac{1}{\beta}Tr\ln(-\mathcal{G}_{\xi_l})\right], \tag{6.95}$$

and

$$\chi_{\text{corr}} = Tr\frac{1}{\beta}\left[\frac{\partial^2\Sigma_{\xi_l}}{\partial H_z^2}\mathcal{G}_{\xi_l} + \frac{\partial\Sigma_{\xi_l}}{\partial H_z}\frac{\partial\mathcal{G}_{\xi_l}}{\partial H_z}\right]. \tag{6.96}$$

It can be easily shown [57–59] that

$$\frac{1}{\beta}Tr\ln(-\mathcal{G}_{\xi_l}) = -\frac{1}{2\pi i}tr\oint_C \varphi(\xi)\mathcal{G}(\xi)d\xi, \tag{6.97}$$

where C encircles the imaginary axis in a counterclockwise direction and

$$\varphi(\xi) = -\frac{1}{\beta}\ln[1 + e^{-\beta(\xi-\mu)}]. \tag{6.98}$$

After some lengthy algebra [50], we obtain

$$\begin{aligned}\chi_{\text{qp}} = &-\sum_{\mathbf{k}\rho}\Big[2\langle\mathbf{k}\rho \mid \Sigma_3 \mid \mathbf{k}\rho\rangle f(\varepsilon_{\mathbf{k}\rho}) + \left(\langle\mathbf{k}\rho \mid \mu_{\text{B}}\sigma_z + \Sigma_1 \mid \mathbf{k}\rho\rangle\right)^2 f'(\varepsilon_{\mathbf{k}\rho})\\ &+ \frac{\chi_{\text{cw}}}{\mu_{\text{B}}}\langle\mathbf{k}\rho \mid \mu_{\text{B}}\sigma_z + \Sigma_1 \mid \mathbf{k}\rho\rangle\langle\mathbf{k}\rho \mid V\sigma_z + \Sigma_2 \mid \mathbf{k}\rho\rangle f'(\varepsilon_{\mathbf{k}\rho})\Big]\\ &-\frac{1}{2\mu_{\text{B}}{}^2}(\chi_{\text{cw}})^2\sum_{\mathbf{k}\mathbf{k}'\rho}\langle\mathbf{k}\rho \mid \sigma_z Ve^{-i(\mathbf{k}-\mathbf{k}')\cdot\mathbf{R}_1} + \Sigma_2 \mid \mathbf{k}'\rho\rangle\langle\mathbf{k}'\rho \mid \sigma_z Ve^{i(\mathbf{k}-\mathbf{k}')\cdot\mathbf{R}_2}\\ &+\Sigma_2 \mid \mathbf{k}\rho\rangle\frac{f(\varepsilon_{\mathbf{k}\rho})}{\varepsilon_{\mathbf{k}\rho} - \varepsilon_{\mathbf{k}'\rho}}.\end{aligned} \tag{6.99}$$

Here, $V = 1/2\mathcal{J}$, $|\mathbf{k}\rho\rangle = \psi_{\mathbf{k}\rho}(\mathbf{r}) = \rho e^{i\mathbf{k}\cdot\mathbf{r}}$ and is a plane wave state for spin ρ. $f(\varepsilon_{\mathbf{k}\rho})$ is the Fermi function and $f'(\varepsilon_{\mathbf{k}\rho})$ denotes its derivative with respect to energy. The free-electron basis is valid only when the bare electron mass is replaced by the effective mass of the system [50]. The last term in equation (6.99) is an oscillatory term and arises from an RKKY type of interaction.

Similarly, we can derive an expression for χ_{corr} from equation (6.96) after some lengthy algebra [50],

$$\begin{aligned}\chi_{\text{corr}} = &\sum_{\mathbf{k}\rho}\Big[2\langle\mathbf{k}\rho \mid \Sigma_3 \mid \mathbf{k}\rho\rangle f(\varepsilon_{\mathbf{k}\rho}) + \langle\mathbf{k}\rho \mid \mu_{\text{B}}\sigma_z + \Sigma_1 \mid \mathbf{k}\rho\rangle\\ &\times\langle\mathbf{k}\rho \mid \Sigma_1 + \frac{1}{2\mu_{\text{B}}}\chi_{\text{cw}}\Sigma_2 \mid \mathbf{k}\rho\rangle f'(\varepsilon_{\mathbf{k}\rho}) + \frac{\chi_{\text{cw}}}{2\mu_{\text{B}}}\langle\mathbf{k}\rho \mid \Sigma_1 \mid \mathbf{k}\rho\rangle\\ &\times\langle\mathbf{k}\rho \mid V\sigma_x + \Sigma_2 \mid \mathbf{k}\rho\rangle f'(\varepsilon_{\mathbf{k}\rho})\Big]\\ &+\frac{1}{2\mu_{\text{B}}{}^2}(\chi_{\text{cw}})^2\sum_{\mathbf{k}\mathbf{k}'\rho}\langle\mathbf{k}\rho \mid \Sigma_2 \mid \mathbf{k}'\rho\rangle\langle\mathbf{k}'\rho \mid \sigma_z Ve^{i(\mathbf{k}-\mathbf{k}')\cdot\mathbf{R}_2}\\ &+\Sigma_2 \mid \mathbf{k}\rho\rangle\frac{f(\varepsilon_{\mathbf{k}\rho})}{\varepsilon_{\mathbf{k}\rho} - \varepsilon_{\mathbf{k}'\rho}}.\end{aligned} \tag{6.100}$$

From equations (6.94), (6.99), and (6.100), after some more algebra, we obtain [50],

$$\chi_S = \chi_S^i + \chi_S^h, \tag{6.101}$$

where

$$\chi_S^i = \frac{1}{1-\alpha}\chi_S^0, \tag{6.102}$$

$$\chi_S^0 = -\sum_{\mathbf{k}\rho}\langle\mathbf{k}\rho \mid \mu_B\sigma_z \mid \mathbf{k}\rho\rangle\langle\mathbf{k}\rho \mid \mu_B\sigma_z \mid \mathbf{k}\rho\rangle f'(\varepsilon_{\mathbf{k}\rho}), \tag{6.103}$$

$$\chi_S^h = \chi_{cw}P_S, \tag{6.104}$$

and

$$\begin{aligned} P_S = {} & \frac{1}{1-\alpha}\sum_{\mathbf{k}\rho}\langle\mathbf{k}\rho \mid \sigma_z \mid \mathbf{k}\rho\rangle\langle\mathbf{k}\rho \mid V\sigma_z \mid \mathbf{k}\rho\rangle f'(\varepsilon_{\mathbf{k}\rho}) \\ & - \frac{1}{(1-\gamma)}\frac{\chi_{cw}}{4\mu_B^2}\sum_{\mathbf{k}\mathbf{k}'\rho}\left[\langle\mathbf{k}\rho \mid \sigma_z V \mid \mathbf{k}'\rho\rangle\langle\mathbf{k}'\rho \mid \sigma_z V \mid \mathbf{k}\rho\rangle e^{i(\mathbf{k}-\mathbf{k}')\cdot(\mathbf{R}_1-\mathbf{R}_2)} + \text{c.c.}\right] \\ & \times\frac{f(\varepsilon_{\mathbf{k}\rho})}{\varepsilon_{\mathbf{k}\rho}-\varepsilon_{\mathbf{k}'\rho}}. \end{aligned} \tag{6.105}$$

where

$$\alpha = -U\sum_{\mathbf{k}'\rho} f'(\varepsilon_{\mathbf{k}'\rho}), \tag{6.106}$$

$$\gamma = -U\sum_{\mathbf{k},\mathbf{k}',\rho}\frac{f(\varepsilon_{\mathbf{k}\rho}) - f(\varepsilon_{\mathbf{k}'\rho})}{\varepsilon_{\mathbf{k}\rho}-\varepsilon_{\mathbf{k}'\rho}}, \tag{6.107}$$

and U is defined through

$$\Sigma(\mathbf{k}) = -\frac{1}{\beta}U\sum_{\xi_l}\mathcal{G}_{\xi_l}. \tag{6.108}$$

The relation between U, Σ_1, and Σ_2 is clearly defined in equations (4.32)–(4.34) of Ref. [50].

Here α and γ are the intraband and interband exchange enhancement parameters, χ_S^i is the spin susceptibility of only interacting itinerant electrons, and χ_S^h the contribution to the spin susceptibility from the c-l hybridization. P_S is the EPR shift [60, 61] which is very similar to the Knight shift [57, 58]. The second term in P_S is due to an RKKY type of oscillatory interaction.

6.5.4. Results and discussion

Introducing dimensionless parameters and the relation $m = \chi_P H_z$ where $\chi_P = \chi_S / N\mu_B$, one obtains from equation (6.77),

$$b^2 = gc^2m^2 + \frac{m}{(m_s - m)}[1 + \{(a - d)m + dm_S\}^2], \tag{6.109}$$

where

$$\begin{aligned} b^2 &= \frac{4\mu_B^2 h^2}{\hbar^2 \gamma_{eh} \gamma_m}, c = \frac{2\mu_B}{\hbar \gamma_{eh} \chi_P}, d = d_f + d_c, \\ d_{f/c} &= \frac{U_{f/c}}{4\hbar \gamma_{eh}}, a = c + v, v = \frac{2V'}{\hbar \gamma_{eh} \chi_P}, \text{ and } g = \frac{\gamma_{eh}}{\gamma_m}. \end{aligned} \tag{6.110}$$

In order to compare the theoretical results with the experimental results observed in $CeRu_2Si_2$, this model has been parameterized [50]. In Figure 6.5, the magnetic moment m (μ_B) versus the magnetic field h (T) has been plotted. The damping constants γ_m and γ_{eh} have been chosen as 5.71×10^7 s^{-1} and 1.5×10^{15} s^{-1}. The values of c and d_f have been chosen to be 1.0×10^{-4} and 2.8 as to give average value of χ_P to be 1.17×10^{-4} in cgs units and U_f to be 11.05 eV. v is chosen to be 0.015 so that $V' = 150\ \mu_B$.

In Figure 6.5, the m versus h curve has been plotted for two values of U_c and the theoretical results come closer to the experiment when U_f and U_c are of the same order which implies that both f electron and conduction-electron correlations are important.

In Figure 6.5, we observe that the sudden rise in magnetization takes place roughly at 6.4 T, while the experimental value is about 7.8 T. In addition, in the low-field regime, the theoretical curve should look linear, in contrast to what it looks now. One of the reasons is that while a detail analysis of the paramagnetic susceptibility has been done in Section 6.5.3, there has been no numerical estimate of it. Instead, χ_P has been used as a parameter even though the itinerant and the local characters of the electrons as well as the electron–electron correlations have been included in the theory. The parameterization for $CeRu_2Si_2$ has been done mainly since different heavy fermion systems show different behavior. In addition, a unified picture for a clear and quantitative understanding of the effective mass as a function of the magnetic field is necessary before χ_P can be calculated for a particular heavy fermion system.

Despite qualitative and phenomenological treatments of some mechanisms, our model [50] explains some basic features of metamagnetism in $CeRu_2Si_2$. The weakness of the model is the MFA and attempts are being made to improve the model by going beyond the MFA.

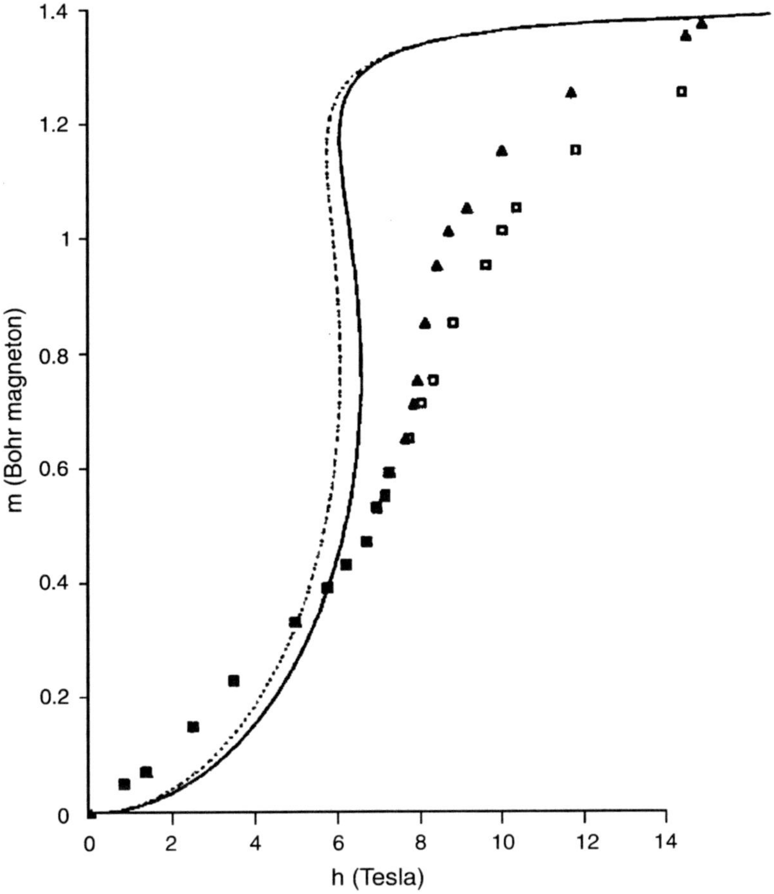

Figure 6.5. Magnetic moment m (μ_B) versus magnetic field h (Tesla). The dotted curve represents the metamagnetic behavior for $U_c = 8.68$ eV and the bold curve for $U_c = 10.48$ eV. The experimental curves are taken from Refs. [5, 12], and represent two temperatures (Δs represent $T = 1.35$ K and □s represent $T = 4.2$ K). Reproduced with the permission of Elsevier from Ref. [50].

References

[1] L. Neel, Ann. Phys. (Paris) **5**, 232 (1936).
[2] L. Neel, Inst. Int. Phys. Solvay Conf. Phys. 10th, 1955, p. 251.
[3] J. Becquerel and J. van der Handel, J. Phys. Radium **10**, 10 (1939).
[4] S.M.M. Evans, Europhys. Lett. **17**, 469 (1992).
[5] P. Haen, J. Flouquet, F. Lapierre, P. Lejay, and G. Remeny, J. Low Temp. Phys. **67**, 391 (1987).
[6] J. Mignot, J. Flouquet, P. Haen, F. Lapierre, L. Puech, and J. Voiron, J. Magn. Magn. Mater. **76–77**, 97 (1988).
[7] C. Sekine, T. Yosida, S. Murrayama, K. Hoshi, and T. Sakakibara, Physica B **186–188**, 511 (1993).

[8] M.B. Suvasini, G.Y. Guo, W.M. Temmerman, and G.A. Gehring, Physica B **186–188**, 860 (1993).
[9] H. Ebert, Phys. Rev. B **38**, 9391 (1988).
[10] G.Y. Guo, W.M. Temmerman, and H. Ebert, Physica B **172**, 61 (1991).
[11] P. Frings and J. Franse, Phys. Rev. B **31**, 4355 (1988).
[12] L. Taillefer, J. Flouquet, and G.G. Lonzarich, Physica B **169**, 257 (1991).
[13] M.R. Norman, Phys. Rev. Lett. **59**, 232 (1987).
[14] G. Aeppli, A. Goldman, G. Shirane, E. Bucher, and M.-Ch. Lux-Steiner, Phys. Rev. Lett. **58**, 808 (1987).
[15] P. Coleman and G.G. Lonzarich, *Narrow Band Phenomena*, p. 31 (NATO ASI series, Plenum, New York, 1988).
[16] D.M. Edwards, *Narrow Band Phenomena*, p. 23 (NATO ASI series, Plenum, New York, 1988).
[17] D.M. Edwards, Physica B **169**, 271 (1991).
[18] J.A. Hertz and D. M. Edwards, J. Phys. F **3**, 2174 (1973).
[19] T. Koyama and M. Tachiki, Phys. Rev. B **34**, 3272 (1986).
[20] J.K. Freericks and L.M. Falicov, Phys. Rev. B **46**, 874 (1992).
[21] P.W. Anderson, Phys. Rev. **124**, 41 (1961).
[22] J. Hubbard, Proc. R. Soc. London, Ser. A **276**, 238 (1963); **277**, 237 (1964); **281**, 401 (1964); **285**, 542 (1965); **296**, 82 (1967); **296**, 100 (1967).
[23] A.B. Harris and R.V. Lange, Phys. Rev. **157**, 295 (1967).
[24] A.H. MacDonald, S.M. Girvin, and D. Yoshioka, Phys. Rev. B **37**, 9753 (1988); **41**, 2565 (1990).
[25] A.M. Oles, Phys. Rev. B **41**, 2565 (1990).
[26] J.C. Parlebas, R.H. Victoria, and L.M. Falicov, J. Phys. (Paris) **47**, 1029 (1986).
[27] J.C. Parlebas, R.H. Victoria, and L.M. Falicov, J. Magn. Magn. Mater. **54–57**, 405 (1986).
[28] A. Reich and L.M. Falicov, Phys. Rev. B **34**, 6752 (1986).
[29] P.K. Misra, D.G. Kanhere, and J. Callaway, Phys. Rev. B **35**, 503 (1987).
[30] J. Callaway, D.P. Chen, D.G. Kanhere, and P.K. Misra, Phys. Rev. B **38**, 2583 (1988).
[31] J. Callaway and D. P. Chen, J. Appl. Phys. **64**, 5944 (1988).
[32] D.P. Chen and J. Callaway, Phys. Rev. B **38**, 11869 (1988).
[33] A. Reich and L.M. Falicov, Phys. Rev. B **37**, 5560 (1988).
[34] J.K. Freericks and L.M. Falicov, Phys. Rev. B **42**, 4960 (1990).
[35] G.S. Tripathi, Phys. Rev. B **52**, 6522 (1995).
[36] G.S. Tripathi, R.L. Hota, B. Mishra, and P.K. Misra, J. Magn. Magn. Mater. **195**, 745 (1999).
[37] Y. Ono, J. Phys. Soc. Jpn. **67**, 6522 (1998).
[38] K. Ohara, K. Hanazawa, and K. Yosida, J. Phys. Soc. Jpn. **68**, 521 (1999).
[39] B. Coqblin and J.R. Schrieffer, Phys. Rev. **185**, 847 (1969).
[40] H. Satoh and F. Ohkawa, Phys. Rev. B **57**, 5891 (1998).
[41] H. Satoh and F. Ohkawa, Phys. Rev. B **63**, 184401 (2001).
[42] F.J. Ohkawa, Phys. Rev. B **46**, 209 (1992).
[43] F.J. Ohkawa, J. Phys. Soc. Jpn. **67**, 525 (1998).

[44] C. Paulsen, A. Lacerda, L. Puech, P. Haen, P. Lejay, J.L. Tholence, J. Flouquet, and A. de Visser, J. Low Temp. Phys. **81**, 317 (1990).
[45] T. Muoto, Phys. Rev. B **62**, 15589 (2000).
[46] D. Meyer, T. Wegner, M. Potthoff, and W. Nolting, Physica B **270**, 225 (1999).
[47] D. Meyer and W. Nolting, Phys. Rev. B **61**, 13465 (2000).
[48] D. Meyer and W. Nolting, Phys. Rev. B **62**, 5657 (2000).
[49] D. Meyer and W. Nolting, Phys. Rev. B **64**, 052402 (2001).
[50] G.S. Tripathi, B.G. Mahanty, and P.K. Misra, Physica B **339**, 164 (2003).
[51] P.K. Misra and L.M. Roth, Phys. Rev. **177**, 1089 (1969).
[52] S.K. Misra, P.K. Misra, and S.D. Mahanti, Phys. Rev. B **26**, 1903 (1982).
[53] N.N. Bogoliubov, *Lectures on Quantum Statistics, Vol. 2* (McDonald, London, 1971).
[54] L.M. Keldysh, Sov. Phys. JETP **20**, 1018 (1965).
[55] J.M. Luttinger and J.C. Ward, Phys. Rev. **118**, 1417 (1960).
[56] F.A. Buot, Phys. Rev. B **14**, 3310 (1976).
[57] G.S. Tripathi, L.K. Das, P.K. Misra, and S.D. Mahanti, Solid State Commun. **38**, 1207 (1981).
[58] G.S. Tripathi, L.K. Das, P.K. Misra, and S.D. Mahanti, Phys. Rev. B **25**, 3091 (1982).
[59] G.S. Tripathi, J. Phys. C **18**, L1157 (1985).
[60] G.S. Tripathi, Phys. Lett. A **115**, 169 (1986).
[61] G.S. Tripathi, Phys. Rev. B **35**, 5247 (1987).

Chapter 7

Heavy Fermion Superconductors (Ce-Based Compounds)

7.1. Overview

Superconductivity and magnetism were considered to be antagonistic phenomena since the local magnetic moments break up the spin-singlet state of the Cooper pairs [1]. In fact, very low concentrations of paramagnetic impurities suppress superconductivity in a classical BCS superconductor. Therefore, the pioneering discovery of superconductivity in $CeCu_2Si_2$ by Steglich et al. [2] in 1979 created a new scenario for several reasons. First, since the effective mass of the electrons in this heavy fermion system is a few hundred times the electron mass below a coherence temperature, the pairing of such heavy quasi-particles through the electron–phonon mechanism, as postulated in the BCS theory [3], was unlikely. Second, since the Ce ion in this heavy fermion compound is trivalent ($4f^1$), it has a local magnetic moment. Usually less than 1 at% of Ce^{3+} ions are sufficient to suppress superconductivity in a classical (BCS) superconductor. However, in $CeCu_2Si_2$, 100 at% of Ce^{3+} ions are necessary to obtain the superconducting state. The reference compound $LaCu_2Si_2$ which is based on non-magnetic La^{3+} ($4f^0$) is not a superconductor [4] and a few at% La^{3+} substituted for Ce^{3+} destroys the superconducting state in $CeCu_2Si_2$. Thus, the non-magnetic impurities are pair breaking in this material in contrast to classical superconductors. Third, the physical properties in the superconducting phase of $CeCu_2Si_2$ at low temperatures have shown power-law temperature dependence which is different from the exponential decay observed in the conventional s-wave superconductors. Thus, the discovery of superconductivity in $CeCu_2Si_2$, which is an even-parity superconductor with line nodes, became the pioneering work which led to the search for unconventional superconductivity in strongly correlated electron systems (SCES).

Since 1979, approximately 25 unconventional superconductors have been discovered in heavy fermion systems. While most of these systems are Ce- and U-based compounds, a few others are quasi-two-dimensional (2D) in nature and filled skutterdites. The multiphase diagrams in UPt_3 [5–7] and U $(Be_{1-x}Th_x)_{13}$ [8–13] indicate unusual superconductivity with multicomponents. In fact, UPt_3 is the first odd-parity superconductor to be discovered in heavy fermion systems [14, 15]. UPd_2Al_3 [16] and UNi_2Al_3 [17] are unconventional superconductors coexisting with the AF phase and are considered to have even- and odd-parity pairing states, respectively [18–20]. There is coexistence of hidden order and unconventional superconductivity in URu_2Si_2 [21]. UPt_3 [5], URu_2Si_2 [22], UNi_2Al_3 [17], and UPd_2Al_3 [16] have the following common features: (a) they order antiferromagnetically below T_N, ranging from 5 to 17 K, (b) they exhibit, well below T_N and coexisting with AF order, a heavy Landau-Fermi-liquid (LFL) state which becomes unstable against a superconducting transition at T_c (ranging between 0.5 and 2 K).

Recently, a variety of heavy fermion Ce-based superconductors have been discovered due to progress in experiments under pressure. These include $CeCu_2Ge_2$ [23–25], $CePd_2Si_2$ [26, 27], $CeRh_2Si_2$ [28], $CeNi_2Ge_2$ [29], and $CeIn_3$ [27, 30]. These materials, which have the same $ThCr_2Si_2$-type crystal structure as $CeCu_2Si_2$ (except $CeIn_3$) are AF metals at ambient pressure, while under high pressures, the AF phases abruptly disappear accompanied by SC transitions.

The family of $CeTIn_5$ (T=Co, Rh, and Ir), which has $HoCoGa_5$-type crystal structure [31–33], has attracted great deal of attention since they possess relatively high transition temperature (T_c) such as T_c = 2.3 K for $CeCoIn_5$, which is the highest among Ce- and U-based heavy fermion superconductors. Another interesting example is the coexistence of superconductivity and ferromagnetism in UGe_2 at high pressure and URhGe at ambient pressure [34, 35]. It is generally believed that a triplet pairing state coexists with the ferromagnetic phase in such superconductors. The unconventional T^3-behavior has been observed in the NQR/NMR-$1/T_1$ [36].

The heavy fermion superconductors which do not have either Ce or U atom include the filled skutterdite compound $PrOs_4Sb_{12}$ (T_c = 1.85 K), where the possibility of double transition [37] was observed. A very high transition temperature, T_c = 18.5 K, has been reported in the Pu-based compound $PuCoGa_5$ which has the same crystal structure as the family of $CeTIn_5$ [38].

There is a great variety of ground states of the heavy fermion superconductors. However, it is not possible to review the superconductivity of each heavy fermion compound. In this chapter, we shall discuss the superconductivity in some prototype Ce-based compounds. We shall discuss the superconductivity in U-based compounds in Chapter 8 and that of other heavy fermion superconductors in Chapter 9. We shall briefly discuss the various theoretical models for these superconductors in Chapter 10 with the warning that this is an area where the experimentalists are way ahead of the theorists like in other areas of heavy fermion systems.

7.2. $CeCu_2Si_2$

As mentioned earlier, Steglich et al. [2] had discovered the first heavy fermion superconductor, $CeCu_2Si_2$, in 1979. Their results for the low-temperature properties of polycrystalline $CeCu_2Si_2$ is shown in Figure 7.1 by plotting the resistivity for $B = 0$ and $B = 0.1$ T and the low-field AC magnetic susceptibility (inset) as a function of temperature. Their results indicated a bulk superconducting transition below $T_c \approx 0.6$ K. The inset of Figure 7.1 clearly shows that the $CeCu_2Si_2$ sample has a non-magnetic low-T state and a Pauli-like susceptibility before turning superconducting. In Figure 7.2, the ratio of the specific heat (C) and temperature (T), C/T, has been plotted as a function of temperature for two polycrystalline samples of $CeCu_2Si_2$.

The linear T-dependence of the resistivity (Figure 7.1) and the negative temperature coefficient in the Sommerfeld coefficient in the normal (n)-state electronic specific heat, $\gamma = \Delta C(T)/T$ (Figure 7.2), indicated the vicinity of a magnetic instability. As mentioned earlier, $CeCu_2Si_2$ is an even-parity superconductor with line nodes. The pair correlation

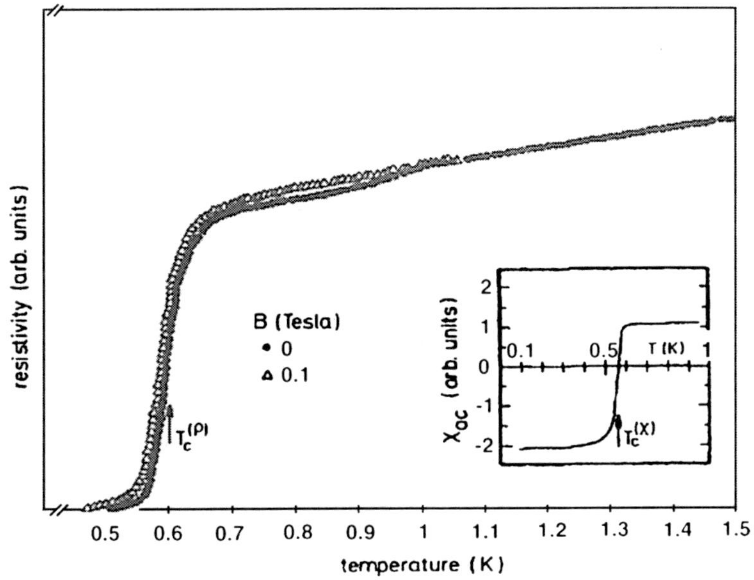

Figure 7.1. Low-temperature properties of polycrystalline $CeCu_2Si_2$: Resistivity (main part) and low-field ac magnetic susceptibility (inset) as a function of temperature. Reproduced with the permission of Elsevier from Ref. [39].

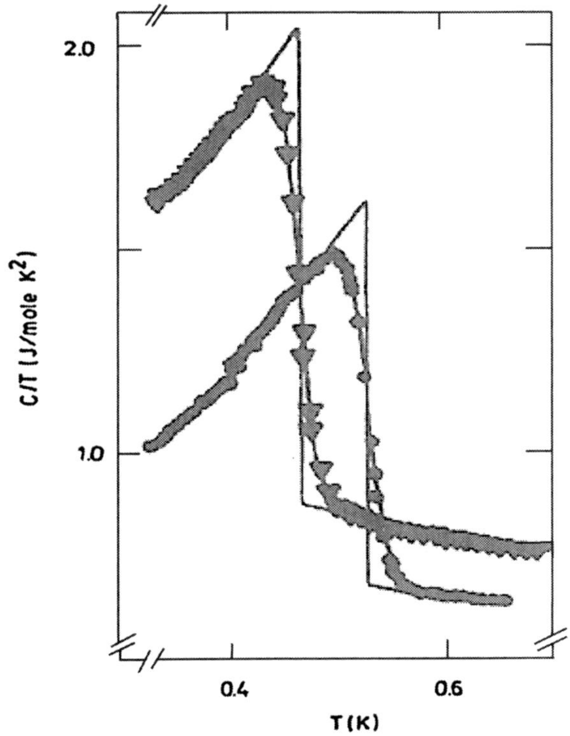

Figure 7.2. Low-temperature C/T versus T for two samples of polycrystalline $CeCu_2Si_2$. Reproduced with the permission of Elsevier from Ref. [39].

function in a superconductor can be written as [40]

$$F_{\alpha\beta}(x) = \langle \psi_\alpha(x)\psi_\beta(0)\rangle, \tag{7.1}$$

where

$$F_{\alpha\beta}(x) = [i\sigma_2(F_s(x) + \sigma \cdot F_p(x)]_{\alpha\beta}. \tag{7.2}$$

$F_{\alpha\beta}$ is expanded in terms of a singlet, even-parity pair function $F_s(x) = F_s(-x)$, and a triplet, odd-parity pair function $F_p(x) = -F_p(-x)$. Either F_s or F_p is zero when parity is a good quantum number.

From Figure 7.2, it may be noticed that the n-state value of $\gamma(T \to 0)$ is about 1 J/K^2 mol which scales with the huge jump height $\Delta C(T_c)/T_c$. This indicates that the quasiparticles with large effective mass ($m^* \approx 10^3 m_e$) form massive Cooper pairs. The renormalized Fermi velocity v_F^* of these heavy fermions (HFs) is correspondingly reduced compared to v_F of the free electrons in a simple metal like Cu. Since the entropy associated with the Fermi liquid (FL) and the superconductor is derived from f-spin entropy, heavy electron quasiparticles are combinations of f-spins and conduction electrons (s-, p-, and d-states). Thus, they represent a type of composite fermions with a dominating local 4f component and a mixture of itinerant conduction electron contributions. However, the quasiparticles carry an elementary charge $|e|$ and a spin 1/2 as concluded from the empirical Kadowaki–Woods relation [41]

$$\frac{A}{\gamma^2} = W \sim 1 \times 10^{-5}\, \mu\Omega\, \text{cm}\, (\text{K mol/mJ})^2, \tag{7.3}$$

between the quadratic "A" coefficient of the resistivity ($\rho = \rho_0 + AT^2$) and the specific heat coefficient $\gamma = C/T$, as well as from the Sommerfeld–Wilson [42] ratios.

The inclusion of this "Kondo physics" into heavy fermion superconductivity is a theoretical challenge. The HFs, severely hindered in their mobility by the strong on-site Coulomb correlations acting on the 4f shells, are formed along with the local Kondo singlets. Thus, v_F^* is of the same order as v_s, the velocity of sound, and the retardation of the electron–phonon interaction which is essential in BCS theory is absent. As discussed in Chapter 3, the rare-earth-based heavy fermion metals exhibit non-Fermi-liquid (NFL) phenomena due to the existence of "nearby" AF quantum phase transition (QPT). The RKKY and the local Kondo interaction, which are the two competing fundamental interactions, cancel each other resulting in "residual interactions" between the quasiparticles.

The sample dependence of $CeCu_2Si_2$, which is clearly evident from Figure 7.2, was resolved by Geibel and co-workers [43] who studied the ternary chemical Ce–Cu–Si phase diagram. Within the narrow homogeneity range of the primary 1:2:2 phase, which allows for an exchange between Cu and Si atoms by not more than 1%, several ground-state properties were obtained. In Figure 7.3, the different phases measured by a general coupling parameter g, which occurs at different strengths of the 4f-conduction electron hybridization, is shown [4]. At the weak-coupling side, it shows coexistence (at sufficiently low T) between SC and AF ordered "A-phase" [44]. From measurements of $CeCu_2Si_2$ single crystals, it was identified as an incommensurate SDW phase with very small ordered moment [45]. Near the true stoichiometry point and at stronger coupling, SC expels the

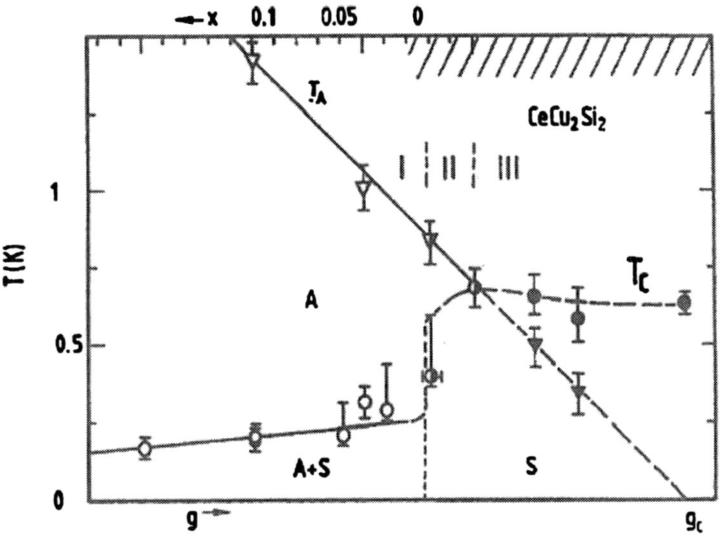

Figure 7.3. Generic phase diagram of $CeCu_2Si_2$ combining data for high-quality polycrystals, undoped ones from the homogeneity range (hatched) and Ge-doped ones, $CeCu_2(Si_{1-x}Ge_x)_2$. The coupling constant g was assumed to be linear in $(1 - T_A)$. Sectors I, II, and III indicate samples "A", "A + S", and "S" type, respectively. Reproduced with the permission of Elsevier from Ref. [4].

A-phase which can be recovered by applying a large magnetic field which can suppress SC. At the QCP, the A-phase smoothly disappears as a function of the 4f-conduction electron hybridization.

The QPT for $CeCu_2Si_2$ at $g = g_c$ (Figure 7.3), where the A-phase disappears continuously as a function of pressure and/or Cu excess, is called an itinerant or SDW-type QPT. In addition, anomalous NFL power laws have been observed in the temperature-dependences of electrical resistivity ($\Delta\rho \sim T^{1.5}$) and the Sommerfeld coefficient ($\gamma = \gamma_0 - \beta T^{0.5}$) [46, 47]. Such a quantum critical state is called a "nearly antiferromagnetic FL" and is typical of three-dimensional critical fluctuations [48–50].

Figure 7.4 shows how the doping of 10 at% Ge on Si sites in the superconducting state of $CeCu_2Si_2$ could be weakened, so that SC occurs only below a reduced T_c, under a narrow "dome" centered around the AF-QCP [51]. A second superconducting "dome" is found upon further increasing the pressure.

The first SC narrow "dome" at low pressure is believed to be due to extended AF spin fluctuations (paramagnons) [27]. In fact, while the quantum critical fluctuations are low lying in energy, high-frequency fluctuations are required to form Cooper pairs [52–55]. The second superconducting dome coincides with a weak valence transition and it is believed that the Cooper pair is due to the exchange of virtual fluctuations of the charge density [56–60].

The origin of valence fluctuation superconductivity in some Ce-based compounds is due to the fact that in metallic Ce, the phase diagram shows a first-order valence discontinuity line [61]. This line separates the γ-Ce with a 4f shell occupation $n_f = 1.0$ from the α-Ce with $n_f \approx 0.9$. The valence transition is isostructural and the line has a critical end point in the vicinity of $p_{cr} = 2$ GPa and $T_{cr} = 600$ K. In cases where p_{cr} is positive,

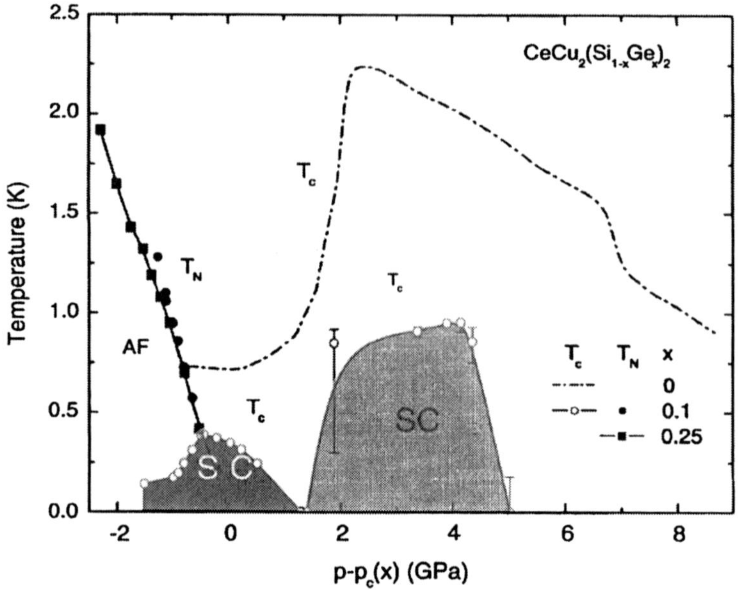

Figure 7.4. T–P phase diagram of $CeCu_2(Si_{1-x}Ge_x)_2$ showing transition temperatures into the "A" phase (T_N, closed symbols) and the superconducting state (T_c, open symbols) versus relative pressure $\Delta p = p - p_c$. The magnetic transition lines for $x = 0.1$, $x = 0.25$, and $x = 1$ coincide since appropriate p_c values are chosen. Reproduced with the permission of Elsevier from Ref. [39].

either T_{cr} is very high [62] or T_{cr} is negative and only a crossover regime is accessible even at $T = 0$. The exceptions are $CeCu_2Si_2$ and $CeCu_2Ge_2$ for which T_{cr} is likely positive although small. In such a situation, the associated low-energy valence fluctuations can mediate superconductivity [63].

Miyake [58] has proposed that the superconducting phase diagram for $CeCu_2(Ge,Si)_2$ (shown in Figure 7.5) exhibits a maximum in the transition temperature in close vicinity to a valence-changing critical point [59]. Miyake has argued that superconductivity may develop around the region where the critical end point is suppressed to zero, to become a QCP.

It is now clear that for $CeCu_2Si_2$, the critical valence fluctuations (VF), between Ce $4f^1$ and $4f^0$ + [5d6s] electronic configurations, form the basis of the pairing mechanism. In Figure 7.6, the resistance of $CeCu_2Si_2$ below 4 K and up to 6.5 GPa in hydrostatic condition is shown [60]. The width of the resistive superconducting transition varies strongly with pressure. It is sharp at the lowest pressure, where T_c is close to the ambient pressure value. Near the maximum of the transition temperature, the transition sharpens again but does not coincide with the narrowest transition in the high-pressure region.

The upper part of the resistance drops in the broad transition at 1.76 GPa could be suppressed by increasing the measurement current, leading to a sharp transition with a width close to the ambient pressure value. The $R = 0$ point could be suppressed more rapidly with a magnetic field than the upper part. The recently discovered heavy fermion superconductor $CeNiGe_3$ [64] displays similarly broad transition widths, along with a large width in residual resistivity, and is the most likely candidate for VF-mediated superconductivity in another compound. The relative position of the magnetic and valence

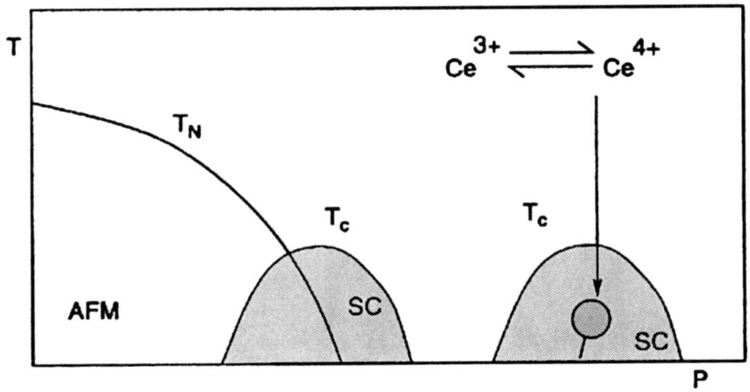

Figure 7.5. Schematic phase diagram for $CeCu_2(Ge,Si)_2$ (Ref. [58]) illustrating possible valence fluctuation critical point beneath the superconducting dome at high pressures. Reproduced with the permission of Elsevier from Ref. [40].

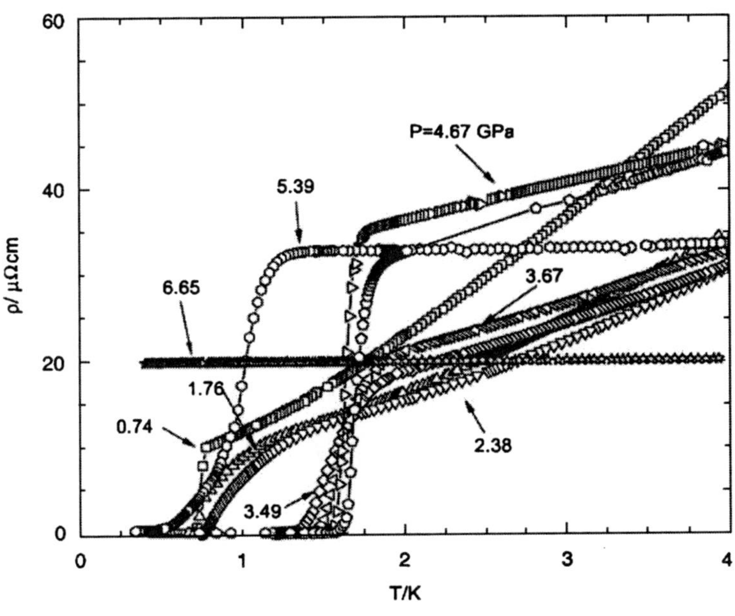

Figure 7.6. Resistive superconducting transition in $CeCu_2Si_2$ in a helium pressure-transmitting medium. The large and pressure-dependent transition widths are evident. Reproduced with the permission of Elsevier from Ref. [60].

instability pressures P_c and P_v plays an important role in the presence or otherwise of VF-mediated superconductivity.

Figure 7.7 shows the superconducting transition in $CeCu_2Si_2$ as detected by AC calorimetry in a helium pressure medium [60]. The beginning of the C_p transition corresponds to the temperature at which the resistivity reaches zero. The jump starts off relatively small and sharp, then grows much bigger as T_c increases. However, the specific heat peak starts to broaden and collapse before T_c reaches a maximum.

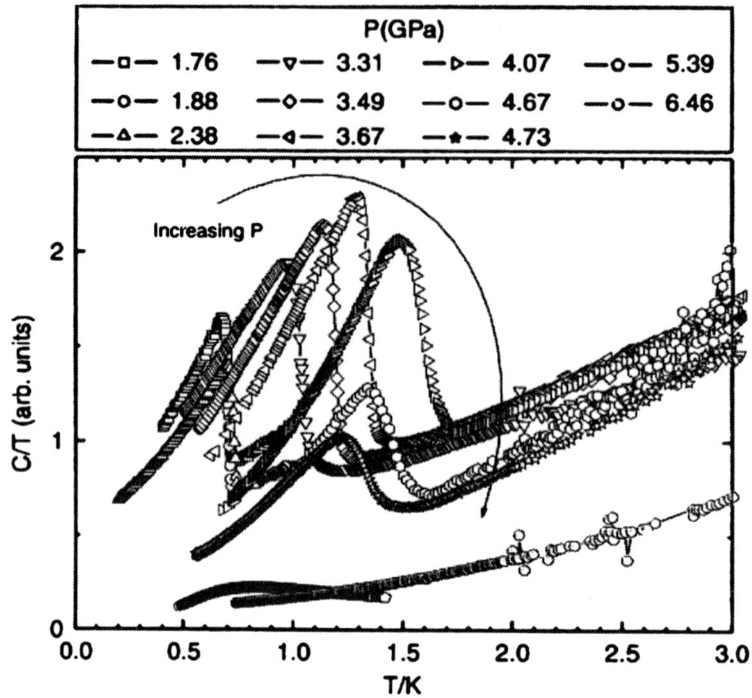

Figure 7.7. Superconducting specific heat jump of $CeCu_2Si_2$ at different pressures extracted from the ac calorimetry signal and simultaneously obtained with ρ (Figure 7.6). The amplitude of the jump seems to be closer to $P_v \approx 4.5$ GPa than the low pressure near P_c. Reproduced with the permission of Elsevier from Ref. [60].

There is a large enhancement of the residual resistivity ρ_0 of $CeCu_2Si_2$ under pressure which is shown in Figure 7.8 for two samples [60]. The coexistence of superconductivity with such enhanced impurity scattering is highly unlikely in a spin-fluctuation mediation scenario. The superconducting transitions are shown both at ambient pressure, and at 4.34 GPa, close to the maximum of ρ_0. The sample labeled # 50 has a slightly lower T_c at ambient pressure than sample # 57, along with a residual resistivity of the order of 70 $\mu\Omega$ cm if the normal state resistivity is extrapolated to $T = 0$.

It may be noted from Figure 7.8 that at 4.34 GPa, the normal state resistivity at T_c has more than doubled in sample # 50, and increased substantially in sample # 57 (which was slightly corrected for experimental difficulties [65]). Similar behavior has been observed in crystalline samples [24]. The enhancement of the impurity scattering is reduced as the temperature increases and the two samples exhibit similar behavior approaching room temperature. These results indicate that there is a VF-enhanced impurity scattering, in addition to the usual electron–electron, phonon, and magnetic scattering terms to the temperature-dependent contributions to the resistivity.

Holmes et al. [59] have performed simultaneous resistivity and ac specific heat measurements under pressure on single-crystalline $CeCu_2Si_2$ to over 6 GPa in a hydrostatic helium pressure medium. Their plot of T_c (K) versus P (GPa) is shown in Figure 7.9 which shows the superconducting phase diagram determined by both resistivity and specific

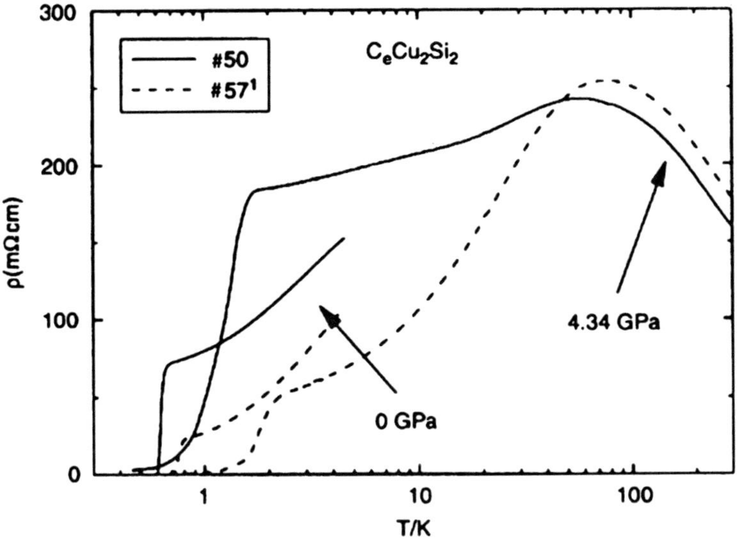

Figure 7.8. Resistivity at ambient pressure and close to P_V of two polycrystalline $CeCu_2Si_2$ samples. The large increase in residual resistivity under pressure is combined with a nearly complete resistive transition. Reproduced with the permission of Elsevier from Ref. [60].

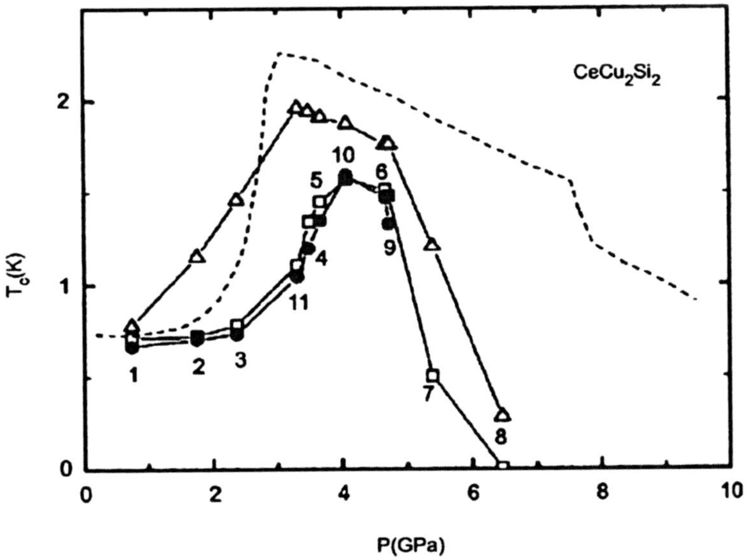

Figure 7.9. $T_c(P)$ in $CeCu_2Si_2$ determined from resistivity and specific heat measurements. The triangles show T_c determined from the onset of the resistive transition (T_c^{onset}), the squares show the completion ($T_c^{R=0}$), and the solid circles show the midpoint of the specific heat jump. The numbers indicate the sequence of pressures. The dotted line shows T_c determined by susceptibility in a different sample, also in a helium pressure medium (Ref. [66]). Reproduced with the permission of the American Physical Society from Ref. [59].

heat, on increasing and decreasing the pressure. Two qualitatively different types of behavior can be seen in the same sample, represented by the onset and completion of the resistive transition. The transition onset T_c^{onset} (P) has sharp kinks which behaves differently from the temperature $T_c^{R=0}$ (P), at which the resistance vanishes. In fact, $T_c^{R=0}$ (P) has a narrower peak.

7.3. $CeCu_2Ge_2$

The RKKY interaction dominates the Kondo effect in $CeCu_2Ge_2$ which orders antiferromagnetically at ambient pressure. The magnetic ordering (T_M) and superconducting transition (T_c) temperature as a function of P has been studied by the high-pressure resistivity of single-crystalline $CeCu_2Ge_2$ [24, 67]. This has also led to the establishment of various links between properties of the normal and superconducting phase. The Kondo temperature (T_K) is small in comparison to the crystal field (CF) splitting $\Delta_{CF} = 191$ K [68]. The magnetic resistivity has two maxima at temperatures T_1^{max} and T_2^{max}, reflecting the Kondo scattering on the ground state and excited CF levels, respectively. In Figure 7.10(a), the P-dependence of these temperatures is plotted along with T_M and T_c [25]. There is a strong P-variation of T_1^{max} whereas T_2^{max} is nearly constant. In the pressure range $7 < P < 25$ GPa, the nearly exponential rise of T_1^{max} of about two orders of magnitude probably corresponds to the enhancement of the Kondo coupling which then dominates the RKKY interaction. At low P, there is evidence of two magnetic transitions. At $P_c \approx 9.4$ GPa, T_M saturates at about 2 K. Close to P_c, superconductivity suddenly emerges and the transition temperature $T_c \approx 0.6$ K remains constant up to ~ 13 GPa. T_c increases above $P = 13$ GPa and reaches a maximum of $T_c = 2$ K at $P = 16$ GPa. At this pressure, T_1^{max} and T_2^{max} merge, indicating an intermediate valence regime when the Kondo temperature T_K becomes the same order as that of the CF splitting.

From Figure 7.10(b), it is evident that the maximum of $T_c(P)$ is correlated to the peak in ρ_0 and to the drop of A. A is related to the effective mass m^* by $A \propto (m^*)^2$. In Figure 7.10(b), the $A(P)$ dependence above P_c can be divided into two parts. In the range $P_c < P < \sim 16$ GPa, a smooth decrease occurs while A drops by more than two orders of magnitude for $P > \sim 16$ GPa. The drop of A is likely due to the change of the ground-state degeneracy when the system crosses from the Kondo lattice state towards the intermediate valence state.

7.4. $CePd_2Si_2$

$CePd_2Si_2$ is isostructural to $CeCu_2Si_2$ ($ThCr_2Si_2$ structure) but differs in the number of d electrons in the d metal constituent and therefore in the nature of the Fermi surface and in the magnetic properties. $CePd_2Si_2$ orders antiferromagnetically below a Neel temperature T_N of about 10 K [69]. The magnetic nature can be pictured as ferromagnetic planes aligned perpendicular to the [1 1 0] direction, stacked with the moment of $\sim 0.7\ \mu_B$ pointing along the [1 1 0] direction. Grosche et al. [26] have investigated the crossover of $CePd_2Si_2$ from a magnetically ordered to a non-magnetic spin liquid state in a series of resistance measurements under hydrostatic pressure up to 30 kbar and at temperatures down to below 200 mK. Near the critical pressure, p_c, of ~ 28 kbar, at which the magnetic

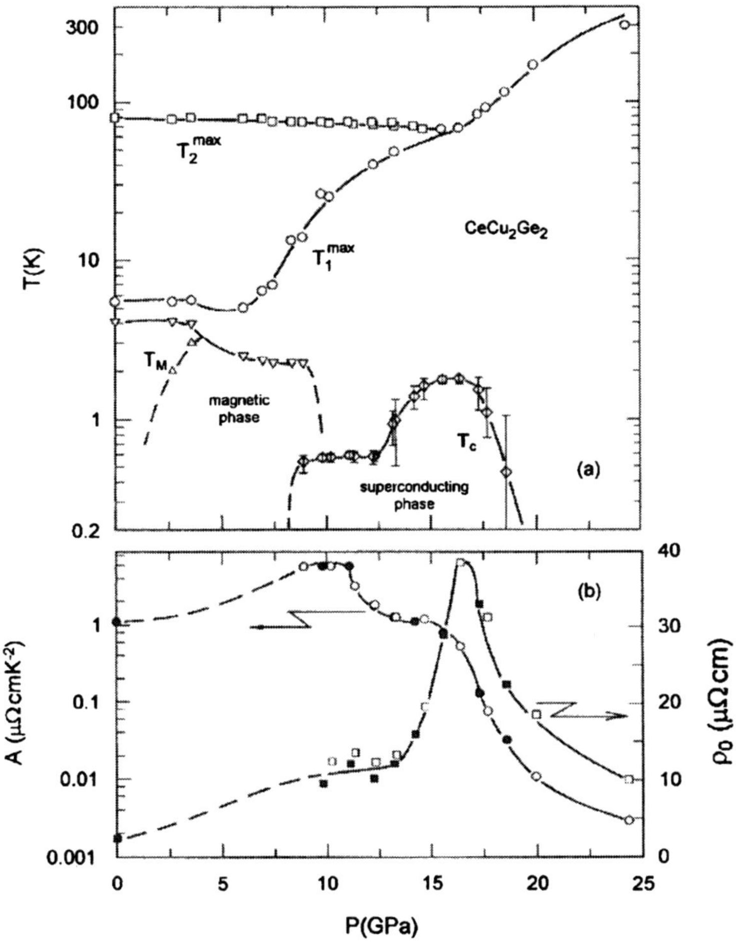

Figure 7.10. (a) Pressure dependence of the temperatures of magnetic ordering T_M, superconducting transition T_c and the maxima of the magnetic resistivity T_1^{max} and T_2^{max} for $CeCu_2Ge_2$. Vertical bars indicate the width of T_c ($\triangle T_c = T_c^{10\%} - T_c^{90\%}$). (b) Pressure dependence of the residual resistivity ρ_0 and the coefficient A in the relation $\rho = \rho_0 + AT^2$ for the same sample. Reproduced with the permission of Elsevier from Ref. [25].

ordering temperature is extrapolated to zero, the electrical resistivity exhibits a quasilinear variation over two orders of magnitude in temperature. This NFL form of $\rho(T)$ extends down to the onset of a new superconducting transition temperature below 430 mK. These results are shown in Figure 7.11.

The striking feature of the experimental result is the transition from an NFL state to a superconducting state at high pressures and low temperatures. The initial slope $\partial B_{c2}/\partial T$ is much higher in $CePd_2Si_2$ at high pressures than in conventional superconductors. The high slope together with the low transition temperature implies a short coherence length, which is consistent with a high effective mass of the quasiparticles forming Cooper pairs. Thus, the electronic interactions in heavy fermion compounds can be influenced in such a way that high pressure favors the Kondo interaction in Ce compounds [27]. Thus, pressure suppresses long-range magnetic order and enhances the screening of the localized

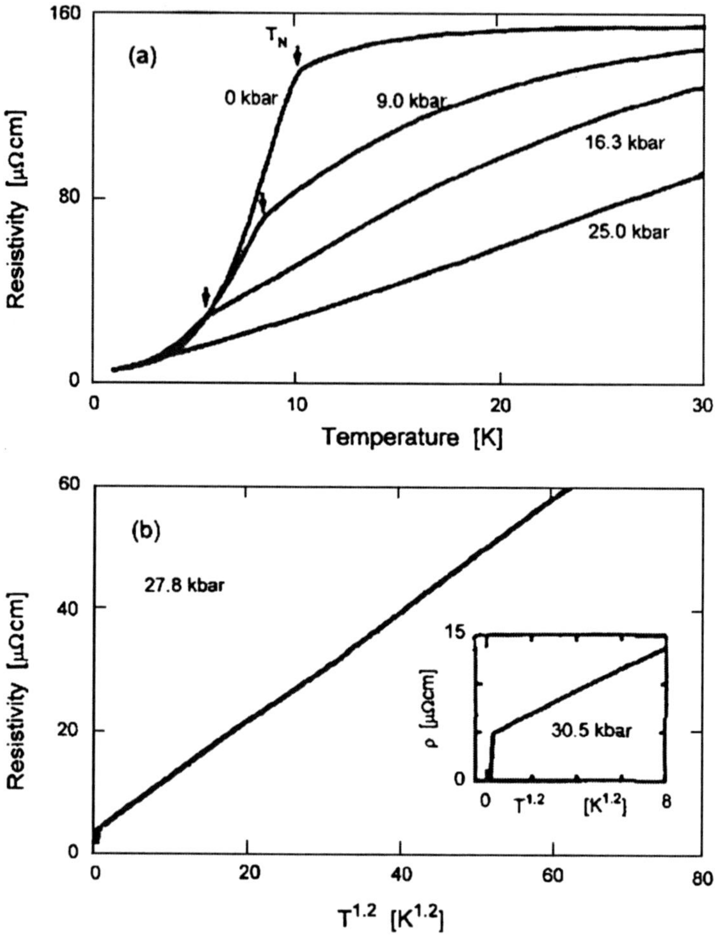

Figure 7.11. (a) The temperature dependence of the resistivity, ρ, of $CePd_2Si_2$ along the a axis at different pressure. The Neel temperature, T_N, marked by arrows, is visible as a sudden change in the slope of $\rho(T)$. (b) Near to and above the critical pressure, p_c, of ~28 kbar, a superconducting transition sets in at about 430 mK (inset, b) below a quasilinear region in $\rho(T)$, which extends over two orders of magnitude in temperature. Reproduced with the permission of Elsevier from Ref. [26].

4f electrons. If both interactions are of similar strength in the vicinity of a critical pressure p_c, a deviation from the FL behavior is observed and some Ce compounds attain a superconducting ground state.

7.5. $CePd_2Ge_2$

$CePd_2Ge_2$ enters an antiferromagnetically ordered phase at $T_N \approx 5.1$ K [70–73] while its Si counterpart $CePd_2Si_2$ exhibits a similar magnetic structure with $T_N \approx 10$ K [69]. Several groups have confirmed the occurrence of a superconducting ground state if the magnetic order is suppressed by external pressure ($p_c \approx 2.8$ GPa) [26, 27, 74–78].

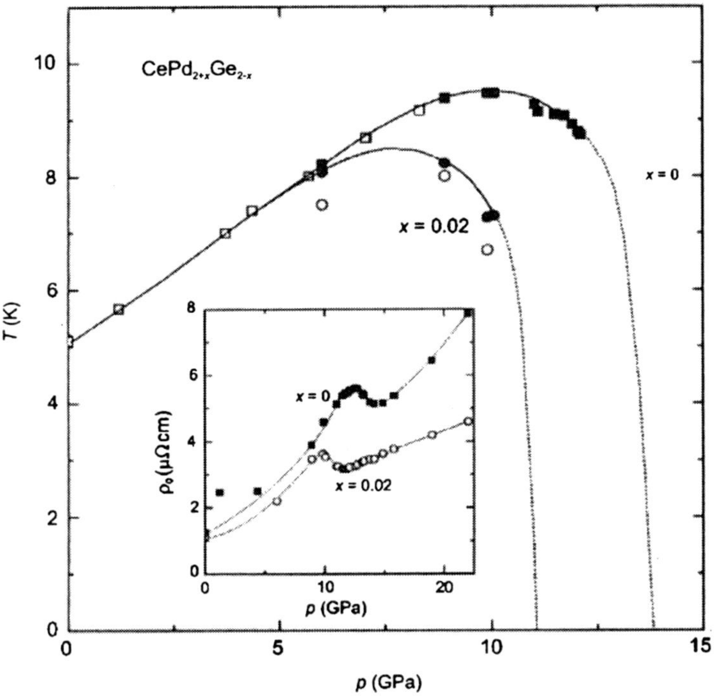

Figure 7.12. Pressure dependence T_N for $CePd_{2+x}Ge_{2-x}$ ($x = 0$ and $x = 0.02$). The T_N values for $x = 0.02$ obtained by the ac calorimetry are represented by open circles. Inset shows that a pronounced variation of the residual resistivity ρ_0 with pressure is observed in both compounds. Reproduced with the permission of the American Physical Society from Ref. [79].

Wilhelm and Jaccard [79] have experimentally obtained the (T, p) phase diagram of $CePd_{2+x}Ge_{2-x}$ (for $x = 0$ and $x = 0.02$) which is shown in Figure 7.12. The extrapolation $T_N \rightarrow 0$ assumes critical pressure $p_c = 11.0$ GPa ($x = 0.2$) and 13.8 GPa ($x = 0$).

The extrapolation $T_N \rightarrow 0$ (dotted lines in Figure 7.12) is based on the assumption that the maximum in the $\tilde{A}(p)$ (defined from $\rho_{mag}(T) = \rho_0 + \tilde{A}T^n$) can be taken as critical pressure. The inset of Figure 7.12 indicates how the verge of magnetism affects the residual resistivity ρ_0. It is very sensitive to small pressure changes and exhibits anomalies around p_c. ρ_0 attains a local maximum just below p_c and passes through a local minimum above p_c.

The (T, p) phase diagram of $CePd_{2+x}Ge_{2-x}$ can be explained by Doniach' model [80]. $T_K \propto \exp[-1/Jn(E_F)]$ and $T_{RKKY} \propto [Jn(E_F)]^2$ where T_K is the Kondo temperature, T_{RKKY} the RKKY interaction, J the exchange coupling between the conduction electrons, and $n(E_F)$ the density of states at the Fermi energy E_F. The RKKY interaction dominates the Kondo effect for small $Jn(E_F)$ values which is enhanced by pressure and forces the system into a non-magnetic state for $p > p_c$.

In Figure 7.13, the transition temperatures of both $CePd_2Ge_2$ and $CePd_2Si_2$ as a function of a calculated unit cell volume V have been plotted [79]. The transformation of pressure into volume has been described in Ref. [79]. The data shown in Figure 7.13 are for samples with their crystallographic c axis parallel to the applied force exerted to the anvils.

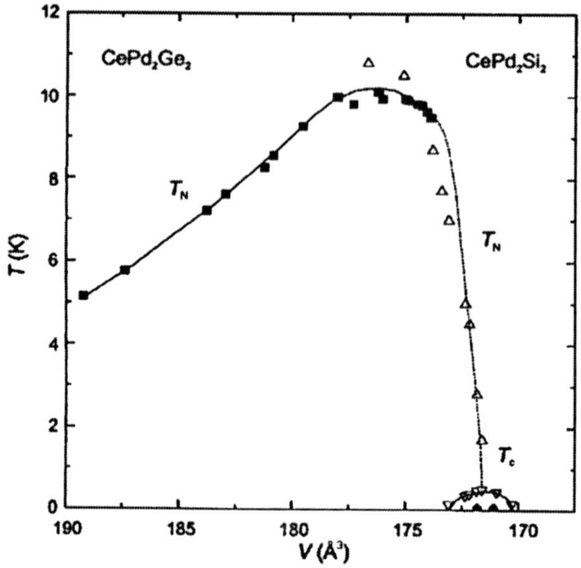

Figure 7.13. Transition temperatures of $CePd_2Ge_2$ and $CePd_2Si_2$ as a function of the calculated cell volume, V. T_N and T_c represent the Neel temperature and superconducting transition temperature, respectively. The data of $CePd_2Si_2$ (open symbols) are taken from Ref. [78]. Reproduced with the permission of the American Physical Society from Ref. [79].

The combined phase diagram in Figure 7.13 clearly shows the analogy between $CePd_2Ge_2$ at high pressure and $CePd_2Si_2$ at moderate pressures. The reduced volume where $\rho_{mag}(T)$ suddenly decreases to 110 mK in $CePd_{2.02}Ge_{1.98}$ and 70 mK in $CePd_2Ge_2$ is similar to the volume where superconductivity occurs in $CePd_2Si_2$. One possible explanation is the incipient superconductivity [79] at this extremely low temperature. Wilhelm and Jaccard [79] have also plotted the variation of $\rho_{mag}(T)$ with pressure and have obtained deviation from FL behavior around p_c, which can be understood within the framework of spin-fluctuation theory [49, 50]. In fact, $\rho(T) = \rho_0 + \tilde{A}T^n$, with $n = 1.5$, for spin fluctuations with three-dimensional character. The minimum values of n obtained by Wilhelm and Jaccard in their resistivity measurement (Figure 4 in Ref. [79]) is close to this value.

7.6. $CeRh_2Si_2$

Movschovich et al. [28] have reported the discovery of a superconducting transition in the heavy fermion compound $CeRh_2Si_2$ under hydrostatic pressure. Superconductivity appears at pressures above 9 kbar which is near the critical pressure required to suppress antiferromagnetic order $[T_N(P = 0) = 36\,K]$. Superconductivity occurs at a temperature of $T_c \approx 350$ mK. The effective mass of $m^* \approx 200m_e$ supports the nature of the heavy fermion nature of superconductivity in $CeRh_2Si_2$.

In Figure 7.14, the P–T phase diagram of $CeRh_2Si_2$ obtained by Movschovich et al. [28] has been reproduced. The superconducting phase appears in the vicinity of where hydrostatic pressure suppresses magnetic order below detectable limits which is also a common feature of pressure-induced superconductivity in $CeCu_2Ge_2$ and $CePd_2Si_2$

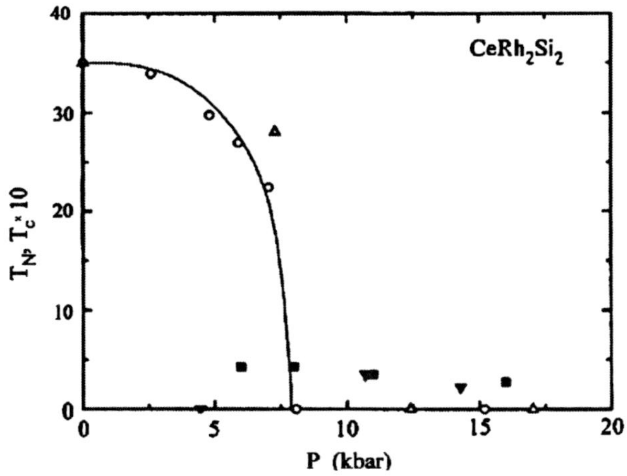

Figure 7.14. *P–T* phase diagram for $CeRh_2Si_2$. Open symbols represent the antiferromagnetic ordering transition $T_N(P)$. Solid symbols are $T_c(P)$ points of the superconducting transitions obtained from the onset of resistivity for several samples that are represented by different symbols. The temperature scales of T_N and T_c are different and the error bars are of the order of the size of the symbols and therefore omitted. Reproduced with the permission of the American Physical Society from Ref. [28].

(as noted earlier). An important criterion for the observation of superconductivity in CeM_2X_2 compounds appears to be their proximity to magnetism, which suggests that spin fluctuations may be important for pairing.

In Figure 7.15, we summarize the superconducting properties of the Ce-based compounds discussed above by reproducing the characteristic (p, T) phase diagrams which has been plotted in Ref. [63]. It shows that in $CeCu_2Si_2$ and $CeCu_2Ge_2$, the compounds with valence-fluctuation-mediated superconductivity, the valence and magnetic critical points (p_v and p_c) are well separated and the characteristic signatures associated with both are distinct. In contrast, the valence and spin-fluctuation regions appear much closer in the other Ce-based compounds.

7.7. $CeNi_2Ge_2$

$CeNi_2Ge_2$ is characterized at ambient pressure by temperature dependence of the specific heat and of the resistivity, which deviate strongly from the standard FL predictions and are reminiscent of the behavior observed in its sibling system $CePd_2Si_2$. The electronic contribution to the resistivity $\Delta\rho = \rho(T) - \rho_0 \sim T^{1+\delta}$ ($\delta \sim 0.5$) over several orders of magnitude in temperature and down to the temperatures of the order of 50 mK at zero pressure [81] although in some samples $\rho_0 \sim T^2$ were later observed [82]. The phase diagram in a series of resistivity measurements under high hydrostatic pressure indicates that at $p > 15$ kbar, a new superconducting transition appears below 220 mK in $CeNi_2Ge_2$. It shifts to higher temperatures with increasing pressure reaching ~400 mK at ~26 kbar.

In Figure 7.16, the temperature dependence of the resistivity, ρ, of $CeNi_2Ge_2$ along the a axis for a long range of pressures has been plotted.

In Figure 7.17, the *P–T* phase diagram of the related materials $CePd_2Si_2$ and $CeNi_2Ge_2$ has been plotted [82].

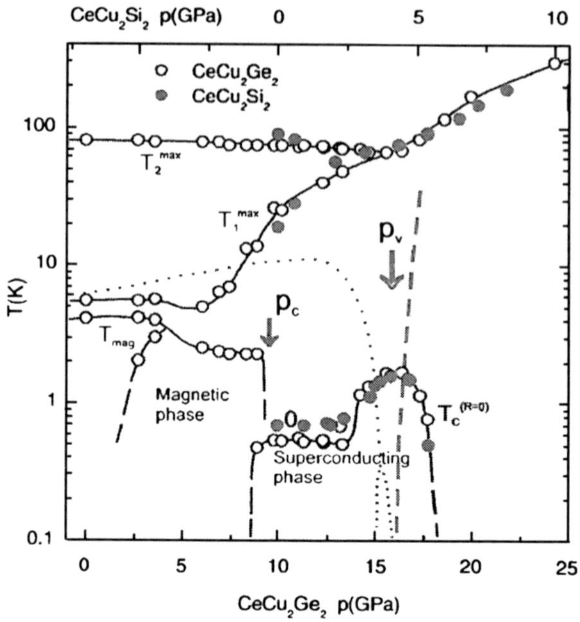

Figure 7.15. (p, T) phase diagram of $CeCu_2(Si/Ge)_2$ showing the two critical pressures p_c and p_v. Around p_v, superconductivity is mediated by valence fluctuations, the temperatures T_1^{max} and T_2^{max} merge and the resistivity is linear in T. The dashed line represents a hypothetical first-order valence discontinuity. The dotted line represents the magnetic ordering temperature of $CePd_2(Si/Ge)_2$ and the narrow T_c domain of the Si compound, plotted for matching T_1^{max}. Reproduced with the permission of Elsevier from Ref. [63].

It is evident from Figure 7.17 that both materials exhibit superconducting instabilities at high pressures, below an anomalous normal state which deviates from conventional FL behavior and defies a quantitative analysis in terms of fluctuations of an antiferromagnetic local order parameter. In $CePd_2Si_2$, the formation of a superconducting phase is linked to the disappearance of magnetism like in all other Ce-based heavy fermion superconductors. In contrast, the superconductivity in $CeNi_2Ge_2$ occurs in a separate region of the phase diagram with no connection to a spin-aligned state.

7.8. $CeIn_3$

Walker et al. [30] have reported that close to the border of antiferromagnetic order at high pressure, the cubic stoichiometric compound $CeIn_3$ displays an unconventional normal-state resistivity followed by superconductivity at low temperatures. Their experimental result can be summarized as follows. At ambient pressure the resistivity versus temperature exhibits a broad peak around 50 K and a cusp near 10 K that marks the onset of antiferromagnetic order (Figure 1 in Ref. [30]). The Neel temperature (T_N), as inferred from the position of this cusp, decreases rapidly with pressure and appears to fall toward absolute zero just above 25 kbar where the cusp becomes ill defined (Figure 7.18). However, the broad peak at a higher temperature T_{sf} appears to move slowly with pressure.

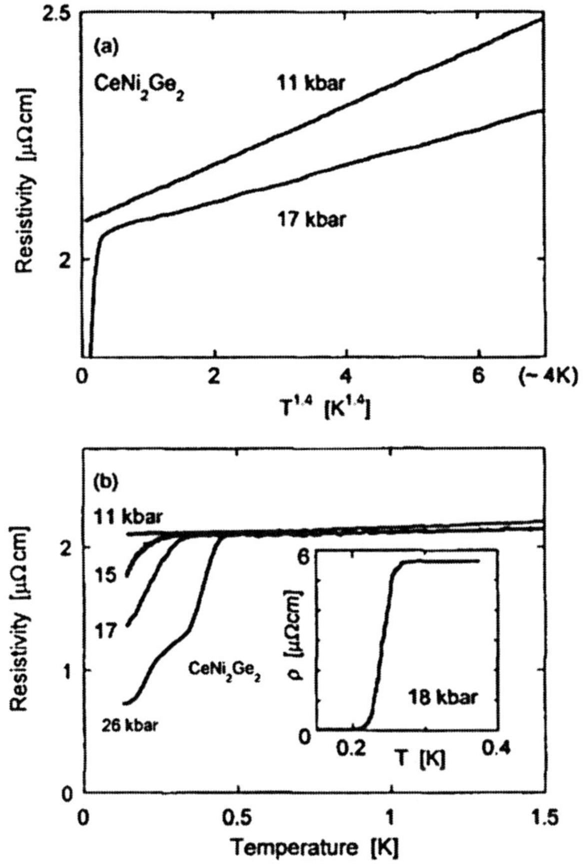

Figure 7.16. The temperature dependence of the resistivity, ρ, of $CeNi_2Ge_2$ along the a axis for a long range of pressures. (a) An NFL form for the resistivity, $\rho(T) \sim T^{1.4}$ is observed over a wide range of temperature at low pressures. With increasing pressure, the pre-factor of the power law decreases, with no major changes in the exponent. At $T < 250$ mK, an anomaly develops in $\rho(T)$ for pressures in excess of 11 kbar. (b) The pressure dependence of the size and position of this phenomenon, as well as measurements on a different sample (inset), point towards the occurrence of superconductivity in $CeNi_2Ge_2$ at high pressure. Reproduced with the permission of Elsevier from Ref. [82].

The superconducting transition temperature (T_s) varies rapidly with pressure near the critical pressure where T_N collapses toward absolute zero. In Figure 7.18, their experimental result for the P–T phase diagram of $CeIn_3$ has been reproduced.

Their experimental results indicate that at ambient pressure, the resistivity varies as T^3 over a wide range below T_N, as expected from the effect of antiferromagnons. ρ varies as T^2 below 10 K for pressures about 30 kbar due to antiferromagnetic spin fluctuations in an FL state well away from the critical pressure p_c where $T_N \to 0$. In the vicinity of p_c, the resistivity exponent increases with decreasing temperature (Figure 3 of Ref. [30]) and remains well below the FL value of 2 over a wide range below 10 K.

Figure 7.19 shows the superconducting transition of this nearly antiferromagnetic normal state in the vicinity of critical pressure and below 200 mK. The superconducting transition

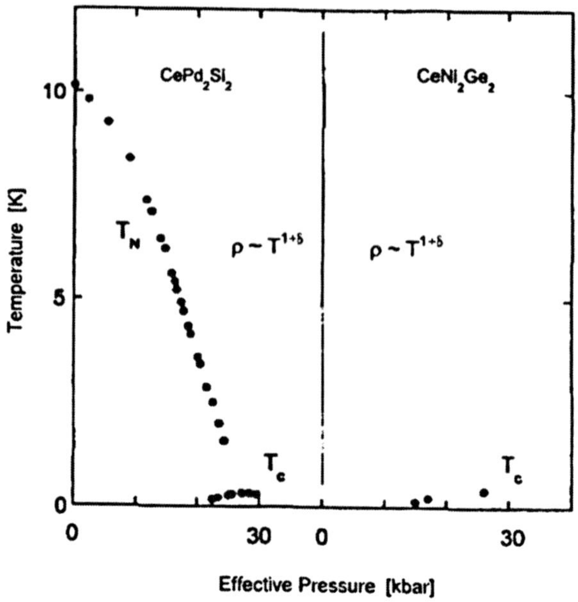

Figure 7.17. Compound phase diagram of the $CePd_2Si_2/CeNi_2Ge_2$ system, spanning the antiferromagnetic QCP as a function of effective pressure. Near the critical point, superconductivity is observed in $CePd_2Si_2$ below a quasilinear form for resistivity which extends over almost two decades of temperature. The resistivity of $CeNi_2Ge_2$ exhibits a similar behavior at ambient pressure. As the pressure is increased, the material undergoes a transition to a superconducting state at high pressures in excess of 15 kbar. Reproduced with the permission of Elsevier from Ref. [82].

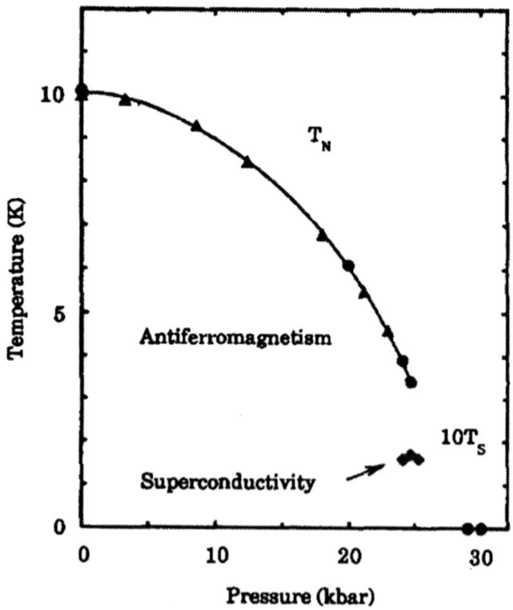

Figure 7.18. The temperature versus pressure phase diagram of $CeIn_3$. The data for T_N represented by filled triangles (all below 25 kbar) is taken from Ref. [83], but with the pressure scaled by a factor of 0.9. The filled circles and diamonds are pressure measurements of Walker et al. [30]. T_S has been scaled by a factor of 10. The solid line represents a polynomial fit to the data for T_N. Reproduced with the permission of Elsevier from Ref. [30].

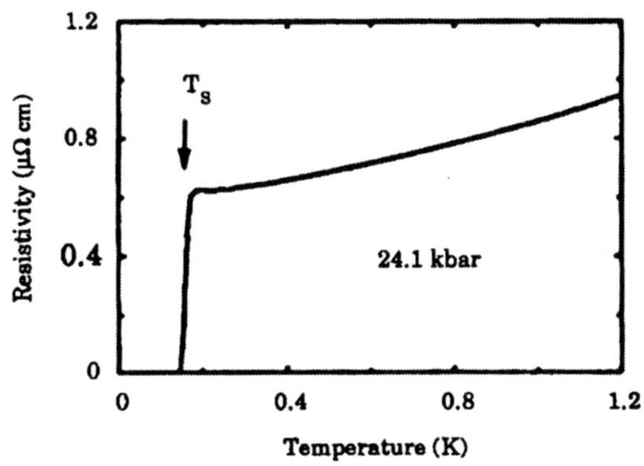

Figure 7.19. The resistivity of $CeIn_3$ near the superconducting transition in the vicinity of the critical pressure near antiferromagnetic order. Reproduced with the permission of Elsevier from Ref. [30].

temperature T_S varies rapidly near the border of antiferromagnetism. As discussed earlier, a similar behavior close to p_c for antiferromagnetic order has also been found for the tetragonal compounds $CePd_2Si_2$ and $CeRh_2Si_2$ although they have different normal-state behaviors than that of $CeIn_3$ near p_c.

7.9. $CePt_3Si$

Bauer et al. [84] have discovered the novel heavy fermion superconductor $CePt_3Si$ which crystallizes in the $CePt_3B$ structure as a tetragonally distorted low symmetry variant of the $AuCu_3$ structure type. It derives from hypothetical $CePt_3$ with cubic $AuCu_3$ structure by filling the void with Si which causes a tetragonal distortion of the unit cell with $c/a = 1.336$. $CePt_3Si$ exhibits antiferromagnetic order at $T_N = 2.2$ K and enters into a heavy fermion superconducting state at $T_c = 0.75$ K. It is the first heavy fermion superconductor without a center of symmetry. In most previous studies on either conventional or unconventional SC, it was assumed that the crystal has an inversion center, which allows separate consideration of the even (spin-singlet) and odd (spin-triplet) components of SC order parameter (equations (7.1) and (7.2)). Since a center is absent in $CePt_3Si$, the novel relaxation behavior found below T_c hints at a new class of SC state.

The zero-field μSR data [85] have demonstrated the bulk character of the antiferromagnetic state in heavy fermion superconductor $CePt_3Si$, suggesting a microscopic coexistence between magnetism and superconductivity. In addition, a slight change of the μSR response upon cooling the sample below T_c can be ascribed to a coupling of the superconducting and magnetic order parameters and/or to the decrease of the hyperfine contribution acting on the muon.

In Figure 7.20(a), the field-dependent specific heat of $CePt_3Si$ at low temperatures and at various external magnetic fields has been shown [86]. In Figure 7.20(b), the temperature dependence of the upper critical field H_{c2} is deduced from the specific heat and resistivity measurements [86].

From Figure 7.20, it is obvious that the application of magnetic fields reduces T_c, giving rise to $H'_{c2} = dH_{c2}/dT \approx -8.5$T/K. An extrapolation of $T_c(H)$ towards zero yields

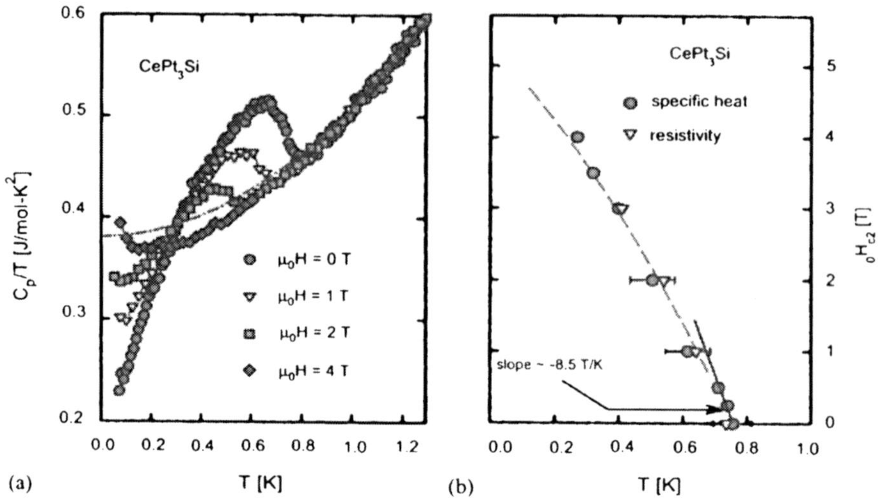

Figure 7.20. (a) Temperature-dependent specific heat of C_p/T of $CePt_3Si$ for various values of applied fields; the dashed line is a T^3 extrapolation of $C_p(T)$ at 0 T. (b) Temperature dependence of the upper critical field H_{c2}. The solid straight line yields $H'_{c2} \approx -8.5$T/K; the dashed line is a guide to the eye. Reproduced with the permission of Elsevier from Ref. [86].

$H_{c2}(0) \approx 5$ T, well above the Paul–Clogston limiting field [84]. An estimate of the Sommerfeld coefficient from the high field data gives 0.36 J/mol K^2, in agreement with the value obtained from an extrapolation of the normal state in the zero field data (Figure 7.20).

Yogi et al. [87] have presented NMR data which confirms the presence of the coexistence of triplet and singlet pairs, exhibiting a small coherence peak characteristic of s-wave superconductivity, slowly crossing over to a T^3 dependence at low temperatures that is characteristic of line nodes around the Fermi surface. Their results are shown in Figure 7.21.

A nice theoretical account, including the prediction of a novel vortex phase [88], has been given by Agterburg et al. [89].

The temperature-dependent electrical resistivity ρ of $CePt_3(Si_{1-x}Ge_x)$ has been plotted in Figure 7.22 for several concentrations x and for temperatures below about 4 K [86].

The samples investigated are characterized by a concentration dependent onset of SC, where the SC/Ge substitution suppresses the SC roughly above $x = 0.1$. At higher temperatures, ρ(T) exhibits a pronounced curvature which can be taken as a signature for long-range magnetic order, in agreement with specific heat measurements. The $d\rho/dT$ plot (right axis of Figure 7.22) shows pronounced magnetic anomalies which indicates broadened magnetic phase transitions and the respective temperatures increase continuously with increasing Ge content.

In order to study the pressure dependence of the superconducting transition temperature, T_c, and of the Neel temperature T_N, Nicklas et al. [90] have plotted the resistivity of $CePt_3Si$ as a function of temperature for different pressures which is shown in Figure 7.23.

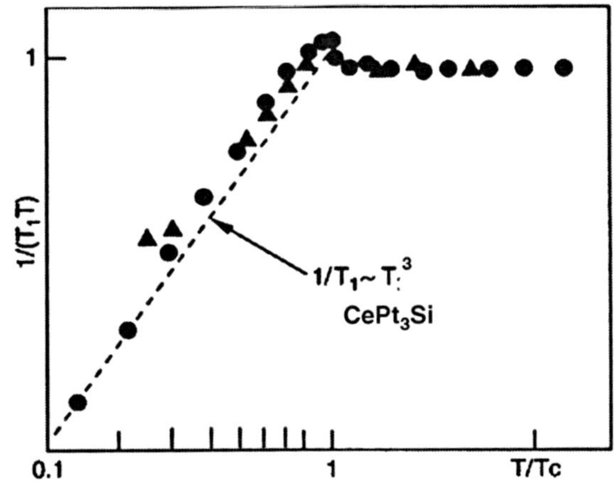

Figure 7.21. Sketch of the temperature dependence of $1/T_1T$ (normalized with respect to its value at T_c) for the parity violating superconductor $CePt_3Si$. Reproduced with the permission of Elsevier from Ref. [40].

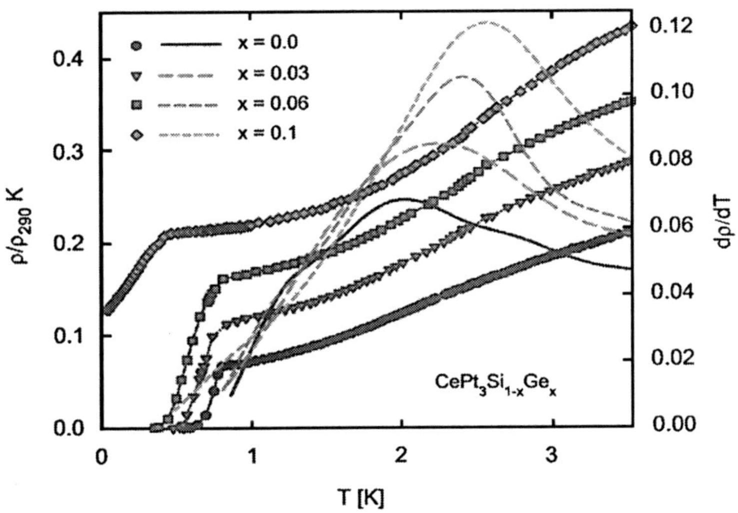

Figure 7.22. Low-temperature electrical resistivity ρ for various concentrations x of $CePt_3Si_{1-x}Ge_x$ (symbols), and temperature derivatives $d\rho/dT$ (lines, right axis). Reproduced with the permission of Elsevier from Ref. [86].

As shown in Figure 7.23, the antiferromagnetic transition at 2.2 K leads to a change in slope of the resistivity data. T_N vanishes non-monotonically in a first-order-like transition at $P_c \approx 0.5$ GPa. In the paramagnetic phase at temperatures right above T_N at atmospheric pressure, the resistivity decreases with increasing pressure, which can be attributed to an increase of the Kondo temperature [87].

The phase diagram shown in Figure 7.24 summarizes the characteristic temperatures obtained for $CePt_3(Si_{1-x}Ge_x)$ at ambient pressure [86] as well as the resistivity studies on $CePt_3Si$ under hydrostatic pressure [90]. The details of conversion of pressure to unit cell volume have been explained in Ref. [86]. The increase of the unit cell volume by the

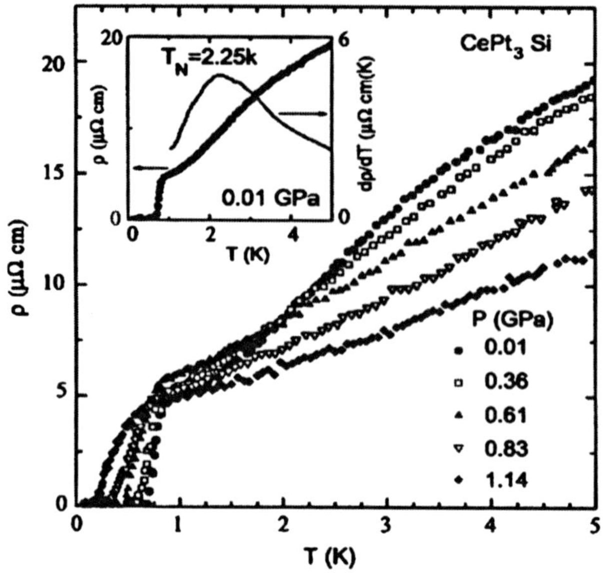

Figure 7.23. Resistivity of $CePt_3Si$ as a function of temperature for different pressures. T_c decreases with increasing pressure accompanied by a broadening of the transition. Inset: Resistivity and its derivative for the initial pressure $P = 0.01$ GPa. The maximum of the derivative indicates T_N. Reproduced with the permission of Elsevier from Ref. [90].

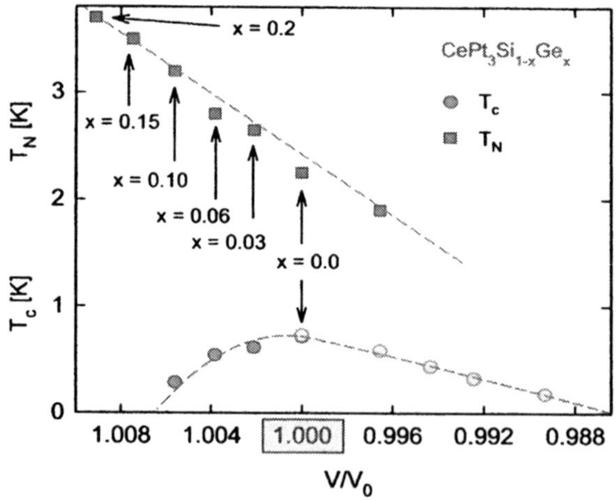

Figure 7.24. Phase diagram of $CePt_3Si_{1-x}Ge_x$. Reproduced with the permission of Elsevier from Ref. [86].

Si/Ge substitution is responsible for a decrease of hybridization. This leads to a decrease of the Kondo interaction and strengthening of the magnetic interaction. The SC transition temperature becomes suppressed and finally vanishes at $x > 0.1$.

From Figure 7.24, it is apparent that increasing hydrostatic pressure is responsible for a decrease of both T_C and T_N. Ternary $CePt_3Si$ thus appears to be at the maximum position of the "superconducting dome". However, it is not known whether this dome is constrained within the magnetic phase [86].

7.10. $CeCoIn_5$

The heavy fermion superconductor $CeCoIn_5$ was discovered recently by Petrovic et al. [32, 33]. $CeCoIn_5$ has the $HoCoGa_5$ crystal structure. The superconducting transition temperature $T_c = 2.3$ K is the highest critical temperature among the heavy fermion compounds. $CeCoIn_5$ is a quasi-2D system and the de Haas–van Alphen effect data indicate a quasi-2D Fermi surface [91, 92]. The specific heat data [93] and the thermal conductivity measurements [94] suggest the presence of nodes in the superconducting gap. $CeCoIn_5$ also exhibits quantum criticality [95–97]. All these properties have led to the possibility of a Fulde–Ferrell–Larkin–Ovchinnikov (FFLO) superconducting state [98, 99] in $CeCoIn_5$.

In the 1960s, Fulde and Ferrell [98] and Larkin and Ovchinnikov [99] had independently developed theories of inhomogeneous superconducting states. These states result from the competition between superconducting condensate energy and the magnetic Zeeman energy which lowers the total energy of the electrons in the normal state. This competition is strong when the superconductivity is of a spin-singlet nature. In this case, the superconducting Cooper pairs form with opposite spins and the electrons cannot lower the total energy of the system by preferentially aligning their spins along the magnetic field. This effect, called Pauli limiting, leads to suppression of superconductivity in magnetic field. The characteristic Pauli field H_P determines the upper limit of the superconducting upper critical field H_{c2} [100]. When Pauli limiting is the dominant mechanism for suppression of superconductivity, a new inhomogeneous superconducting FFLO state would appear at high fields between the normal and the mixed, or the vortex, state below the critical temperature T_{FFLO} [101], with planes of normal electrons that can take advantage of Pauli susceptibility. In the FFLO state, pair breaking due to the Pauli paramagnetic effect is reduced by the formation of a new pairing state ($\mathbf{k}_\uparrow$, $-\mathbf{k} + \mathbf{q}_\downarrow$), with $|\mathbf{q}| \sim 2\mu_B H/\hbar v_F$ (v_F is the Fermi velocity) between the Zeeman split parts of the Fermi surface. In addition, an inhomogeneous superconducting phase with a spatially order parameter $\Delta(\mathbf{r}) = \Delta_0 \cos(\mathbf{q} \cdot \mathbf{r})$ and spin polarization shows up in the vicinity of the upper critical field H_{c2} [102]. It has been shown that in the presence of a substantial orbital effect, planar nodes appear periodically perpendicular to the Abrikosov vertex lattice [101, 103–105].

One of the intriguing features is the T and H phase dependence of the phase boundary between the FFLO and non-FFLO superconducting state (hereafter referred to as the BCS state). $H^{\parallel}_{FFLO}$ ($H \parallel ab$) exhibits an unusually large shift to higher fields at higher temperatures [97, 106]. The results of Bianchi et al. [97] is shown in Figure 7.25.

The complete H–T phase diagram of $CeCoIn_5$ based on the specific heat measurements for field $H \perp [0\ 0\ 1]$ is shown in Figure 7.26. The second-to-first order change is indicated by the arrow at $T_0 = 1.1 \pm 0.1$ K for $H \parallel [1\ 1\ 0]$, which is about 10% higher than T_0 for $H \parallel [1\ 0\ 0]$.

It may be noted that $T_{FFLO} = 130$ mK for $H \parallel [0\ 0\ 1]$ which is approximately half of the value for $H \parallel [1\ 0\ 0]$ indicates that the FFLO state is more stable when the magnetic

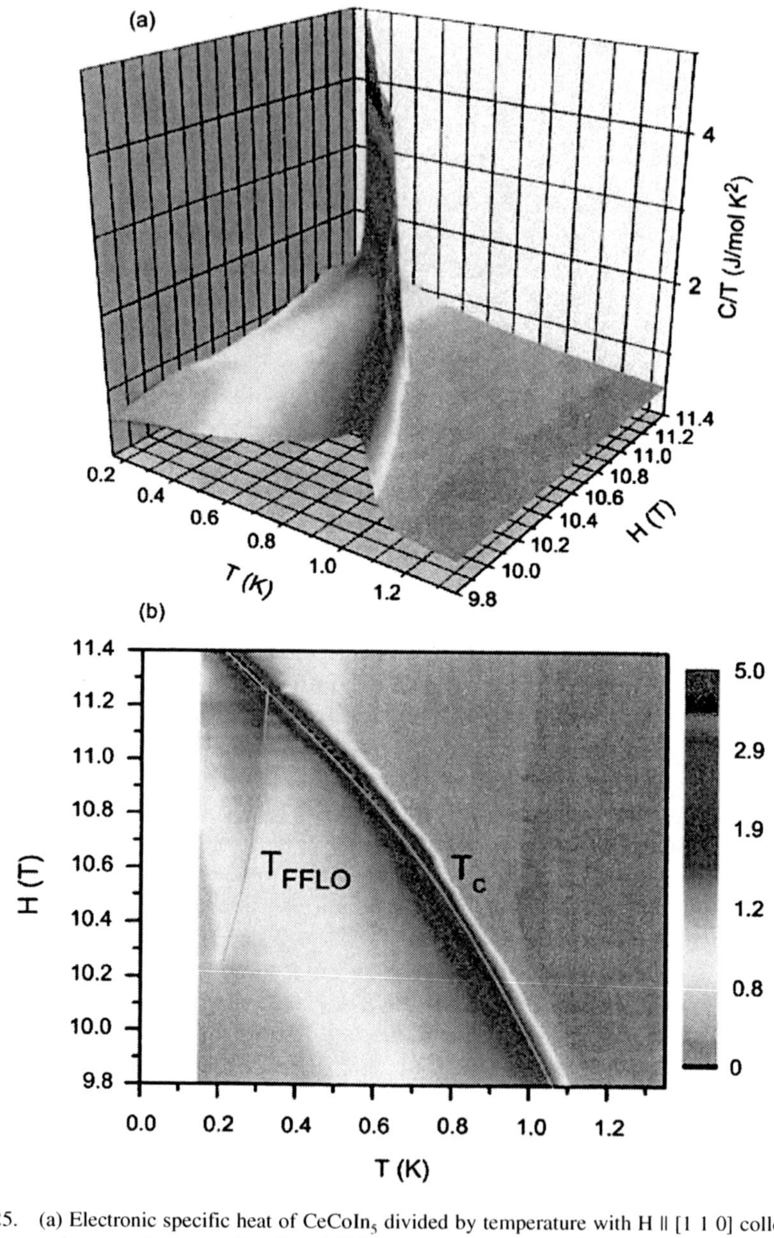

Figure 7.25. (a) Electronic specific heat of $CeCoIn_5$ divided by temperature with H ∥ [1 1 0] collected with the temperature decay method, as a function of field and temperature. (b) Contour plot of the data in (a) in the H–T plane. Gray lines indicate the superconducting phase transition T_C and the FFLO-mixed state T_{FFLO} anomaly. The color scale is same in (a) and (b). Reproduced with the permission of the American Physical Society from Ref. [97].

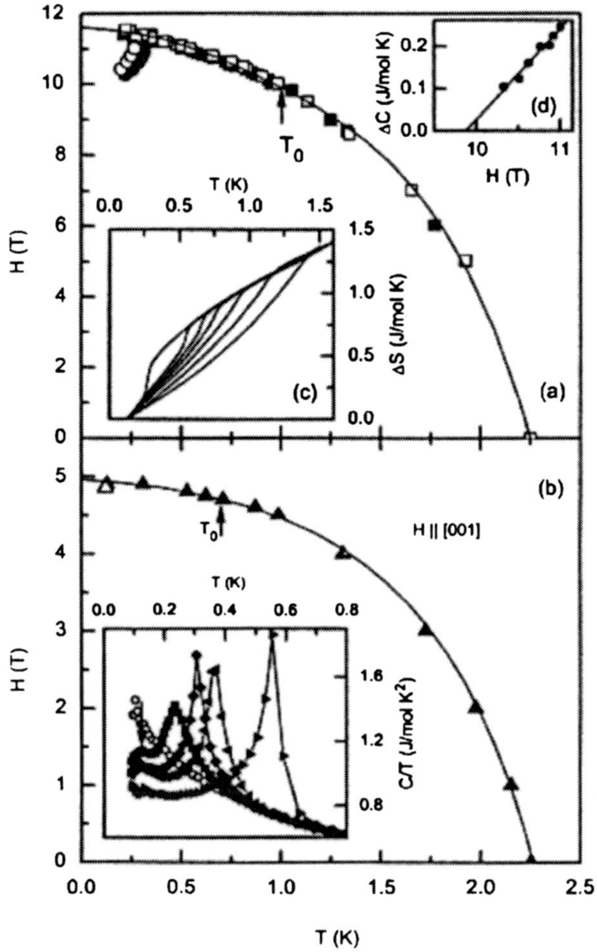

Figure 7.26. (a) H–T phase diagram of $CeCoIn_5$ with both $H \parallel [1\ 1\ 0]$ (filled symbols) and $H \parallel [1\ 0\ 0]$ (open symbols). (O) and (●) indicate the T_{FFLO} anomaly for $H \parallel [1\ 0\ 0]$ and $H \parallel [1\ 1\ 0]$, respectively. Inset (c) shows entropy gain from $T = 0.13$ K for fields of 11.4, 11, 10.8, 10.22, 9.5, and 8.6 T from left to right. Inset (d) shows the T_{FFLO} anomaly obtained from equal area concentration. (b) H–T phase diagram for $H \parallel [0\ 0\ 1]$. (Δ) indicate the T_{FFLO} anomaly. Inset: Sommerfeld coefficient; (O) 5 T, solid symbols 4.9, 4.875, 4.85, and 4.8 T from left to right. Arrow indicates T_{FFLO} anomaly at 4.9 T. Solid line in (a) and (b) are guides to the eye for superconducting phase boundaries. Reproduced with the permission of the American Physical Society from Ref. [97].

field is in the *a–b* plane of this quasi-2D compound. The work of Radovan et al. [106] has a similar phase diagram for $H \perp [0\ 0\ 1]$ but when the field is tilted out of the *a–b* plane, the FFLO phase disappears supporting the 2D nature of the FFLO phenomenon in $CeCoIn_5$ in contrast to the 3D picture emerging from the results of Bianchi et al. [97].

The experimental results of Bianchi et al. [97] as well as that of Radovan et al. [106] show a remarkable T and H dependence of the phase boundary between the FFLO and non-FFLO superconducting state (hereafter referred to as the BCS state). $H^{\parallel}_{FFLO}$ exhibits an

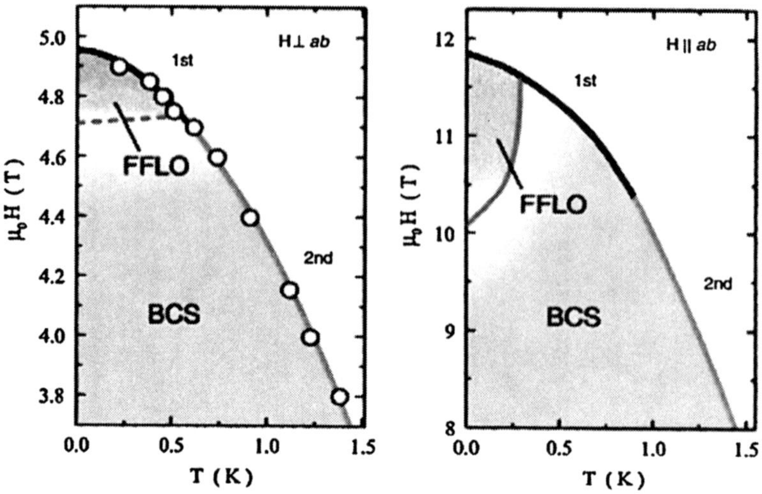

Figure 7.27. *H–T* phase diagrams at low temperatures and high fields for $H \perp ab$ (left panel) and $H \parallel ab$ (right panel). The colored portions display the FFLO and BCS regions. The open circles are $H_{c2}^{\perp}$ determined by the NMR experiments. The black and green (gray) lines represent the upper critical fields which are in the first and the second order, respectively. The red or dark gray dashed and solid lines represent the phase boundary separating the FFLO and BCS states. Precise determination of the phase boundary between the FFLO and BCS states is difficult in the present experiments for $H \perp ab$. Reproduced with the permission of the American Physical Society from Ref. [102].

unusually large shift to higher fields with higher temperatures. In addition, it is believed that the 2D nature of $CeCoIn_5$ is essential for the formation of FFLO state since the strong reduction of the orbital pair breaking in a parallel field and the nesting properties of the quasi-2D Fermi surface are expected to stabilize the FFLO state.

Kumagai et al. [102] have conducted ^{115}In NMR study for $H \perp ab$ in order to resolve the differences between the two contradictory results of Refs. [97] and [106]. Their result of *H–T* phase diagram is shown in Figure 7.27(a). The *H–T* phase diagram of Bianchi et al. [97] as well as that of Radovan et al. [106] for $H \parallel ab$ is shown in Figure 7.27(b).

In order to establish the existence of a possible FFLO state, it is important to elucidate the structure of the flux line lattice (FLL). Ultrasound velocity measurements of $CeCoIn_5$ [107] provide strong evidence of the presence of FFLO phase in which the order parameter is spatially modulated and has planar nodes perpendicular to the FLL. The results of ultrasonic measurements [108] combined with the heat capacity measurements [97] has led to an interesting experimental *H–T* phase diagram below 0.7 K for $CeCoIn_5$ which is shown in Figure 7.28.

There are a large number of experimental studies of the superconducting state in $CeCoIn_5$ which indicate the presence of line nodes in the superconducting gap [93, 109–111]. The presence of a fourfold anisotropy in the thermal conductivity suggests an order parameter with $d_{x^2-y^2}$ symmetry [94]. However, specific heat measurements [112] show a fourfold anisotropy which points to a d_{xy} nodal gap structure.

In order to resolve these anomalies, Tanatar et al. [113] have measured the specific heat and thermal conductivity in the superconducting state of $Ce_{1-x}La_xCoIn_5$. They have

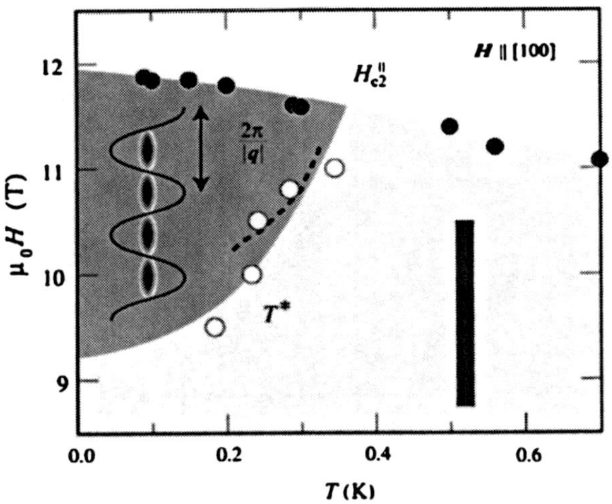

Figure 7.28. Experimental H–T phase diagram for $CeCoIn_5$ below 0.7 K for $H \parallel [1\ 0\ 0]$. The transition temperatures T^*, indicated by open circles, were obtained from ultrasonic measurements (from Figure 1 of Ref. [108]). The solid circles are the upper critical field $H_{c2}^{\parallel}$ determined by the ultrasonic experiments. The dotted line is the second-order transition line determined by the heat capacity reported in Ref. [97]. The schematic figures are sketches of a flux line above and below transition. Reproduced with the permission of Elsevier from Ref. [108].

observed the suppression of T_c with increasing impurity concentration of x. The suppression of T_c is accompanied by the increase in residual electronic specific heat expected of a d-wave superconductor, but it occurs in parallel with decrease in residual electronic thermal conductivity. Their result for specific heat is shown in Figure 7.29.

Based on their contrasting results for specific heat and thermal conductivity, Tanatar et al. [113] have proposed the presence of uncondensed electrons coexisting with nodal quasiparticles. They have also proposed that superconductivity in $CeCoIn_5$ originates on the cylindrical sheets of the Fermi surface, with a negligible gap on the light three-dimensional pockets. This scenario of uncondensed electrons would resolve the apparent contradiction in the angular field dependence of the thermal conductivity [94] and the specific heat [112].

7.11. $CeRhIn_5$

In the $CeRhIn_5$ structure, there are two inequivalent In sites: a single In(1) site, analogous to the single In site in $CeIn_3$, and four In(2) sites, two on each of the lateral faces of the unit cell that are equidistant above and below the Rh layer [114]. The T–P phase diagram of $CeRhIn_5$ is shown in Figure 7.30(b). As the pressure is gradually increased, at 14.5 kbar, the resistivity slowly decreases with temperature and at 16.3 kbar, there is a broad transition beginning near 2 K to a zero resistance state. The transition width sharpens with increasing pressure to less than 0.05 K at 21 kbar where the onset of superconductivity is at 2.17 K. The abrupt loss of a signature for T_N, the sudden appearance of superconductivity, and the rapid sharpening of the

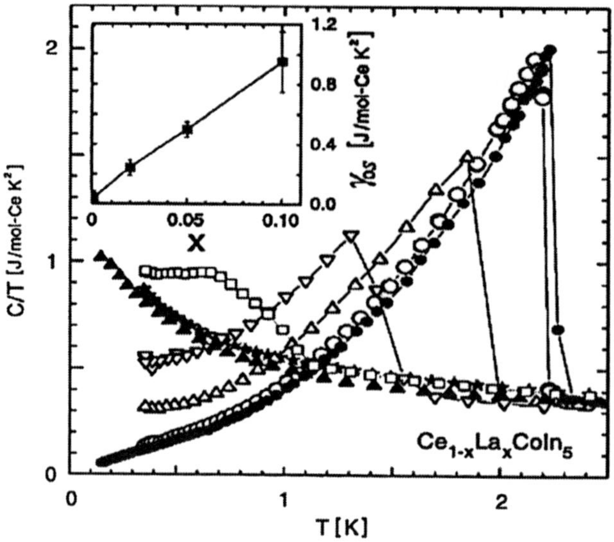

Figure 7.29. Temperature dependence of specific heat C of $Ce_{1-x}La_xCoIn_5$ for $x = 0$ (open circles), 0.02 (open triangles), 0.05 (open down triangles), 0.1 (open squares), and 0.15 (solid stars) in zero magnetic field plotted as C/T versus T. The data of Petrovic et al. [33] for pure $CeCoIn_5$ in zero field (solid circles) and 5 T (solid up triangles) are shown for comparison. Inset shows $\gamma_{0s} \equiv C_{0s}/T\ (T \to 0)$. Reproduced with the permission of the American Physical Society from Ref. [113].

transition width suggests a first-order-like transition at a critical pressure between 14.5 and 16.3 kbar.

The specific heat measurements on $CeRhIn_5$ at 19 kbar are shown in Figure 7.30(a). C/T begins to increase more rapidly below 5 K, reaches plateau near 2.5 K, and shows a well-defined feature at T_c. $C \propto T^2$, a power law expected for a superconductor in which there are line nodes in the gap function [115]. The state reflected in the plateau of C/T near 2.5 K is also found as a weak feature in resistivity measurements, and its pressure dependence is shown as the transition labeled $T_?$ in Figure 7.30(b). This phase transition first appears in resistivity measurements at pressures near 10 kbar and persists in the pressure regime of superconductivity. The transition $T_?$ and the superconducting transition appear to gap the entire Fermi surface. At a field of 9 T, the $T_?$ transition remains unchanged while the superconductivity is suppressed below 0.3 K. This $T_?$ transition is probably one due to a charge- or spin-density wave state that gaps part of the Fermi surface [114] which coexists with superconductivity.

7.12. $CeIrIn_5$

$CeIrIn_5$ belongs to a family of "115" family of compounds [32] and changes its ground state from antiferromagnetic to superconducting at ambient pressure and temperature $T_c = 0.4$ K. Its quasi-2D tetragonal structure can be viewed as layers of $CeIn_3$ as the parent compound (like $CeCoIn_5$). This layered crystal structure appears to be important for superconductivity since the relatively high T_cs in this compound scales with structural

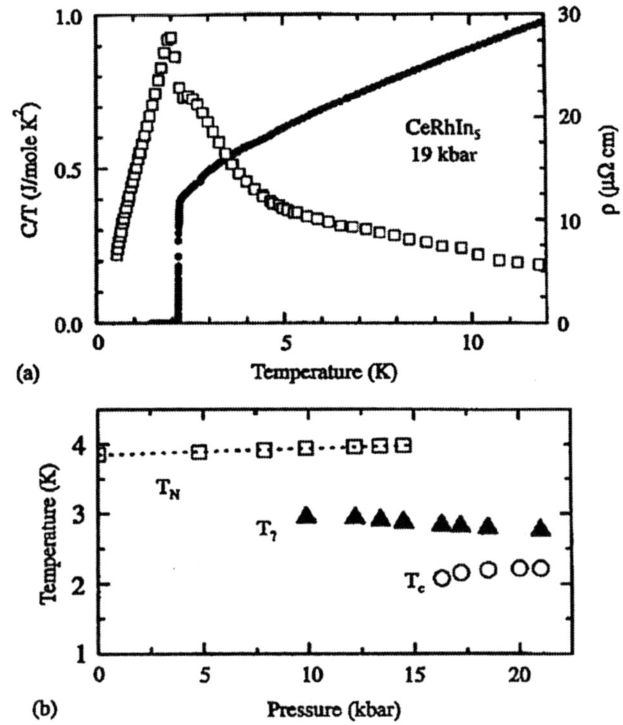

Figure 7.30. (a) Magnetic contribution to the specific heat C divided by temperature (open squares) and electrical resistivity (solid squares) for $CeRhIn_5$ at 19 kbar. The magnetic contribution to specific heat was estimated by subtracting the specific heat of $LaRhIn_5$ at atmospheric pressure from the total specific heat of $CeRhIn_5$ at pressure. (b) The temperature–pressure phase diagram determined from resistivity measurements. T_N, T_c, and $T_?$ correspond to the Neel temperature, onset temperature for superconductivity, and the transition state to an unknown state, respectively. Reproduced with the permission of Elsevier from Ref. [114].

anisotropy [115]. Figure 7.31 shows the specific heat– and magnetic field–temperature phase diagrams [116] of $CeIrIn_5$.

Figure 7.31(a) shows specific heat of $CeIrIn_5$ as a function of temperature for several magnetic fields, with **H** ∥ **(a,b)**. Magnetic field suppresses T_c to zero by $H = 1\ T$, and the size of the specific heat anomaly at T_c gradually decreases due to the suppression of the superconducting fraction of the sample in the vortex state. Figure 7.31(b) shows ac magnetic susceptibility data both as a function of temperature at fixed field (main body) as well at a fixed temperature $T = 80$ mK (inset). The midpoint of χ_{AC} anomaly was taken as T_c or H_{c2} respectively.

Figure 7.32(a) shows the magnetic field–temperature phase diagrams for $CeIrIn_5$. Figure 7.32(b) shows the phase diagram of $CeCoIn_5$ (for comparison) to focus on the Pauli limiting analysis in order to determine whether $CeIrIn_5$ undergoes an FFLO transition like $CeCoIn_5$. From Figure 7.32(a), $T_c = 0.4$ K, $H_{c2}^{\perp}$ $(T = 0) \approx 5.3$kG, $\partial H_{c2}^{\perp}/\partial T|_{T_c} = -2.5$T/K, $H_{c2}^{\parallel}$ (T = 0) = 9.3kG, and $\partial H_{c2}^{\parallel}/\partial T|_{T_c} = -4.8$T/K, where ⊥ and ∥ are with respect to the c axis. There is a magnetic field anisotropy in H_{c2} of about a factor of 2 between **H** ⊥ **c** and **H** ∥ **c**. For $CeIrIn_5$, the Clogston limit on H_{c2} [100]

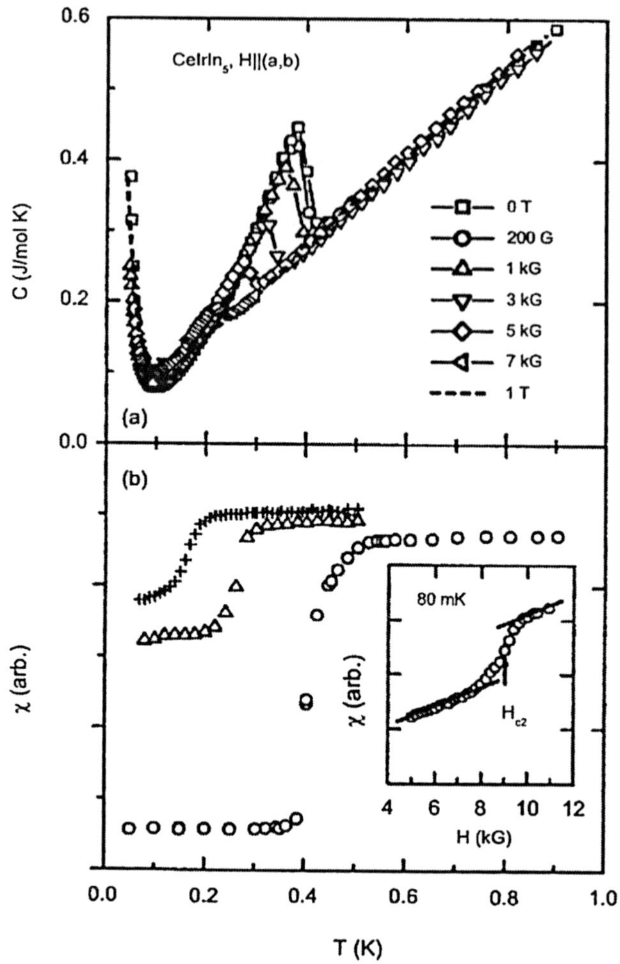

Figure 7.31. (a) Specific heat of $CeIrIn_5$ as a function of temperature for various magnetic fields in the plane **H** || **(a,b)**. (b) Magnetic susceptibility of $CeIrIn_5$ in magnetic field parallel **H** || **(a,b)**, (O) 0 kG, (Δ) 5 kG, (+) 8 kG. Inset: Susceptibility as a function of field at 80 mK. Solid lines are linear fit to the data above and below the transition regime. Reproduced with the permission of Elsevier from Ref. [116].

(for a BCS superconductor with non-interacting electrons) is $H_P = 1.84$ T/K $\times\ T_c =$ 0.74 T. For **H** || **c**, the measured value $H_{c2}^{\perp} = 5.3$ kG for $CeIrIn_5$ is less than H_P. However, there is no strong experimental evidence for or against the Pauli limiting effects in $CeIrIn_5$ [116].

The temperature dependence of the specific heat of $CeIrIn_5$ in zero field is plotted in Figure 7.33(a) and that of $CeCoIn_5$ is plotted in Figure 7.33(b) for comparison.

The vertical intercepts of the solid lines in the main body of Figure 7.33 gives the coefficients of the T-linear contribution to the specific heat, $\gamma_0 = 0.110 \pm 0.010$ J/mol K^2 for $CeIrIn_5$ and $\gamma_0 = 0.035 \pm 0.005$ J/mol K^2 for $CeCoIn_5$. The T^2 terms in electronic specific heat of $CeIrIn_5$ and $CeCoIn_5$ indicate the presence of lines of nodes in the energy gap, as in a number of other unconventional superconductors.

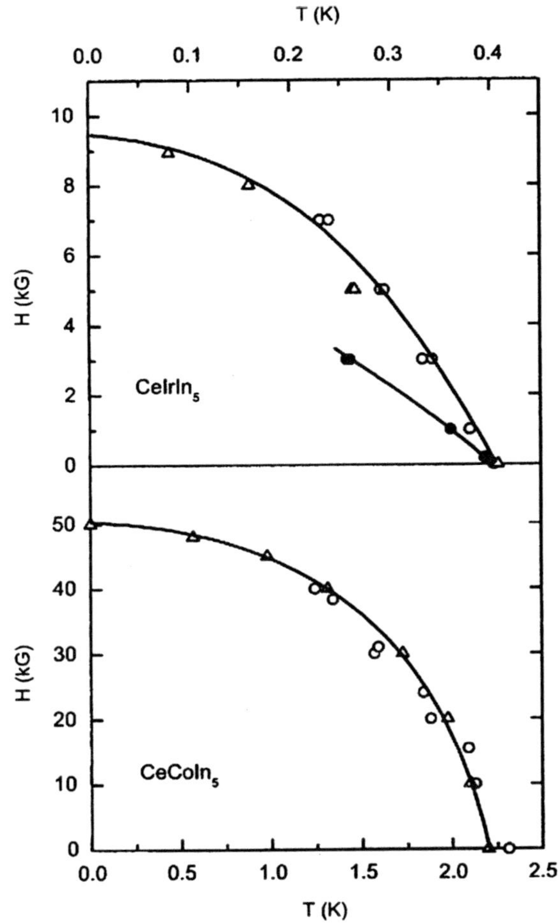

Figure 7.32. (a) Magnetic field–temperature phase diagram for CeIrIn$_5$: (O) from specific heat for **H** ∥ **(a,b)**, (Δ) from magnetic susceptibility for **H** ∥ **(a,b)**, (●) from specific heat for **H** ∥ **c**. (b) Phase diagram of CeCoIn$_5$ for **H** ∥ **c**, (Δ) from specific heat, (O) from resistivity. Solid lines in (a) and (b) are guides to the eye. Reproduced with the permission of Elsevier from Ref. [116].

7.13. CeNiGe$_3$

Recently, it was reported that CeNiGe$_3$, which shows an antiferromagnetic (AF) ordering below $T_N = 5.5$ K at ambient pressure and exhibits a QCP at $P_C \sim 6-7$ GPa, shows the superconductivity in wide pressure range of 1.7–9.3 GPa [64, 117]. In addition to the superconductivity around P_C, the sharp superconducting transition is also observed by both the susceptibility and resistivity measurements at pressures far below P_C, where 4f-localized picture is still dominant. This type of superconductivity appears to be unusual for f-electron systems and is probably the first of its kind.

The experimental results of Kotegawa et al. [117] is shown in Figures 7.34 and 7.35. In Figure 7.34, the temperature dependences of the resistivity at several pressures measured

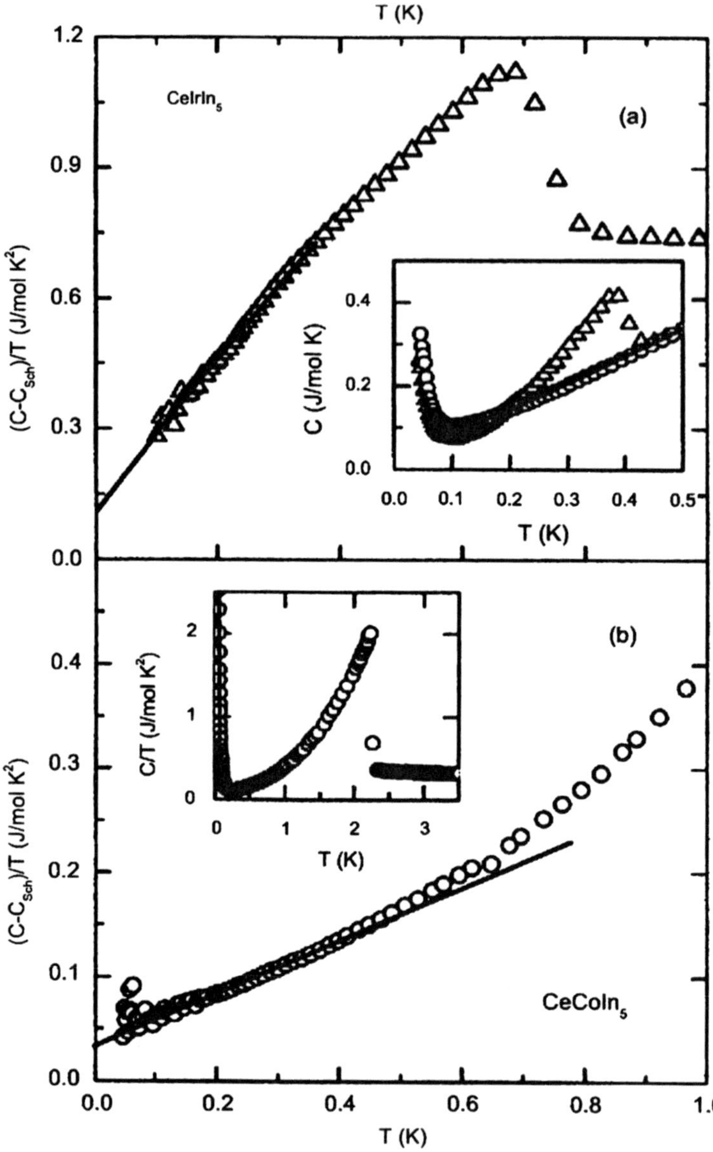

Figure 7.33. (a) Specific heat of $CeIrIn_5$, as $(C - C_{Sch})/T$ versus T, where T_{Sch} is the nuclear Schottky contribution. The solid line is a fit to zero-field data for $T < 0.2$ K. Inset: $C(T)$ for $CeIrIn_5$: (Δ) $H = 0$ kG, (O) $H = 5$ kG. The solid line is the sum of γT and nuclear Schottky C_{Sch} for $H = 5$ kG data. (b) $(C - C_{Sch})/T$ of $CeCoIn_5$ at zero field. The solid line is a fit for $0.1\ K < T < 0.4$ K. Inset: Specific heat of $CeCoIn_5$ below 3.5 K; the solid line is the In nuclear quadrupolar contribution C_{Sch}. Reproduced with the permission of Elsevier from Ref. [116].

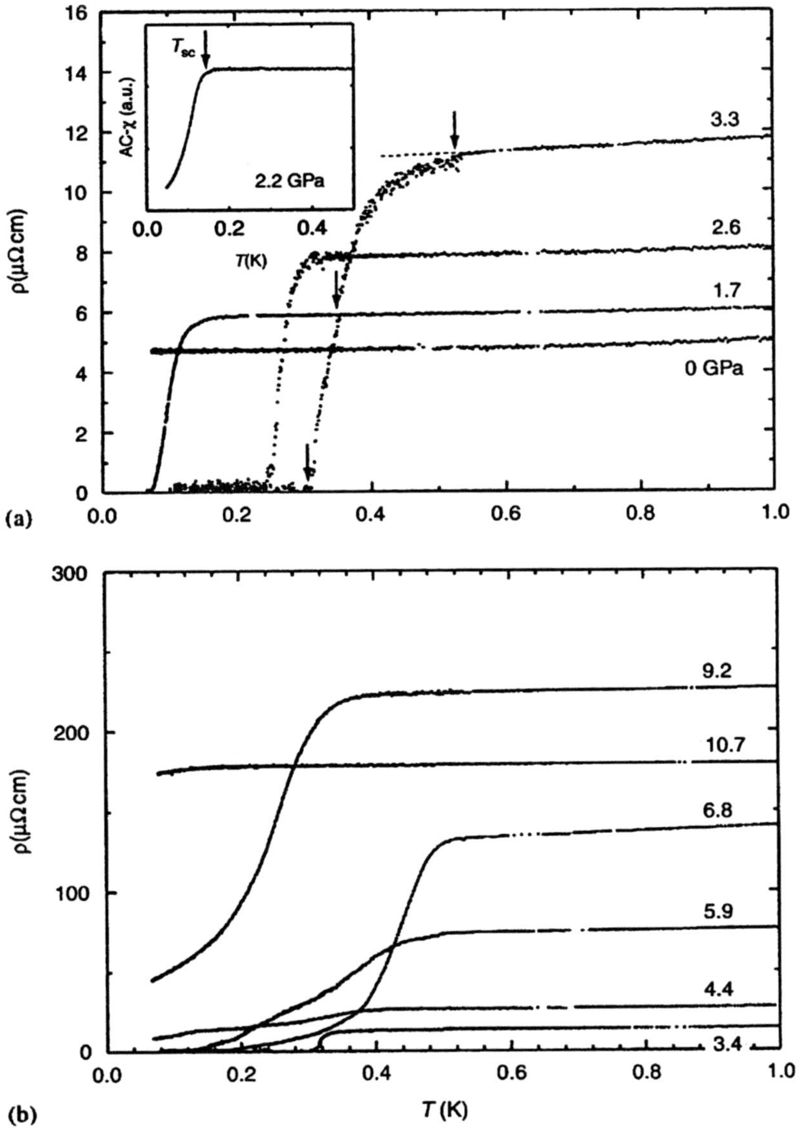

Figure 7.34. The temperature dependences of the resistivity of $CeNiGe_3$, at several pressures, measured by (a) the indenter cell up to 3.3 GPa, and (b) the diamond-anvil-cell up to 10.7 GPa. The inset shows the result of ac susceptibility at 2.2 GPa. Reproduced with the permission of Elsevier from Ref. [117].

by (a) the indenter cell up to 3.3 GPa and (b) the diamond anvil cell (DAC) up to 10.7 GPa is shown. The inset of Figure 7.34(a) shows the result of AC-susceptibility at 2.2 GPa.

As shown in the inset of Figure 7.34(a), the clear Meissner effect supports the view that the superconductivity does not arise from some impurity phase.

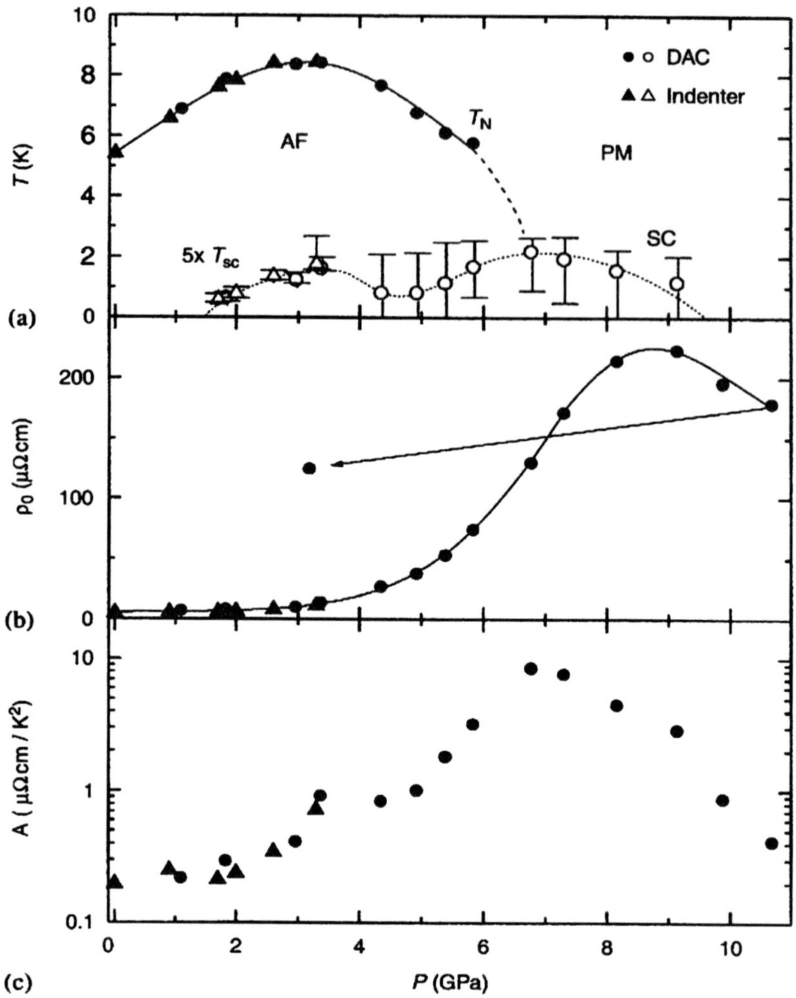

Figure 7.35. (a) The pressure–temperature phase diagram of $CeNiGe_3$. (b) The pressure dependence of ρ_0. (c) The pressure dependence of A. Reproduced with the permission pf Elsevier from Ref. [117].

Kotegawa et al. [117] have plotted the pressure–temperature phase diagram in Figure 7.35(a), the pressure dependence of ρ_0 in Figure 7.35(b), and the pressure dependence of A in Figure 7.35(c) (where $\rho = \rho_0 + AT^2$).

Figure 7.35(a) shows the pressure–temperature phase diagram of $CeNiGe_3$. Each T_{sc} was determined by a midpoint where the resistivity becomes 50%. The onset and offset temperatures of the SC transition are noted as the error bars. ρ_0 and A were determined by fitting the data at the lowest temperature to $\rho(T) = \rho_0 + AT^2$. The data indicates that applying pressure causes significant damage to the sample. Further, superconductivity is realized only in the clean limit of $\xi < l$, where l and ξ are the mean free path and the coherence length, respectively. Further, T_N increases with increasing pressure up to 3 GPa, suggesting that RKKY interaction is dominant rather than the Kondo effect below 3 GPa.

In such a region, the superconductivity is probably due to some exotic mechanism and not due to spin fluctuations which normally plays an important role in most heavy fermion superconductors.

7.14. $Ce_2Ni_3Ge_5$

$Ce_2Ni_3Ge_5$ is an antiferromagnet with a Neel temperature $T_{N1} = 4.8$ K and undergoes a successive transition $T_{N2} = 4.2$ K [118]. The Neel temperature of $Ce_2Ni_3Ge_5$ decreases with increasing pressure P and becomes zero at a critical pressure $P_C \approx 4.0$ GPa. Nakashima et al. [119] have studied the pressure dependences of T_N, A, and ρ_0 which are shown in Figure 7.36. They measured the temperature dependence of the electrical resistivity under pressure by using a cubic anvil cell up to 8 GPa. They have used an indenter cell in the temperature range 40–300 mK under pressures up to 3.9 GPa.

From Figure 7.36(a), the Neel temperature is found to become zero around 4.0 GPa ($=P_c$). The pressure dependences of A and ρ_0 values ($\rho = \rho_0 + AT^2$) indicate that both increase sharply above 3 GPa, and become maximum around P_c.

At ambient pressure the electrical resistivity has two peaks around 100 and 5 K, indicating the characteristic behavior of the Kondo compound [120]. Figure 7.37 shows the temperature dependence of the electrical resistivity at 3.6 GPa. The overall resistivity is very similar to that of a heavy fermion superconductor $CeCu_{2.1}Si_2$, which is shown in Figure 7.38 for comparision. The value of $A = 8.5$ μΩ cm/K^2 at 3.6 GPa in $Ce_2Ni_3Ge_5$ is comparable to $A = 10$ μΩ cm/K^2 in $CeCu_{2.1}Si_2$. These results indicate that $Ce_2Ni_3Ge_5$ forms a heavy fermion state at P_c around which superconductivity is also observed. The inset of Figure 7.37 shows the low-temperature resistivity at a typical pressure of 3.6 GPa, indicating superconductivity below $T_{sc} = 0.26$ K (shown by an arrow) which is compared to $T_{sc} \approx 0.78$ K in $CeCu_{2.1}Si_2$ shown in the inset of Figure 7.38.

7.15. Summary and conclusion

To summarize, hitherto only three Ce-based compounds ($CeCu_2Si_2$, $CeCoIn_5$, and $CeIrIn_5$) have been discovered which become superconductors at ambient pressure and at sufficiently low temperature. $CeCu_2Si_2$ has two superconducting domes (at high pressures), one of which is due to spin fluctuations and the other presumably due to valence fluctuations. The P–T phase diagram in $CeRhIn_5$ and the phase diagram of the alloy system $CeRh_{1-x}Ir_xIn_5$ indicate that the superconductivity appears in the neighborhood of the AF phase. Both $CeCoIn_5$ and $CeIrIn_5$ are quasi-2D systems like the high-T_c cuprates. However, since the Pauli limitings are very different, $CeCoIn_5$ has presumably a transition to the FFLO superconducting phase while the transition of $CeIrIn_5$ to the FFLO phase is still an open question. The power-law behavior of the resistivity in the normal phase with $T^{1.3}$ in $CeIrIn_5$ ($T_c = 0.4$ K) and T in $CeCoIn_5$ ($T_c = 2.3$ K) implies the strong quasi-2D AF spin fluctuations. The fourfold symmetry in the thermal conductivity in $CeCoIn_5$ [94] strongly suggests the $d_{x^2-y^2}$-wave singlet pairing. The pseudogap phenomenon has been reported in the ^{115}In-NQR measurement in $CeRhIn_5$ in the range of pressures ($P = 1.53$–1.75 GPa) under which the coexistent phase exists [121]. These similarities,

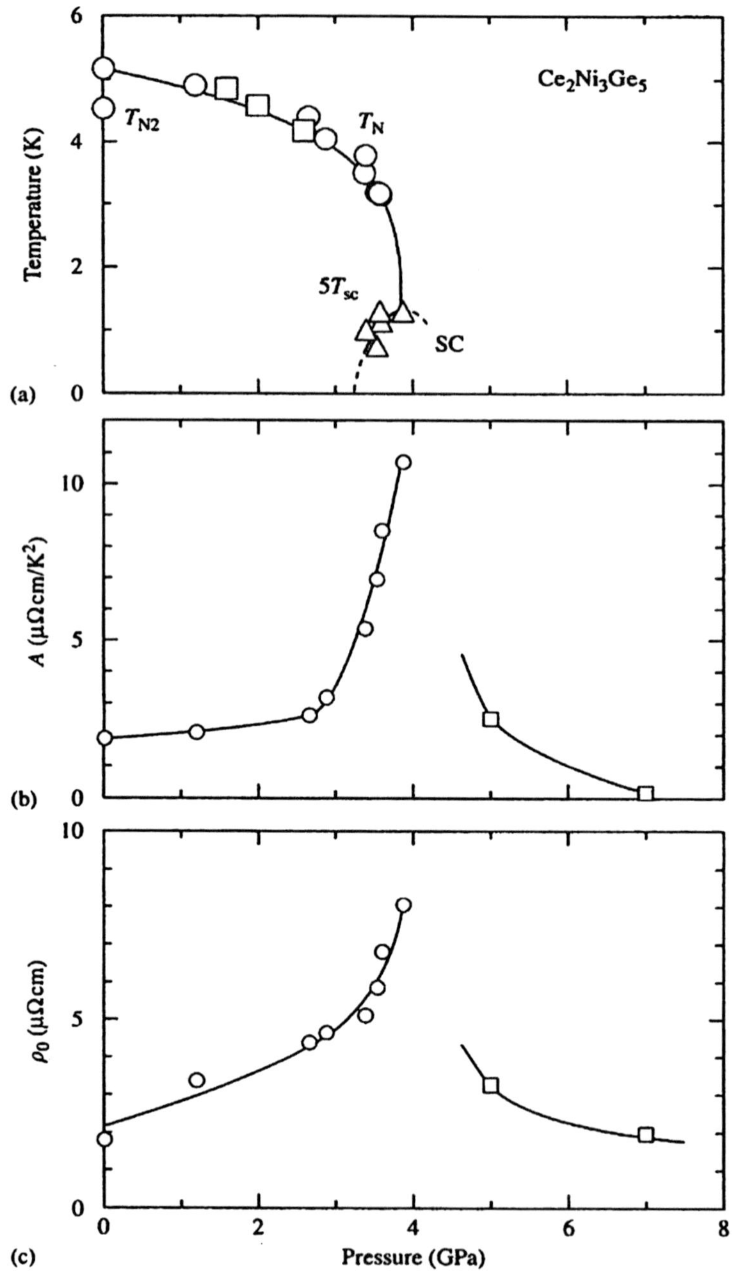

Figure 7.36. Pressure dependences in $Ce_2Ni_3Ge_5$ of (a) T_N and T_{sc}, (b) A, (c) ρ_0. The data shown by circles and squares were obtained by using the indenter and cubic anvil cells, respectively. Reproduced with the permission of Elsevier from Ref. [119].

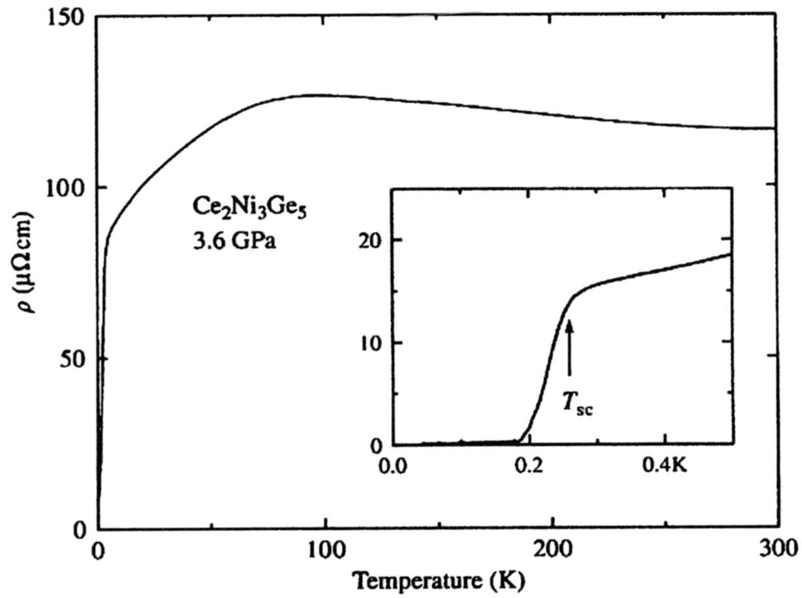

Figure 7.37. Temperature dependence of the electrical resistivity of $Ce_2Ni_3Ge_5$ at 3.6 GPa. An arrow in inset indicates onset of superconductivity. Reproduced with the permission of Elsevier from Ref. [119].

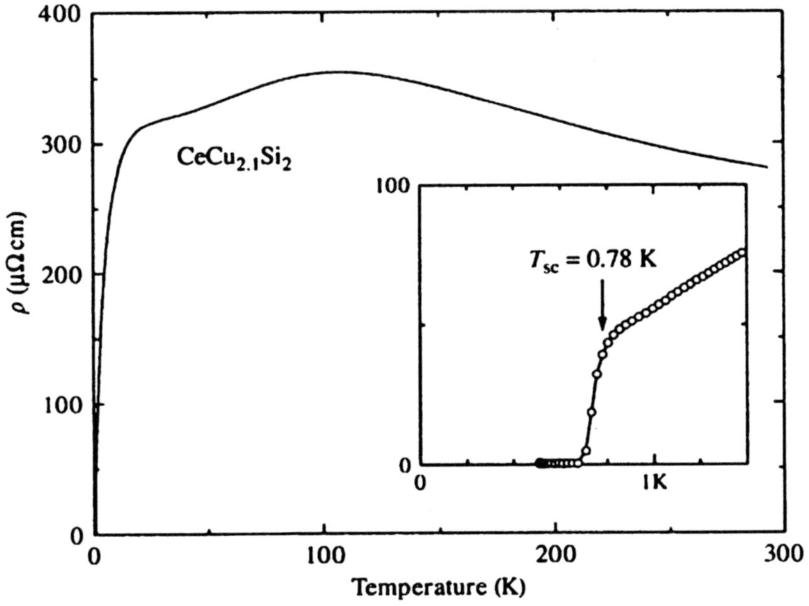

Figure 7.38. Temperature dependence of the electrical resistivity in $CeCu_{2.1}Si_2$. An arrow in inset indicates onset of superconductivity. Reproduced with the permission of Elsevier from Ref. [119].

specially the $d_{x^2-y^2}$-wave pairing and the AF spin fluctuations, indicate that the physics in the superconductivity in these Ce-based heavy fermion superconductors is in common with the cuprate superconductors.

The other Ce-based compounds become superconductors only at high pressures. A detail analysis of the theoretical models for pairing of heavy fermion superconductors will be presented in Chapter 10 with the caveat that the theories are still unfolding and there is no unanimity about the pairing.

References

[1] A.A. Abrikosov and L.P. Gorkov, Zh. Exsp. Teor. Fiz. **39**, 1781 (1960); Sov. Phys. JETP **12**, 1243 (1961) .

[2] F. Steglich, J. Aarts, C.D. Bredl, W. Lieke, D. Meschee, W. Franz, and H. Schafer, Phys. Rev. Lett. **43**, 1892 (1979).

[3] J. Bardeen, L. Cooper, and J.R. Schrieffer, Phys. Rev. **108**, 1175 (1957).

[4] F. Steglich, Physica B **359–361**, 326 (2005).

[5] G.R. Stewart, Z. Fisk, J.O. Willis, and J.L. Smith, Phys. Rev.Lett. **52**, 679 (1984).

[6] G.R. Stewart, Rev. Mod. Phys. **56**, 755 (1984).

[7] K. Hasselbach, L. Taillefer, and J. Flouquet, Phys. Rev. Lett. **63**, 93 (1989).

[8] M. Sigrist and K. Ueda, Rev. Mod. Phys. **63**, 239 (1991).

[9] H.R. Ott, H. Rudigier, P. Delsing, and Z. Fisk, Phys. Rev. Lett. **50**, 1595 (1983).

[10] H.R. Ott, H. Rudigier, E. Felder, Z. Fisk, and J.L. Fisk, Phys. Rev. B **33**, 126 (1986).

[11] H.R. Ott, in *Progress in Low Temperature Physics, Vol. XI,* ed. D.F. Brewer, p. 215 (North-Holland, Amsterdam, 1987).

[12] H.R. Ott, Helv. Phys. Acta **60**, 62 (1987).

[13] Y. Kuramoto and Y. Kitaoka, in *Dynamics of Heavy Electrons*, eds. J. Birman, S.F. Edwards, R. Friend, C.H. Llewellyn Smith, M. Rees, D. Sherrington, and D. Veneziano (Oxford, New York, 2000).

[14] H. Tou, Y. Kitaoka, K. Asayama, N. Kimura, and Y. Onuki, Phys. Rev. Lett. **77**, 1374 (1996).

[15] H. Tou, Y. Kitaoka, K. Ishida, K. Asayama, N. Kimura, Y. Onuki, E. Yamamoto, Y. Haga, and K. Maezawa, Phys. Rev. Lett. **80**, 3129 (1998).

[16] C. Geibel, C. Schank, S. Thies, H. Kitazawa, C.D. Bredl, A. Bohm, M. Rau, A. Grauel, R. Caspary, R. Helfrich, U. Ahleim, G. Weber, and F. Steglich, Z. Phys. B **84**, 1 (1991).

[17] C. Geibel, S. Thies, D. Kaczororowski, A. Mehner, A. Grauel, B. Seidel, U. Ahlheim, R. Helfrich, K. Peterson, C.D. Bredl, and F. Steglich, Z. Phys. B **83**, 305 (1991).

[18] M. Kyogaku, Y. Kitaoka, K. Asayama, C. Geibel, C. Schank, and F. Steglich, J. Phys. Soc. Jpn. **62**, 4016 (1993).

[19] H. Tou, Y. Kitaoka, T. Kamatsuka, H. Tou, K. Asayama, C. Geibel, F. Steglich, S. Sullow, and J.A. Mydosh, Physica B **230–232**, 360 (1997).

[20] K. Ishida, D. Ozaki, T. Kamatsuka, H. Tou, M. Kyogaku, Y. Kitaoka, N. Tateiwa, N.K. Sato, N. Aso, G. Geibel, and F. Steglich, Phys. Rev. Lett. **89**, 037002 (2002).

[21] T.T.M. Palstra, A.A. Menovsky, J. van den Berg, A.J. Dirkmaat, P.H. Kes, G.J. Nieuwenhuys, and J.A. Mydosh, Phys. Rev. Lett. **55**, 2727 (1985).

[22] W. Schlabitz, J. Baumann, B. Polit, U. Rauschwalbe, H.M. Mayer, U. Ahlheim, and C.D. Bredl, Z. Phys. B **62**, 671 (1986).

[23] D. Jaccard, K. Behnia, and J. Sierro, Phys. Lett. A **163**, 275 (1992).

[24] D. Jaccard, E. Vargoz, K. Alami-Yadri, and H. Wilhelm, Rev. High Pressure Sci. Technol. **7**, 412 (1998).

[25] D. Jaccard, H. Wilhelm, K. Alami-Yadri, and E. Vargoz, Physica B **259–261**, 1 (1999).

[26] F.M. Grosche, S.R. Julian, N.D. Mathur, and G.G. Lonzarich, Physica B **223–224**, 50 (1996).

[27] N.D. Mathur, F.M. Grosche, S.R. Julian, I.R. Walker, D.M. Freye, R.K.W. Haselwimmer, and G.G. Lonzarich, Nature **394**, 39 (1998).

[28] R. Movschovich, T. Garf, D. Mandrus, J.D. Thompson, J.L. Smith, and Z. Fisk, Phys. Rev. B **53**, 8241 (1996).

[29] S.J.S. Lister, F.M. Grosche, F.V. Carter, R.K.W. Haselwimmer, S.S. Saxena, N.D. Mathur, S.R. Julian, and G.G. Lonzarich, Z. Phys. B **103**, 263 (1997).

[30] I.R. Walker, F.M. Grosche, D.M. Freye, and G.G. Lonzarich, Physica C **282–287**, 303 (1997).

[31] H. Hegger, C. Petrovic, E.G. Moshopoulou, M.F. Hundley, J.L. Sarrao, Z. Fisk, and J.D. Thompson, Phys. Rev. Lett. **84**, 4986 (2000).

[32] C. Petrovic, R. Movshovich, M. Jaime, P.G. Pagliuso, M.F. Hundley, J.L. Sarrao, Z. Fisk, and J.D. Thompson, Europhys. Lett. **53**, 354 (2001).

[33] C. Petrovic, P.G. Pagliuso, M.F. Hundley, R. Movshovich, J.L. Sarrao, Z. Fisk, and P. Monthoux, J. Phys. Condens. Matter **13**, L337 (2001).

[34] S.S. Saxena, P. Agarwal, K. Ahilan, F.M. Grosche, R.K.W. Haselwimmer, M.J. Steiner, E. Pugh, I.R. Walker, S.R. Julian, P. Monthoux, G.G. Lonzarich, A. Huxley, I. Sheiken, D. Braithwaite, and J. Flouquet, Nature **406**, 587 (2000).

[35] D. Aoki, A. Huxley, E. Ressouche, D. Braithwaite, J. Flouquet, J.-P. Brison, E. Lhotel, and C. Paulsen, Nature **413**, 613 (2001).

[36] H. Kotegawa, S. Kawasaki, A. Harada, Y. Kawasaki, K. Okamoto, G.-Q. Zheng, Y. Kitaoka, E. Yamamoto, Y. Haga, Y. Onuki, K.M. Itoh, and E.E. Haller, J. Phys. Condens. Matter **15**, S2043 (2003).

[37] E.D. Bauer, N.A. Frederick, P.-C Ho, V.S. Zapf, and M.B. Maple, Phys. Rev. B **65**, 100506(R) (2002).

[38] J.L. Sarrao, L.A. Morales, J.D. Thompson, B.L. Scott, G.R. Stewart, F. Wastin, J. Rebizant, P. Boulet, E. Colineau, and G.H. Lander, Nature **420**, 297 (2002).

[39] F. Steglich, Physica B **378–380**, 7 (2006).

[40] P. Coleman, Physica B **378–380**, 1160 (2006).

[41] K. Kadowaki and S.B. Woods, Solid State Commun. **58**, 507 (1986).

[42] W. Lieke, U. Rauchschwalbe, C.D. Bredl, F. Steglich, J. Aarts, and F.R. de Boer, J. Appl. Phys. **53**, 2111 (1982).

[43] F. Steglich, P. Gegenwart, C. Geibel, P. Hinze, M. lang, C. Langhammer, G. Sparn, T. Tayama, O. Trovarelli, N. Sato, T. Dahm, and G. Varelogiannis, in *More is Different—Fifty Years of Condensed Matter Physics* eds. N.P. Ong and R.N. Bhatt, p. 191 (Princeton University Press, Princeton, NJ, 2001).

[44] G. Bruls, B. Wolf, D. Finsterbusch, P. Thalmeier, I. Kouroudis, W. Sun, W. Assmus, B. Luthi, M. Lang, K. Gloos, F. Steglich, and R. Modler, Phys. Rev. Lett. **72**, 1754 (1994).

[45] O. Stockert, E. Faulhaber, G. Zwicknagel, N. Stuber, H.S. Jeevan, M. Deppe, R. Borth, R. Kuchler, M. Lowenhaupt, C. Geibel, and F. Steglich, Phys. Rev. Lett. **92**, 136401 (2004).
[46] P. Gegenwart, C. Longhammer, C. Geibel, R. Helfrich, M. Lang, G. Sparn, F. Steglich, R. Horn, L. Donnevert, A. Link, and W. Assmus, Phys. Rev. Lett. **72**, 1501 (1998).
[47] G. Sparn, L. Donnevert, P. Hellemann, R. Horn, F. Laube, A. Link, S. Thomas, P. Gegenwart, B. Buschinger, C. Geibel, and F. Steglich, Rev. High Pressure Sci. Technol. **7**, 431 (1998).
[48] J.A. Hertz, Phys. Rev. B **14**, 1165 (1976).
[49] A.J. Millis, Phys. Rev. B **48**, 7183 (1993).
[50] T. Moriya and T. Takimoto, J. Phys. Soc. Jpn. **64**, 960 (1995).
[51] H.Q. Yuan, F.M. Grosche, M. Deppe, C. Geibel, G. Sparn, and F. Steglich, Science **302**, 2104 (2003).
[52] A.J. Millis, S. Sachdev, and C.M. Varma, Phys. Rev. B **37**, 4975 (1988).
[53] T. Moriya and K. Ueda, J. Phys. Soc. Jpn. **63**, 1871 (1994).
[54] G.G. Lonzarich, in *The Electron,* ed. M. Springfield, Chapter 7, (Cambridge University Press, Cambridge, UK, 1997).
[55] P. McHale and P. Monthoux, Phys. Rev. B **67**, 214512 (2003).
[56] K. Miyake, O. Narikiyo, and Y. Onishi, Physica B **259–261**, 676 (1999).
[57] Y. Onishi and K. Miyake, J. Phys. Soc. Jpn. **69**, 3955 (2000).
[58] K. Miyake et al., SCES '05, Abstract Booklet Th-P-1, p. 95.
[59] A.T. Holmes, D. Jaccard, and K. Miyake, Phys. Rev. B **69**, 024508 (2004).
[60] A.T. Holmes and D. Jaccard, Physica B **378–380**, 339 (2006).
[61] D.C. Kosenmanki and K.A. Gschneidner Jr., in *Handbook on the Physics and Chemistry of Rare Earths, Vol. 1,* p.1 (Elsevier, Amsterdam, 1978).
[62] N. Mori, Y. Okayama, H. Takahasi, Y. Haga, and T. Suzuki, Jpn. J. Appl. Phys. **32**, 300 (1993).
[63] D. Jaccard and A.T. Holmes, Physica B **359–361**, 333 (2005).
[64] M. Nakashima, K. Tabata, A. Kobayashi, T.C. Hedo, V. Ubatoko, K. Shimizu, and Y. Onuka, J. Phys. Condens. Matter **16**, L255 (2004).
[65] A.T. Holmes, D. Jaccard, H.S. Jeevan, C. Geibel, and M. Ishikawa, J. Phys. Condens. Matter **17**, 5423 (2005).
[66] F. Thomas, C. Ayache, I.A. Fomine, J. Thomasson, and C. Geibel, J. Phys. Condens. Matter **8**, L51 (1996).
[67] E. Vergoz and D. Jaccard, J. Magn. Magn. Mater. **177–181**, 294 (1998).
[68] G. Knopp, H. Spille, A. Loidl, K. Knorr, U. Rauchschusal, R. Felten, G. Weber, F. Steglich, and P. Murrani, J. Magn. Magn. Mater. **63–64**, 88 (1987).
[69] A. Loidl, K. Knorr, G. Knopp, A. Krimmel, R. Casparry, A. Boln, G. Spam, C. Geibel, F. Steglich, and A.P. Murani, Phys. Rev. B **46**, 9341 (1992).
[70] B. Fak, E. Ressouche, G. Knebel, J. Flouquet, and P. Lejay, Solid State Commun. **115**, 407 (2000).
[71] T. Fukuhara, S. Akamaru, T. Kuwai, J. Sakurai, and K. Maezawa, J. Phys. Soc. Jpn. **67**, 1998 (1998).
[72] H. Iwaski, N. Kobayashi, and Y. Muto, Physica B & C **148**, 64 (1987).
[73] M.J. Besnus, A. Essaihi, G. Fischer, N. Hamdaoui, and A. Meyer, J. Magn. Magn. Mater. **104–107**, 1387 (1992).

[74] S. Raymond, D. Jaccard, H. Wilhelm, and R. Cerny, Solid State Commun. **112**, 617 (1999).
[75] A. Demuer, D. Jaccard, I. Shekin, S. Raymond, B. Salce, J. Thomasson, D. Braithwaite, and J. Flouquet, J. Phys. Condens. Matter **13**, 9335 (2001).
[76] I. Shekin, D. Braithwaite, J.-P. Brison, S. Raymond, D. Jaccard, and J. Flouquet, J. Low Temp. Phys. **122**, 591 (2001).
[77] S. Raymond and D. Jaccard, Phys. Rev. B **61**, 8679 (2000).
[78] A. Demuer, A.T. Holmes, and D. Jaccard, J. Phys. Condens. Matter **14**, L529 (2002).
[79] H. Wilhelm and D. Jaccard, Phys. Rev. B **66**, 064428 (2002).
[80] S. Doniach, Physica B&C, **91**, 231 (1977).
[81] S.R. Julian, C. Pfleiderer, F.M. Grosche, N.D. Mathur, G.J. McMullan, A.J. Diver, I.R. Walker, and G.G. Lonzarich, J. Phys. C **8**, 9675 (1996).
[82] F.M. Grosche, S.J.S. Lister, F.V. Carter, S.S. Saxena, R.K.W. Haselwimmer, N.D. Mathur, S.R. Julian, and G.G. Lonzarich, Physica B **329**, 62 (1997).
[83] J. Flouquet, P. Haen, P. Lejay, P. Morin, D. Jaccard, J. Scweizer, C. Vettier, A.R. Fischer, and N.E. Phillips, J. Magn. Magn. Mater. **90–91**, 377 (1990).
[84] E. Bauer, G. Hilscher, H. Michor, Ch. Paul, E.F. Scheidt, A. Gribanov, Yu Seropegin, H. Noel, M. Sigrist, and P. Rogi, Phys. Rev. Lett. **92**, 027003 (2004).
[85] A. Amato, E. Bauer, and C. Baines, Phys. Rev. B **71**, 092501 (2005).
[86] E. Bauer, G. Hilscher, H. Michor, M. Seiberer, E.W. Scheidt, A. Gribanov, Yu. Seropegin, P. Rogl, A. Amato, W.Y. Song, J.-G. Park, D.T. Adroja, M. Nicklas, G. Sparn, M. Yogi, and Y. Kitaoka, Physica B **359–361**, 360 (2005).
[87] M. Yogi, Y. Kiyota, S. Hashimoto, T. Yasuda, R. Settai, T.D. Matsuda, Y. Haga, Y. Onuki, P. Rogl, and E. Bauer, Phys. Rev. Lett. **87**, 027003 (2004).
[88] R.P. Kapur, D.F. Agterberg, and M. Sigrist, Phys. Rev. Lett. **94**, 137002 (2005).
[89] D.F. Agterburg, R.P. Kapur, M. Sigrist, and A. Koga, Physica B **378–380**, 351 (2006).
[90] M. Nicklas, G. Sparn, R. Lackner, E. Bauer, and F. Steglich, Physica B **359–361**, 386 (2005).
[91] D. Hall, E. Palm, T. Murphy, S. Tozer, Z. Fisk, U. Alver, R.G. Goodrich, J.L. Sarrao, P.G. Pagliuso, and T. Ebihara, Phys. Rev. B **64**, 212508 (2001).
[92] R. Settai, H. Sishido, S. Ikeda, Y. Murakawa, M. Nakashima, D. Aoki, Y. Haga, H. Harima, and Y. Onuki, J. Phys. Condens. Matter **13**, L627 (2001).
[93] R. Movshovich, M. Jaime, J.D. Thompson, C. Petrovic, Z. Fisk, P.G. Pagliuso, and J.L. Sarrao, Phys. Rev. Lett. **86**, 5152 (2001).
[94] K. Izawa, H. Yamaguchi, Y. Matsuda, H. Sishido, R. Settai, and Y. Onuki, Phys. Rev. Lett. **87**, 057002 (2001); K. Izawa, H. Yamaguchi, Y. Matsuda, T. Sasaki, R. Settai, and Y. Onuki, J. Phys. Chem. Solids **63**, 1055 (2002).
[95] V.A. Sidorv, M. Nicklas, P.G. Pagliuso, J.L. Sarrao, Y. Bang, A.V. Balatsky, and J.D. Thompson, Phys. Rev. Lett. **89**, 157004 (2002).
[96] J. Paglione, M.A. Tanatar, D.G. Hawthorn, E. Boaknin, R.W. Hill, F. Ronning, M. Sutherland, and L. Taillefer, Phys. Rev. Lett. **91**, 246405 (2003).
[97] A. Bianchi, R. Movshovich, C. Capan, P.G. Pagluso, and J.L. Sarrao, Phys. Rev. Lett. **91**, 187004 (2003).
[98] P. Fulde and R.A. Ferrell, Phys. Rev. **135**, A550 (1964).
[99] A.I. Larkin and Y.N. Ovchinnikov, J. Exptl. Theoret. Phys. USSR **47**, 1130 (1964); A.I. Larkin and Y.N. Ovchinnikov, Sov. Phys. JETP **20**, 762 (1965).

[100] A.M. Clogston, Phys. Rev. Lett. **2**, 9 (1962).
[101] L.W. Gruenberg and L. Gunther, Phys. Rev. Lett. **16**, 996 (1966).
[102] K. Kumagai, M. Saitoh, T. Oyaizu, Y. Furukawa, S. Takashima, M. Nohara, H. Takagi, and Y. Matsuda, Phys. Rev. Lett. **97**, 227002 (2006).
[103] M. Tachiki, S. Takahasi, P. Gegenwart, M. Weiden, M. Lang, C. Geibel, F. Steglich, R. Modler, C. Paulsen, and Y. Onuki, Z. Phys. B **100**, 369 (1996).
[104] H. Adachi and R. Ikeda, Phys. Rev. B **68**, 184510 (2003).
[105] T. Mizushima, K. Machida, and M. Ichioka, Phys. Rev. Lett. **95**, 117003 (2005).
[106] H.A. Radovan, N.A. Fortune, T.P. Murphy, S.T. Hannahs, E.C. Palm, S.W. Tozer, and D. Hall, Nature (London) **425**, 51 (2003).
[107] T. Watanabe, T. Kasahara, K. Izawa, T. Sakakibara, Y. Matsuda, C.J. van der Beek, T. Hanoguri, H. Sishido, R. Settai, and Y. Onuki, Phys. Rev. B **70**, 020506 (2004).
[108] Y. Matsuda, K. Izawa, T. Watanabe, Y. Kasahara, Y. Nakajima, T. Sakakibara, C.J. van der Beek, M. Nohara, T. Hanaguri, H. Takagi, J. Goryo, K. Maki, P. Thalmeier, S. Osaki, H. Sugawara, H. Sato, H. Shishido, R. Settai, and Y. Onuki, J. Phys. Chem. Solids **66**, 1365 (2005).
[109] Y. Kohori, Y. Yamato, Y. Iwamoto, T. Kohora, E.D. Bauer, M.B. Maple, and J.L. Sarrao, Phys. Rev B **64**, 134526 (2001).
[110] S. Ozcan, D.M. Broun, B.Morgan, R.K.W. Hasselwimmer, J.L. Sarrao, S. Kamal, C.P. Bidinosti, P.J. Turner, M. Raudsepp, and J.l. Waldram, Europhys. Lett. **62**, 412 (2003).
[111] P.M.C. Rourke, M.A. Tanatar, C.S. Turel, J Berdeklis, C. Petrovic, and J.Y.T. Wei. Phys. Rev. Lett. **94**, 107005 (2005).
[112] H. Aoki, T. Sakakibara, H. Sishido, R. Settai, Y. Onuki, P. Miranovic, and K. Michida, J. Phys. Condens. Matter **16**, L13 (2004).
[113] M.A. Tanatar, J. Paglione, S. Nakatsuji, D.G. Hawthorn, E. Boaknin, R.W Hill, F. Ronning, M. Sutherland, L. Taillefer, C. Petrovic, P.C. Canfield, and Z. Fisk, Phys. Rev. Lett. **95**, 067002 (2005).
[114] J.D. Thompson, R. Movshovich, Z. Fisk, F. Bouquet, N. J. Curro, R.A. Fisher, P.C. Hammel, H. Hegger, M.F. Hundley, M. Jaime, P.G. Pagliuso, C. Petrovic, N.E. Phillips, and J.L. Sarrao, J. Magn. Magn. Mater. **226–230**, 5 (2001).
[115] P.G. Pagliuso et al., cond-mat/010575.
[116] R. Movschovich, A. Bianchi, M. Jaime, MF. Hundley, J.D. Thompson, N. Curro, P.C. Hammel, Z. Fisk, P.G. Pagliuso, and J.l. Sarrao, Physica B **312–313**, 7 (2002).
[117] H. Kotegawa, T. Miyoshi, K. Takeda, S. Fukushima, H. Hidaka, K. Tabata, T.C. Kobayashi, M. Nakashima, A. Thamizhavel, R. Settai, and Y. Onuki, Physica B **378–380**, 419 (2006).
[118] L. Durivault, F. Bourie, B. Chevalier, G. Andre, and J. Etourneau, J. Magn. Magn. Mater. **246**, 366(2002).
[119] M. Nakashima, H. Kohara, A. Thamizhavel, T.D. Matsuda, Y. Haga, M. Hedo, Y. Uwatoko, R. Settai, and Y. Onuki, Physica B **378–380**, 402 (2006).
[120] Z. Hossain, H. Hamashima, K. Umeo, T. Takabataka, C. Geibel, and F. Steglich, Phys. Rev. B **62**, 8950 (2000).
[121] H. Kawasaki, T. Mito, G.-Q Zheng, C. Thessieu, Y. Kawasaki, K. Ishida, Y. Kitaoka, T. Muramatsu, T.C. Kobayashi, M. Nakashima, D. Aoki, S. Araki, Y. Haga , R. Settai, and Y. Onuki, Phys. Rev. B 65, 020504 (2002)

Chapter 8

U-Based Superconducting Compounds

8.1. Overview

The U-based heavy fermion superconductors have multiple f-band structures with two or three f electrons per U site, unlike the Ce-based superconductors which have only one f electron per site. Due to the multi-f orbitals and relatively large hybridization terms with the conduction electrons, the U-based compounds do not display clear log T dependence in the resistivity. The multiband structures are complicated and the Fermi surfaces have dominant f character. The first two U-based heavy fermion superconductors, UBe_{13} ($T_c = 0.9$ K) and UPt_3 ($T_c = 0.54$ K), were discovered in 1983 by Ott et al. [1] and in 1984 by Stewart et al. [2], respectively. An elegant review of these early discoveries is available in Ref. [3]. It was evident within a few years that UPt_3 had three superconducting phases [4, 5] which created great impetus for further study of this unusual heavy fermion superconductor. This interest increased after the discovery of multiphase diagrams in $U(Be_{1-x}Th_x)_{13}$ [6–8].

The superconducting and magnetic phase transition in URu_2Si_2 ($T_c = 0.8$ K) was discovered by Palstra et al. [9] in 1985. Susceptibility, magnetization, and specific heat measurements on single-crystal samples indicate both a magnetic phase transition at 17.5 K and a superconducting transition at 0.8 K. The magnetism is related to a confined-moment type of magnetism while the superconductivity is bulk and exhibits abnormal critical behavior. The magnetic and superconducting properties are highly anisotropic with the *c* axis strongly magnetic and the *a* axis favorable for superconductivity.

Geibel et al. [10] discovered superconductivity in UNi_2Al_3 in 1991, which orders antiferromagnetically below T_N. Well below T_N, coexisting with the antiferromagnetic (AF) order, UNi_2Al_3 exhibits a heavy Landau-Fermi-liquid (LFL) state which becomes unstable against an HF-superconducting transition at T_c. Soon after Geibel et al. [11] discovered superconductivity in UPd_2Al_3 ($T_N = 14.3$ K, $T_c \approx 2$ K) which has very similar properties as UNi_2Al_3. In fact, unconventional, i.e., non-phonon-mediated Cooper pairing, could so far be demonstrated convincingly only for UPd_2Al_3. Sato et al., by combining quasiparticle tunneling [12] and INS [13] measurements, showed that in this strong-coupling superconductor, it is the acoustic magnon in the center of the AF Brillouin zone which acts as the "exchange boson" and replaces the optical phonons in classical strong-coupling superconductors such as Pb and Hg [13].

Since these early years, several U-based heavy fermion superconductors have been discovered. In this chapter, we shall make a brief review of the properties of these actinide systems.

8.2. UBe_{13}

UBe_{13} was the first actinide-based heavy fermion compound which was found to be a bulk superconductor below approximately 0.9 K [1]. The cubic UBe_{13} is also one of the most fascinating HF superconductors because superconductivity develops out of a highly unusual normal state characterized by a large and strongly T-dependent resistivity [1, 14].

In addition, upon substituting a small amount of Th for U in $U_{1-x}Th_xBe_{13}$, a non-monotonic evolution of T_c and a second phase transition of T_{c2} below T_{c1}, the superconducting one, is observed in a critical concentration range of x [15, 16].

It was also shown that the superconducting state is formed by heavy mass quasiparticles. This was demonstrated by plotting C_p/T versus T (at low temperatures) which is shown in Figure 8.1 [17]. The anomaly at T_c (Figure 8.1) is compatible with the large γ parameter in the normal state at this temperature.

Immediately after the discovery of superconductivity in UBe_{13}, both Ott et al. [1] and Anderson [18] suspected that an unconventional superconductivity state had been formed. It was unconventional in the sense that the superconducting transition was accompanied by a symmetry reduction, such as breaking time-reversal symmetry, in addition to breaking-gauge symmetry.

The temperature dependence of the specific heat of UBe_{13} well below T_c was the first indication for unconventional superconductivity. Figure 8.2 [17] shows the non-exponential but power-law-type decrease of $C_p(T)$ which was interpreted [19] as being the consequence of nodes in the gap of the electronic excitation spectrum.

NMR techniques were used to probe the Be nuclei of UBe_{13} and confirmed this conclusion by showing a power-law-type decrease of the NMR spin-lattice-relaxation rate $T_1^{-1}(T)$ with decreasing temperature below T_c [20]. Their experiment also showed the

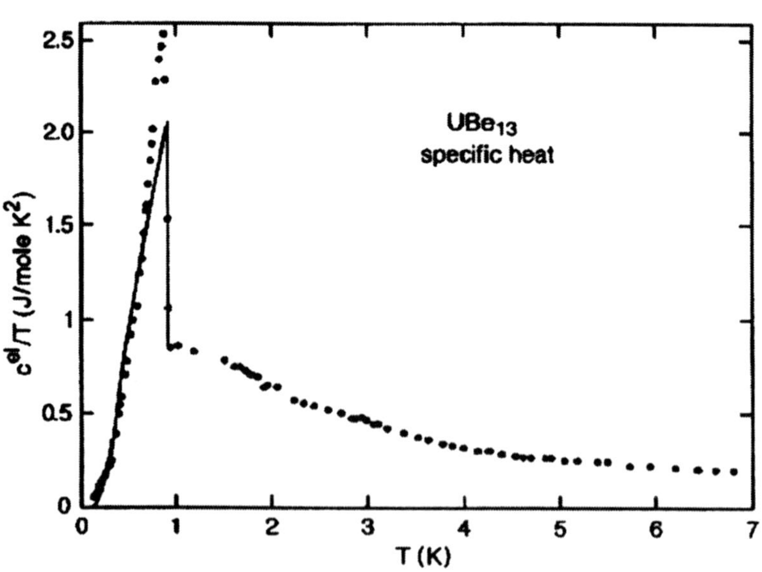

Figure 8.1. Electronic specific heat of UBe_{13} below 7 K. The solid line represents the BCS approximation of the anomaly at and below T_c. Reproduced with the permission of Elsevier from Ref. [17].

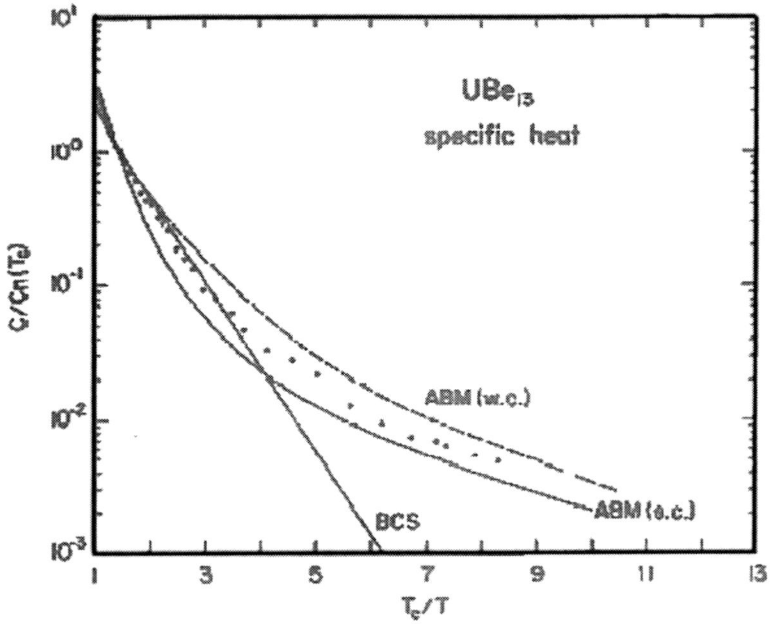

Figure 8.2. Normalized electronic specific heat of UBe_{13} below T_c, plotted versus T_c/T. The solid and broken lines represent calculations assuming point nodes in the gap. Reproduced with the permission of Elsevier from Ref. [17].

absence of coherence effects since there was no enhancement of T_1^{-1}. Experimental results of the temperature dependence of ultrasound absorption [21] were consistent with the persistent of gap nodes at low temperatures and confirmed the absence of coherence effects. The temperature dependence of the upper critical field $H_{c2}(T)$ of UBe_{13} was highly anomalous and indicated robust superconductivity with respect to an external magnetic field [22]. A direct comparison of the temperature dependence of the magnetic field penetration depths λ_L of UBe_{13} and Sn [23] showed that $\lambda_L(T)$ is very different for the two materials. Figure 8.3 [17] shows the temperature-induced variation of the two quantities on the same reduced temperature scale. A power law in T is observed for $\lambda_L(T)$ of UBe_{13} while $\lambda_L(T)$ of Sn varies exponentially with temperature.

It was also found that when small amounts of U atoms in UBe_{13} were replaced by other elements, there was a substantial reduction of the critical temperature [24]. T_c is also first substantially reduced with the alloys $U_{1-x}Th_xBe_{13}$ [15] as x is increased. However, when $x > 0.018$, T_c increases again until it passes over a shallow maximum at $x = 0.033$ and gradually decreases with a reduced slope when x is further increased. Further, in the range $0.019 < x < 0.05$, a second transition at T_{c2} below T_c was discovered by measuring the specific heat of these alloys at very low temperatures. Measurements of $\rho(T)$ and $\chi(T)$ confirmed that the phase at temperatures below the second anomaly of $C_p(T)$ was superconducting.

The phase diagram of superconductivity of $U_{1-x}Th_xBe_{13}$, from these observations as well as from thermodynamic arguments, is shown in Figure 8.4 [17]. One can identify three different superconducting phases, F, L, and U.

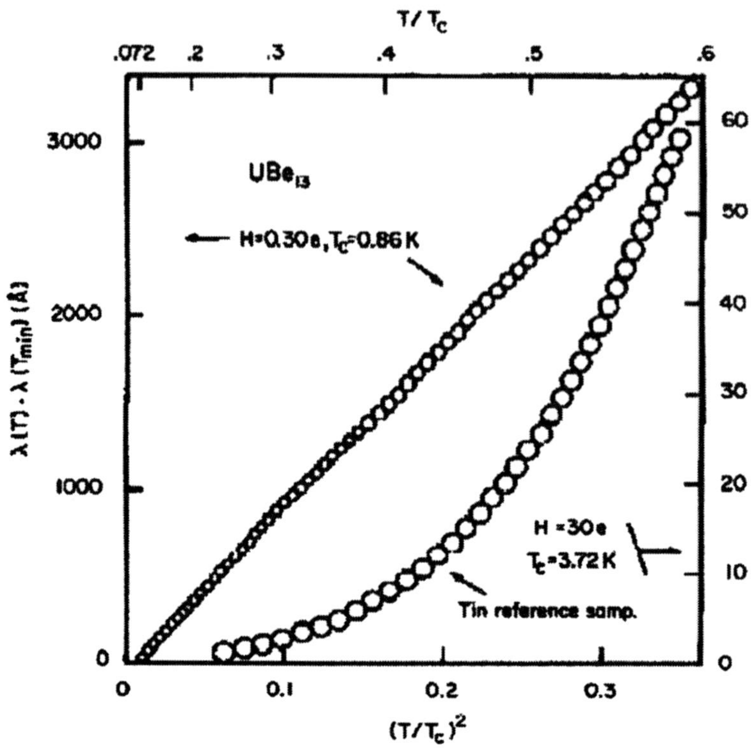

Figure 8.3. Incremental magnetic field penetration depth of superconducting UBe_{13} and the corresponding data for Sn reference sample. Reproduced with the permission of Elsevier from Ref. [17].

From experiments probing the pressure dependence of T_c [16], significantly different values for $\partial T_c/\partial p$ for x larger or smaller than 0.018 were obtained, proving that the phases F and U are different. Further, above a pressure of about 9 kbar, the phases F and U are separated by a normal conducting phase close to $T = 0$ K.

A theoretical Ginzburg–Landau type analysis, assuming an unconventional pairing of quasiparticles and aiming at reproducing the phase diagram of Figure 8.4, concluded that the phase termed L would break time-reversal symmetry [25]. Subsequent experiments employing μSR techniques [26] showed the onset of a spontaneous magnetic field gradient at T_{c2} which grew in a mean-field type manner with decreasing temperature. The formation of spontaneous orbital moments in this type of superconducting phases is consistent with the experimental results. Recent experiments [27], which probed the vortex motion in the mixed state of a sample with approximately 3% Th content, support the fact that the pairing in the L phase breaks time-reversal symmetry.

The replacement of small amounts of U by Th [28] or B for Be [29] affects the superconducting state of UBe_{13} as well as its electronic normal state properties. The temperature variation of the ratio C_p/T for such alloys at low temperature was studied by Ott et al. [30]. As Figure 8.5 [17] indicates, the necessary entropy balance across the transition and below can be achieved only if, for T approaching 0 K, C_p/T grows in an unusual manner.

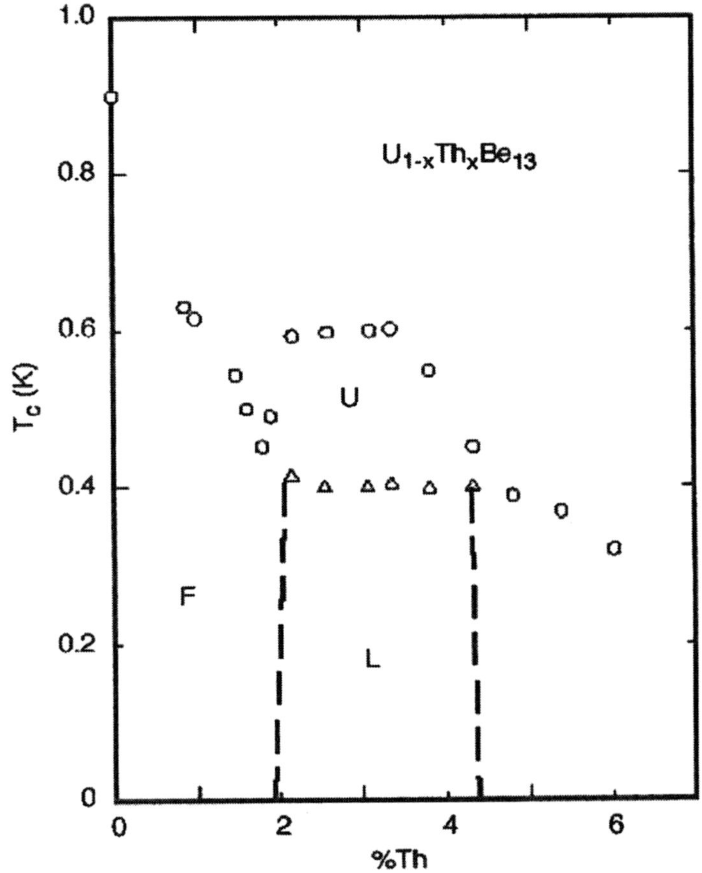

Figure 8.4. x, T phase diagram for superconducting $U_{1-x}Th_xBe_{13}$ as derived from the measurements of the specific heat. The letters F, L, and U denote three superconducting phases. Reproduced with the permission of Elsevier from Ref. [17].

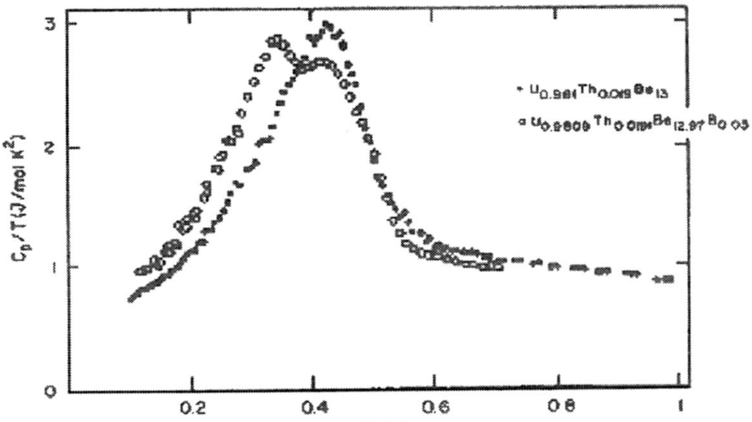

Figure 8.5. Specific heat anomalies of the superconducting transition of UBe_{13} with small concentrations of Th for U and B for Be. Reproduced with the permission of Elsevier from Ref. [17].

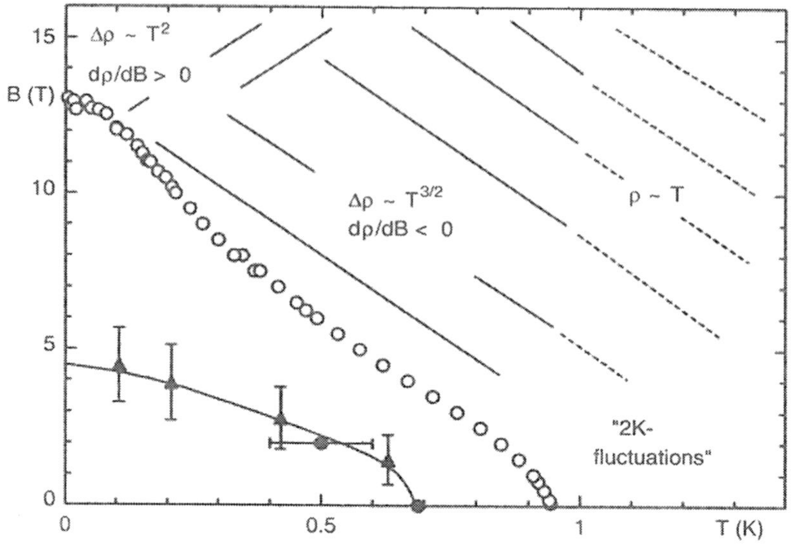

Figure 8.6. *B–T* phase diagram for UBe_{13}. Open circles indicate superconducting phase boundary, defined by $\rho = 0$, as derived from both *T*- and *B*-dependent measurements of the electrical resistivity. Solid symbols indicate "line of anomalies" $B^*(T)$ obtained from temperature-dependent thermal expansion (circles) and magnetic-field-dependent specific heat (triangles) measurements, respectively [32]. Reproduced with the permission of Elsevier from Ref. [31].

Since $T_F \ll 1$ K is unlikely, this data indicates non-Fermi-liquid (NFL) behavior of the underlying normal state.

Gegenwart et al. [31] have investigated the low-temperature normal-state NFL behavior of a high-quality single crystal of UBe_{13}. The temperature dependences of the specific heat and electrical resistivity resemble those of Ce-based heavy fermion systems near the AF quantum critical point (QCP). Thus, the NFL behavior in UBe_{13} is probably caused by a QCP at a finite magnetic field slightly above 4 T, covered by the superconducting (SC) phase. Studies of the thermal expansion and specific heat have revealed a "line of anomalies" $B^*(T)$ in the *B–T* phase diagram shown in Figure 8.6.

$B^*(T)$, which is of short-range AF origin, starts at $T_L = 0.7$ K ($B = 0$) and ends at $B^*(0) \approx 4.5$ T [32]. A clear enhancement of the slope $dT_c(B)/dB$ is observed near 4 T. It is likely that the stabilization of this superconducting state might be related to the "short-range AF" QCP.

8.3. UPt_3

In 1984, Stewart et al. [2] discovered bulk superconductivity in UPt_3 at 0.54 K based on specific heat, resistance, and ac susceptibility measurements. In addition, they presented evidence which indicated that UPt_3 is a spin-fluctuation system. Since then, there has been a large amount of work on superconductivity of UPt_3 because of its unique nature.

Some of these include the thermodynamic evidence for a multiple superconducting regime [4, 5] denoted as phases A, B, and C, as well as the specific heat data which were shown to fit the law [33]

$$C_V = \gamma T + \beta T^3 + \delta T^3 \ln\left(\frac{T}{T_{SF}}\right), \tag{8.1}$$

where $\delta \approx 2$ mJ/mole-K^4 and is a measure of the fluctuations with the characteristic energy T_{SF}. The spin-fluctuation term in the specific heat is unique to UPt_3 since this term is absent in other heavy fermion superconductors. UPt_3 has other unique features some of which will be discussed in this chapter. The large number of excellent reviews on superconductivity of UPt_3 include that of Sigrist and Ueda [6], Taillefer et al. [34], Joynt [35], Taillefer [36], Lohneysen [37], Heffner and Norman [38], and Joynt and Taillefer [39].

The dispersion of bands of UPt_3 is due to the hybridization of the U 5f shell with the Pt 5d shell. Direct overlap of the U atoms is negligible since the minimum U–U distance is 4.13 Å well below the Hill limit of 3.4 Å at which 5f orbitals cease to overlap. The bandwidth is generally about 20 mRy which is greater than the crystal-field splitting. UPt_3 can be classified as an extreme type II superconductor as the magnetic penetration length of $\lambda \approx 6000$ Å is relatively large compared to the superconducting coherence length of $\xi \approx 120$ Å.

One of the noteworthy features of UPt_3 is the absence of a peak in electrical resistivity unlike other heavy fermion compounds, a feature attributed to the single-impurity Kondo effect. The electrical resistivity of UPt_3 is shown as a function of temperature up to 300 K in Figure 8.7.

In Figure 8.7(b), the resistivity of UPt_3 is shown as a function of temperature up to 300 K for a current along each of the high-symmetry directions, $J \| c$ and $J \perp c$. There is no difference between $J \| a$ and $J \| b$. In addition, at all temperatures $\rho_{a,b} \approx 2\rho_c$. In fact, $\rho_{a,b}(300\text{ K}) \approx 230$ μΩ cm and $\rho_c(300\text{ K}) \approx 130$ μΩ cm.

In Figure 8.7(a), the low-temperature behavior of resistivity is shown. It obeys the $\rho(T) = \rho_0 + AT^2$ up to 1.5 K. $A = 1.55 \pm 0.1$ μΩ cm/K^2 for $J \| a, b$ and $A = 0.55 \pm 0.05$ μΩ cm/K^2 for $J \| c$ independent of ρ_0. In the low low-temperature region (below ~1 K), the anisotropy in conduction is almost independent of temperature [40]. The fact that the anisotropy is the same for the elastic and inelastic terms (ρ_0 and A) suggests that it arises from the Fermi velocities rather than from the scattering rates.

A few months after the discovery of superconductivity in UPt_3 by Stewart et al. [2], the evidence of a second superconducting phase was found from a kink in the upper critical field to the hexagonal c axis [41]. The anomalies in ultrasonic attenuation [42, 43] provided evidence for a transition in finite fields within the superconducting state and also within the superconducting phase. Fischer et al. [4] found two successive transitions to superconductivity at T_c^+ and T_c^- in the specific heat which is shown in Figure 8.8.

In Figure 8.8, the zero-field specific heat of UPt_3 is shown as C/T versus T. It is seen that out of the Fermi-liquid regime (well-defined up to 1.5 K), a first discontinuity appears at $T_c^+ = 490$ mK, signaling the onset of superconductivity and a second discontinuity is observed at $T_c^- = 434$ mK.

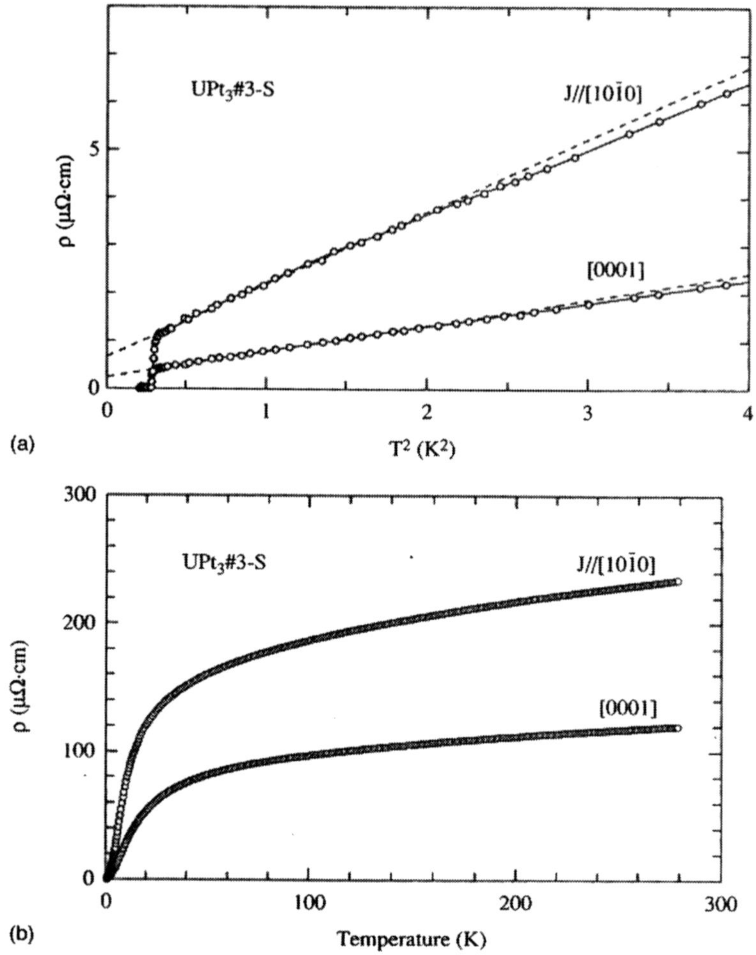

Figure 8.7. Temperature dependence of the electrical resistivity of a high-quality single crystal of UPt_3 for current directions parallel [0 0 0 1] and perpendicular to the hexagonal c axis. Reproduced with the permission of the American Physical Society from Ref. [39].

In order to resolve the presence of a second superconducting order parameter (OP) (T_c^-), already well inside the superconducting regime of UPt_3, the use of the ionic hexagonal lattice of UPt_3 as an indirect probe of its superconducting state has been done. In Figure 8.9, the thermal expansion coefficient $\alpha = L^{-1}dL/dT$ [44], for the two main directions of the lattice response (parallel and perpendicular to the c axis), has been plotted as a function of temperature.

In Figure 8.10, the velocity of longitudinal sound (propagating along the a axis), measured as deviations Δv from its value v at $T = 0.53$ K [45], has been plotted as a function of temperature.

We note that there are several thermodynamic quantities including the specific heat, the thermal expansion, and sound velocity, in which the two transitions T_c^+ and T_c^- are evident. The measured values of T_c^+ and T_c^- are in excellent agreement with each other.

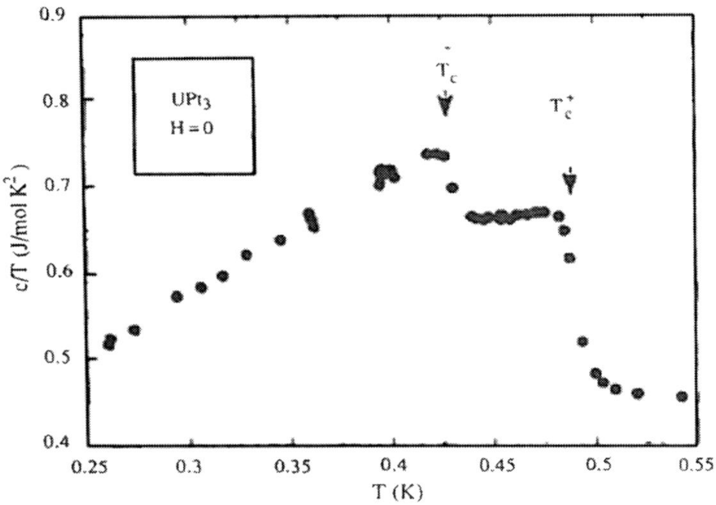

Figure 8.8. Specific heat of UPt_3 in the vicinity of the transition to superconductivity, in the absence of a magnetic field. It reveals the presence of a second (lower) transition at $T_c^- = 434$ mK, within the superconducting regime the onset of which is at $T_c^+ = 490$ mK. Reproduced with the permission of Elsevier from Ref. [34].

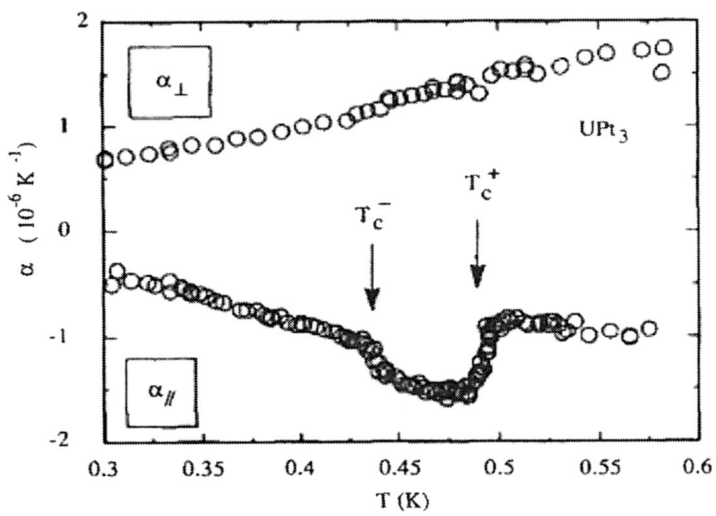

Figure 8.9. Coefficient of linear thermal expansion (α) of UPt_3, parallel and perpendicular to the hexagonal axis as a function of temperature, in zero magnetic field. The two discontinuities in $\alpha_\parallel$ are at $T_c^- = 442$ mK and $T_c^+ = 494$ mK. Reproduced with the permission of Elsevier from Ref. [34].

Hasselbach et al. [5] measured the specific heat of UPt_3 at different magnetic fields. Figure 8.11 shows their result and how the specific-heat peaks merge in a magnetic field.

Based on the thermodynamic evidences, Hasselbach et al. [5] have proposed a (B, T) phase diagram with three different superconducting critical phases (A, B, and C) which exist below an upper critical field line $B_{c2}(T)$ that separates them from the normal state.

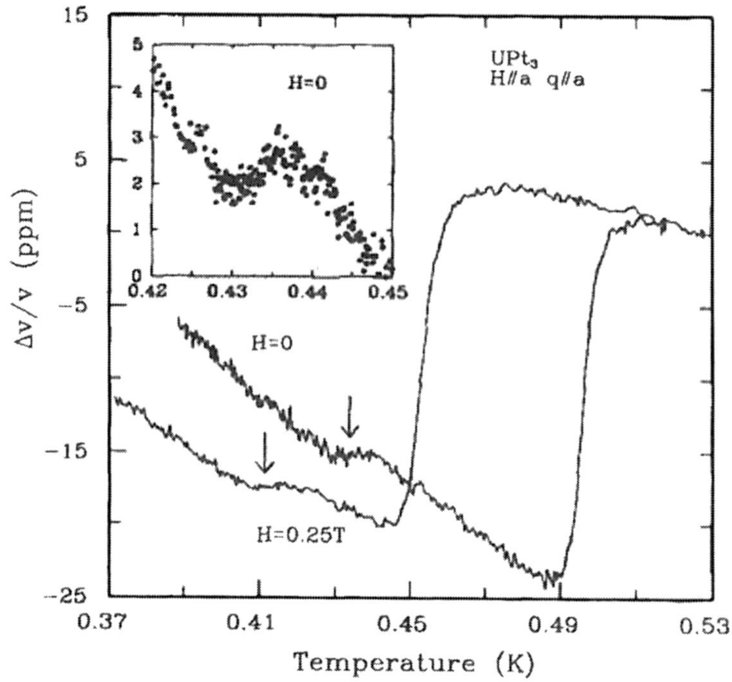

Figure 8.10. Temperature dependence of the longitudinal ultrasonic velocity at $H = 0$ and 2.5 kOe. The inset shows the result of a very slow temperature sweep through the anomaly centered at $T = 434$ mK. The onset of superconductivity is at $T = 495$ mK ($H = 0$). Reproduced with the permission of Elsevier from Ref. [34].

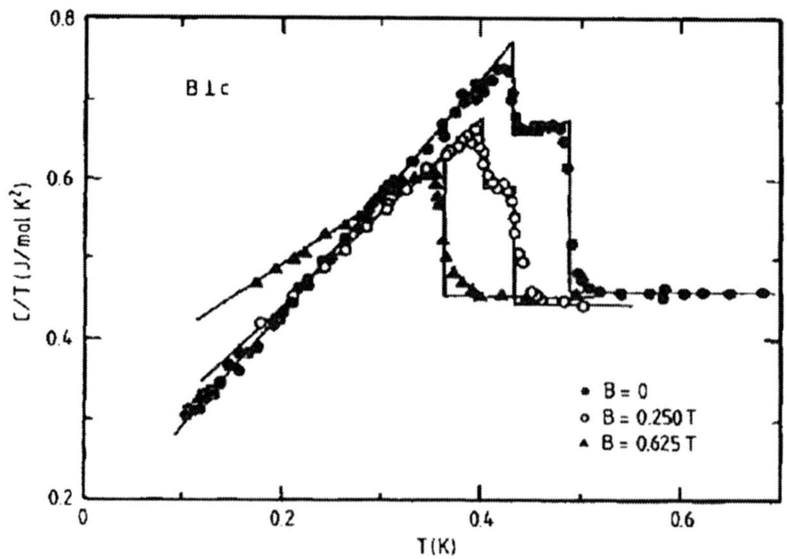

Figure 8.11. Specific heat C of UPt_3 plotted as C/T versus T for magnetic fields B applied perpendicular to the c direction. Reproduced with the permission of Elsevier from Ref. [37].

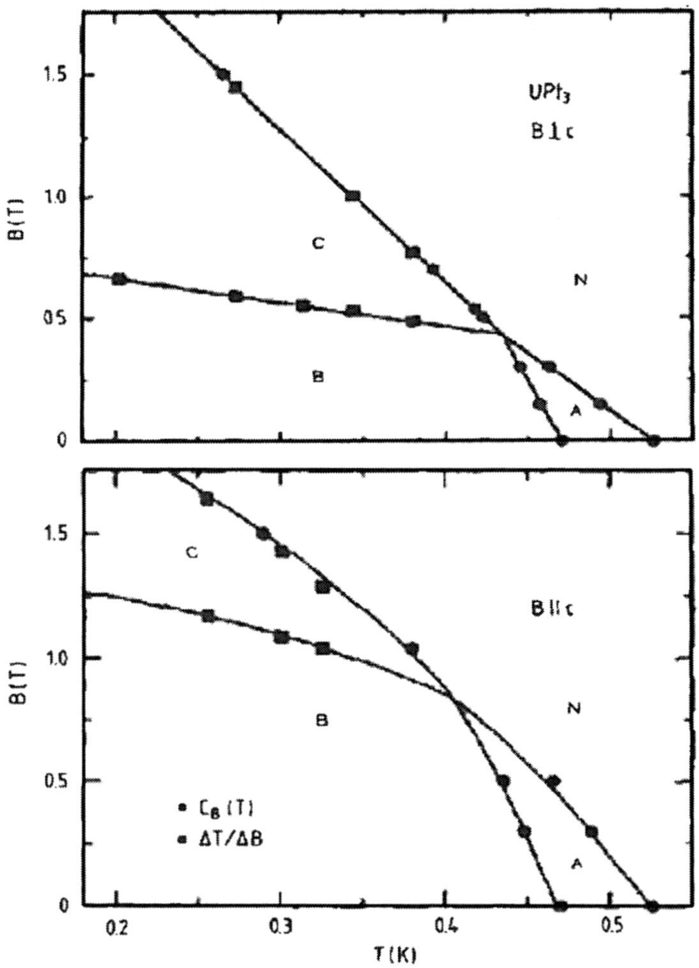

Figure 8.12. (B, T) phase diagram of UPt_3 for a field B applied perpendicular (a) and parallel (b) to the c direction as determined from the specific heat (●) and magneto-caloric effect (■). Solid lines have slopes consistent with a tetracritical point [46]. Reproduced with the permission of Elsevier from Ref. [37].

They have also shown that these phases all meet at a tetracritical point (T^*, H^*). Their (B, T) phase diagram is shown in Figure 8.12.

By analyzing the various thermodynamic data in a magnetic field, Bogenberger et al. [46] have concluded that the BC transition line is most likely of second order. This can also be inferred from the shape of the anomaly in the magnetocaloric effect when crossing the BC line. The large number of experimental data provide strong evidence that the multiple phases are due to superconductivity. For example, while $T_{c+} \approx 0.5$ K can differ by ~0.1 K for different samples, $T_c^+ - T_c^-$ is always of the order of 50 mK. In addition, the repeated observation of only one broad transition in the specific heat of width ~0.1 K can be interpreted as a smeared double transition.

The lower critical point B_{c1} is for $T \to 0$ between 6 and 10 mT which depends on the criteria of determination [47]. It also exhibits a kink [48] similar to B_{c2}, indicating that the lower transition is indeed a thermodynamic phase transition leading to a lowering of the condensation energy. The Meissner effect is very weak, accounting to a few percent of the diamagnetic-shielding signal [48].

The multiplicity of superconducting phases can be interpreted in various ways. All these scenarios postulate an unconventional OP. The possible OP representations allowed for the hexagonal point group of UPt_3 can be classified under the assumption of strong spin-orbit interaction [6]. The OP can be classified according to parity where even parity corresponds to "singlet" or "d-wave" superconductivity when turning on the spin-orbit interaction adiabatically, and odd parity corresponds to "triplet" or "p-wave" superconductivity. One possibility is that the two (zero-field) phases belong to different OP representations which are nearly degenerate [48]. The other possibility is that the two transitions arise from a splitting of an otherwise degenerate state within a single degenerate representation. The theory for these different scenarios will be discussed in detail in Chapter 10.

Behnia et al. [49] showed that the kink in $B_{c2}(T)$ curve (Figure 8.12), which is associated with the tetracritical point, vanishes when hydrostatic pressure is applied. Trappmann et al. [50] have studied the specific heat of UPt_3 under pressure. They have shown that the two sharp transitions observed at ambient pressure merge into a single and equally sharp transition at a critical pressure $p^* \approx 3.7$ kbar. Their results of specific heat measurements for different hydrostatic pressures plotted as C/T versus T is shown in Figure 8.13.

The pressure–temperature phase diagram of UPt_3 for hydrostatic pressure, as determined from specific heat studies by Loheneysen et al. [51], is shown in Figure 8.14.

Figure 8.14 shows that the two transitions observed at ambient pressure collapse into a single transition thereby confirming that the two transitions at low p are both superconducting transitions. The earlier speculations [52] that the weak AF is suppressed on the same pressure scale on which the two transitions merge were confirmed by a neutron-scattering study under pressure [53]. This strengthened the case of a coupling between superconducting and AF OPs.

Some of the unresolved questions raised by Lohneysen [37], in his excellent review, are summarized as follows: (i) T_c^+ is rather isotropic when a magnetic field is applied within the basal plane which can be understood if the magnetic moments rotate under the influence of the magnetic field and always are oriented perpendicular to the field. However, the isotropy with respect to field direction within or perpendicular to the basal plane are more difficult to understand. (ii) The magnetic correlation length is of the order of the superconducting coherence length ξ. It is not clear how this can produce a T_c splitting. (iii) Although the magnetic order breaks the hexagonal symmetry and thus is a prime candidate for symmetry-breaking field (SBF), it is unclear how the tiny AF magnetic moments (0.02 μ_B/U atom) can produce a sizable T_c splitting. We will address some of these issues in the following.

Elastic neutron scattering [54] and magnetic X-ray diffraction [55] measurements at low temperatures revealed that UPt_3 shows a weak AF order below $T_N = 6$ K with an ordered moment of only 0.02 μ_B/U atom [56]. The AF order was found to coexist with the superconductivity at low temperature. In most of the scenarios proposed [57–61] to explain the exotic superconducting phase diagram, the double transition is caused by a SBF which lifts the degeneracy of the components of a vector OP within a single-symmetry

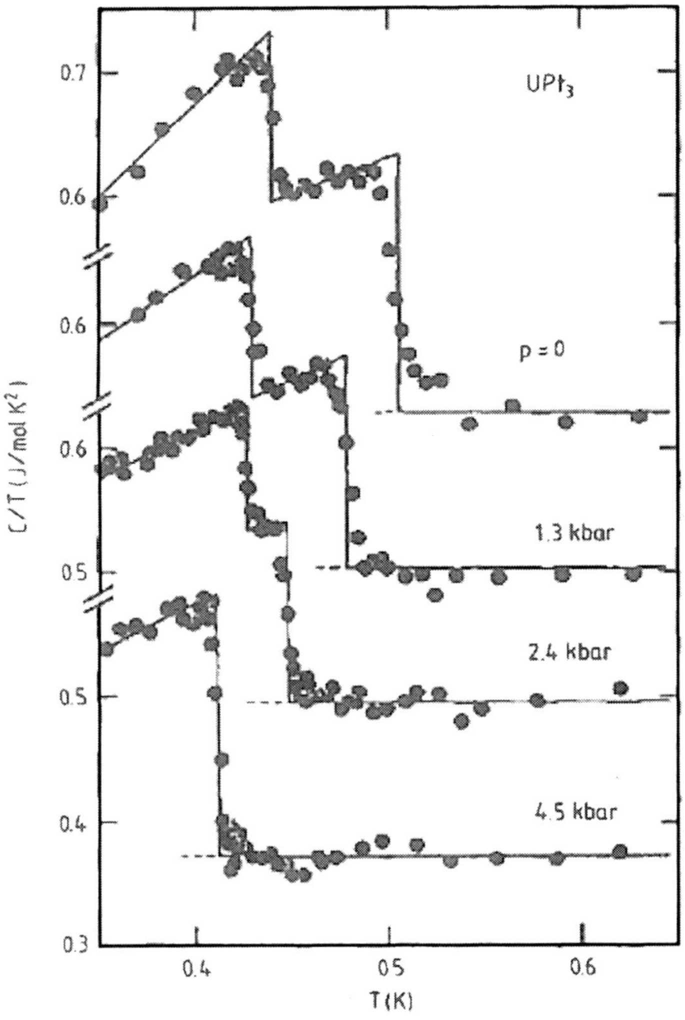

Figure 8.13. Specific heat C of UPt_3 under hydrostatic pressure p plotted as C/T versus T. Reproduced with the permission of Elsevier from Ref. [37].

representation. The most likely candidate for the SBF is the weak AF order since the magnetic order lowers the symmetry of the system [56].

Some of the unusual magnetic and superconducting properties of UPt_3 can be understood by analyzing the neutron-diffraction and μSR experiments in UPt_3 doped with either Pd [62] or with Th [63]. In pure UPt_3, a small-moment antiferromagnetic (SMAF) phase develops below $T_N \sim 6$ K [64]. The size of the ordered moment is unusually small, $m = 0.02 \pm 0.01$ μ_B/U atom. Neutron-diffraction experiments were carried out on annealed $U(Pt_{1-x}Pd_x)$ single crystals with $x = 0.001$, 0.002, 0.005, and 0.01 [64], and the temperature variation of m^2 was plotted to demonstrate that small-moment magnetism is robust

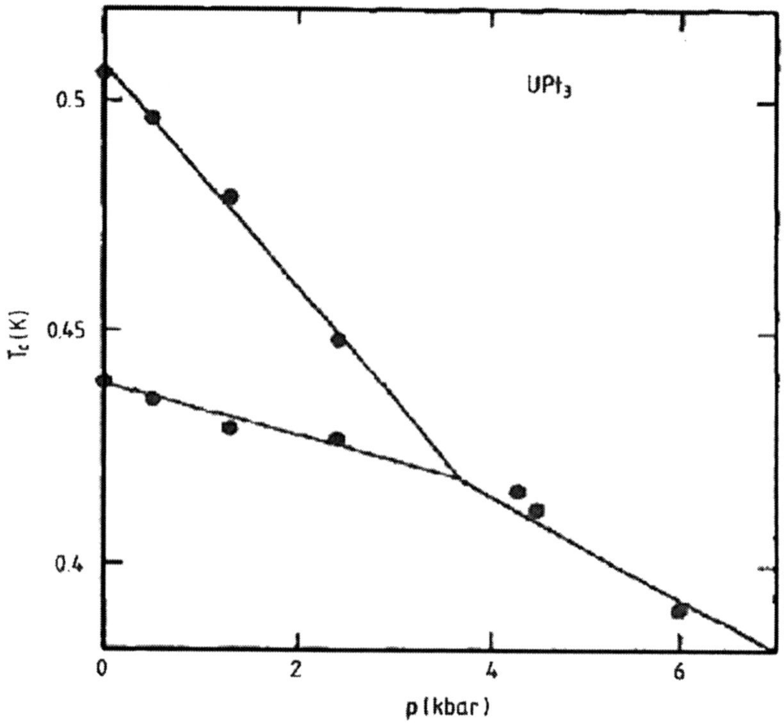

Figure 8.14. (p, T_c) phase diagram of UPt_3 for hydrostatic pressure p. Reproduced with the permission of Elsevier from Ref. [37].

upon alloying with Pd. The size of the ordered moment increases gradually with Pd concentration, but SMAF invariably sets in near $T_N \sim 6$ K for $x \leq 0.01$. This suggests that the transition to SMAF is not a true phase transition. In addition, the fact that SMAF does not show up in the zero-field μSR signal strongly suggests that the small moment fluctuates with time [62].

Neutron-diffraction experiments were carried out on annealed single crystals of $U(Pt_{1-x}Pd_x)$, with $x = 0.01$, 0.02, and 0.05 [64] where a large-moment antiferromagnetic (LMAF) phase developed with $T_N = 1.8$, 3.8, and 5.8 K, respectively. In Figure 8.15, the temperature variation of m^2 measured at the magnetic Bragg peak $\mathbf{Q} = (1/2, 0, 1)$ for annealed $U(Pt_{1-x}Pd_x)_3$ is shown.

The temperature dependence of the Bragg intensity of the sample with $x = 0.01$ shows that $m^2(T)$ starts to rise slowly below $T_N \sim 6$ K, grows rapidly below $\sim$2 K, and then saturates below $\sim$0.5 K. The rapid rise near 2 K suggests a crossover from the small-moment to the large-moment state, with an estimate of $T_N \sim 1.8$ K for the LMAF phase. For $T \to 0$ K, m reaches a value of 0.11 ± 0.03 μ_B/U atom.

The suppression of the double superconducting transition ($T_c^+ = 0.56$ K and $T_c^- = 0.51$ K for UPt_3) as a function of Pd doping, which has been studied in specific heat [65] and electrical resistivity [66, 67] experiments for a large number of crystalline and polycrystalline samples. Figure 8.16 shows the specific heat divided by T versus T of $U(Pt_{1-x}Pd_x)_3$ for various samples.

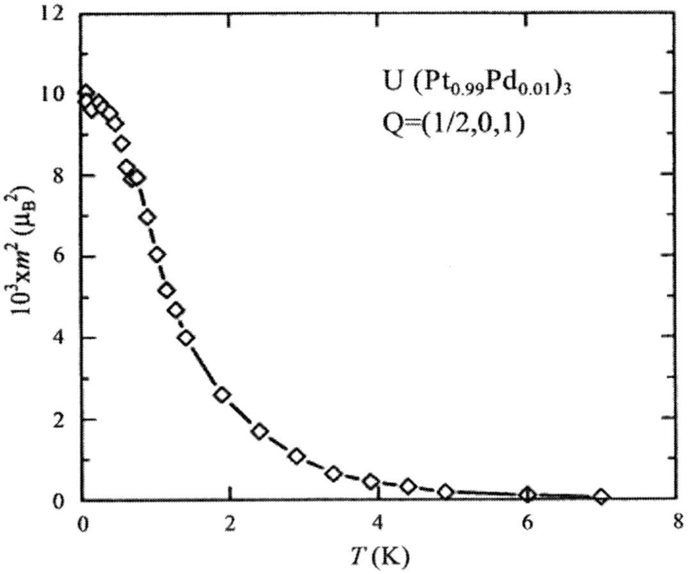

Figure 8.15. Temperature variation of m^2 measured at the magnetic Bragg peak $\mathbf{Q} = (1/2, 0, 1)$ for annealed $U(Pt_{1-x}Pd_x)_3$ with $x = 0.01$. The sharp increase in the intensity near 1.8 K indicates a crossover from SMAF to LMAF. Reproduced with the permission of Elsevier from Ref. [62].

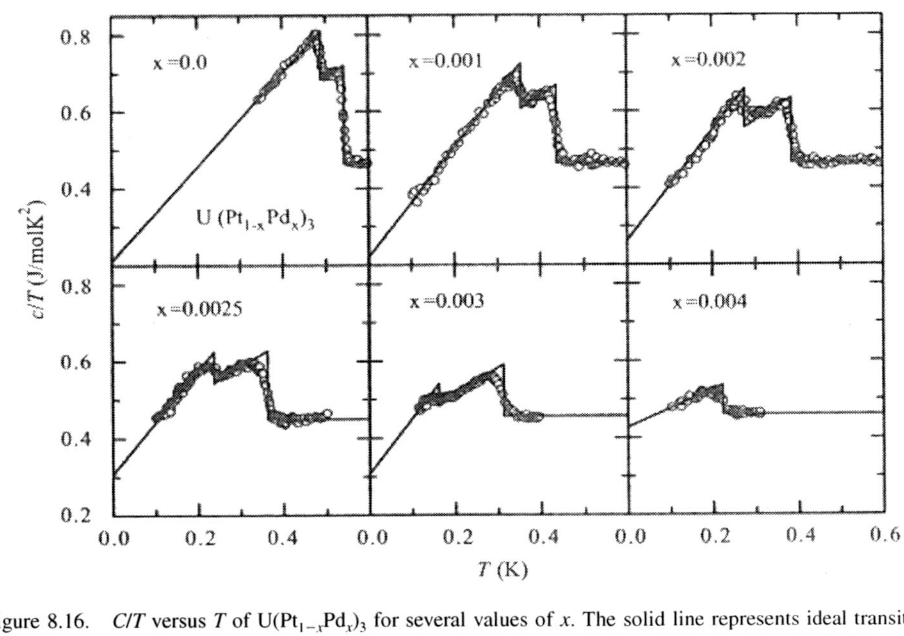

Figure 8.16. C/T versus T of $U(Pt_{1-x}Pd_x)_3$ for several values of x. The solid line represents ideal transition from an equal entropy construction. Reproduced with the permission of Elsevier from Ref. [62].

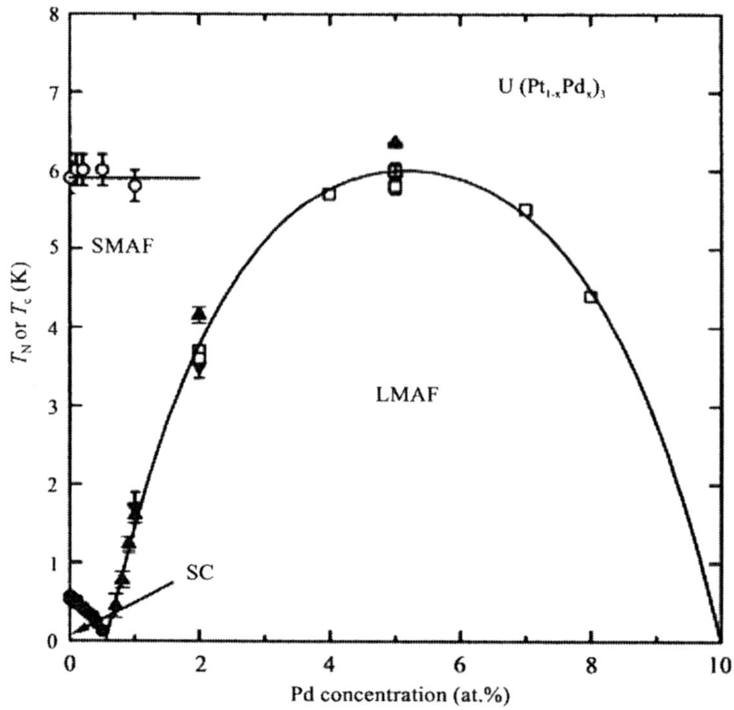

Figure 8.17. Magnetic and superconducting phase diagram for $U(Pt_{1-x}Pd_x)_3$ alloys. The Neel temperatures T_N are measured by neutron diffraction (O and ▼) [64], specific heat (□) [67], and μSR (▲) [68, 69]. Resistively determined superconducting (SC) temperatures T_c^+ (●) are taken from Ref. [66]. The solid lines are guide to the eye. Reproduced with the permission of Elsevier from Ref. [62].

The main findings of Figure 8.16 can be summarized as follows: (i) T_c^+ is suppressed linearly with Pd content at a rate of 0.79 ± 0.04 K/at% Pd, (ii) T_c^- is suppressed at a faster rate of 1.08 ± 0.06 K/at% Pd. Thus, ΔT_c increases at a rate of 0.30 ± 0.02 K/at% Pd. The high-temperature low-field A phase gains stability at the expense of the low-field *B* phase. The increase in ΔT_c is accompanied by an increase in the size of the SMAF-ordered moment [65]. This provides additional evidence that the SMAF acts as the SBF.

Figure 8.17 shows the magnetic and superconducting phase diagram for $U(Pt_{1-x}Pd_x)_3$ alloys [62, 64, 66–69].

The differences between SMAF and LMAF can be summarized as follows: (i) the OP for the LMAF is conventional and conforms a real phase transition while the OP of SMAF is unusual and grows quasilinearly, indicating a crossover phenomenon; (ii) T_N for the SMAF does not change with Pd content (or pressure), while T_N for the LMAF compounds follows a rather Doniach-type phase diagram; (iii) SMAF is not observed in zero-field μSR (and NMR) experiments in contrast to LMAF. The failure to detect the SMAF phase in the zero-field μSR experiments indicates that the moment is fluctuating in time with a frequency >10 MHz. The summary of other experimental results [62] also confirms the fact that it is the LMAF, and not the SMAF phase, which presents the magnetic instability in $U(Pt, Pd)_3$.

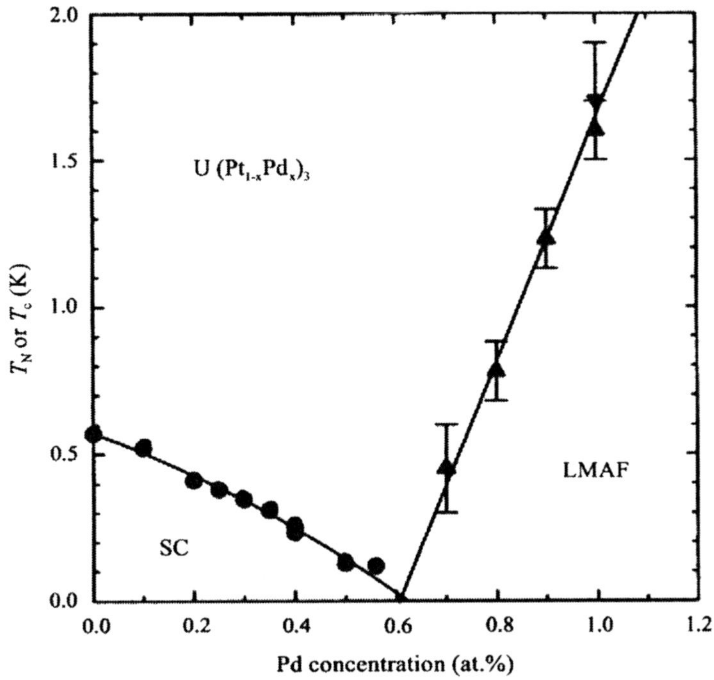

Figure 8.18. Magnetic and superconducting phase diagram for $U(Pt_{1-x}Pd_x)_3$ alloys for $x < 0.012$. The solid lines are guide to the eye. Reproduced with the permission of Elsevier from Ref. [62].

The magnetic and superconducting phase diagram for $x < 0.012$ is shown in Figure 8.18.

The superconducting (T_c^+) phase-transition temperatures have been taken from Ref. [66]. From Figure 8.18, it is obvious that the critical concentration for the suppression of superconductivity coincides with the concentration for the emergence of LMAF, $x_{c,s} = x_{c,A} \approx 0.006$. This suggests that superconductivity is not mediated by AF interactions, but rather by ferromagnetic fluctuations [62] which cannot coexist with long-range AF order.

Graf et al. [70] have studied the onset of LMAF in UPt_3 via Th substitution by muon spin spectroscopy. At low-Th content ($x \leq 0.002$) magnetic ordering on the time scale of the μSR experiment (10^{-8} s) is not detected, as in the case of UPt_3. For $0.006 \leq x \leq 0.05$, spontaneous oscillations in the μSR spectra signal the presence of the LMAF phase. The data are well described by a two-component depolarization function, combining the contribution of a polycrystalline antiferromagnet and a Kubo–Lorenzian response. For $x = 0.01$ and 0.02 the magnetic signal extends up to ~7 K, which suggests that the broadening may be a result of the effects of disorder on the time scale associated with the anomalous AF state (SMAF). These results imply that SMAF-to-LMAF is not a true phase transition but a crossover behavior [70].

Another interesting comparison between $U_{1-x}Th_xPt_3$ and $U(Pt_{1-x}Pd_x)_3$ shown in Figure 8.19 is the transition widths for samples with $x = 0.01$, 0.02, and 0.05, as illustrated by the temperature-dependent amplitude associated with magnetism.

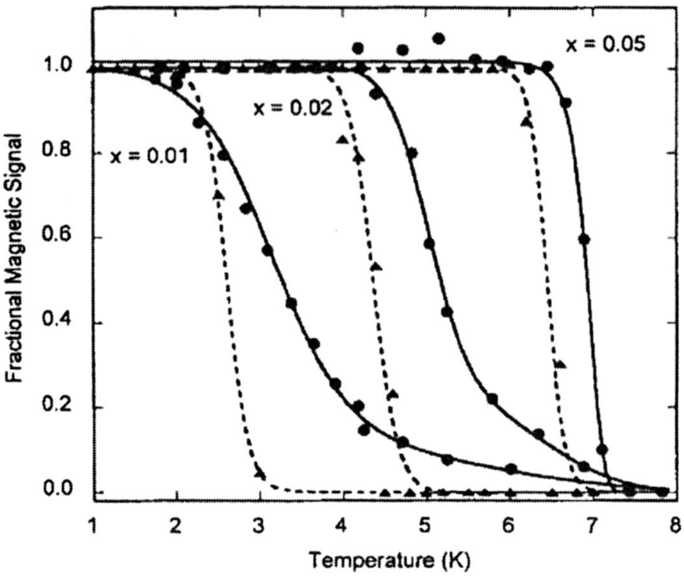

Figure 8.19. Transition widths as illustrated by the temperature-dependent fractional amplitude associated with magnetism. Circles are $U_{1-x}Th_xPt_3$ while triangles are for $U(Pt_{1-x}Pd_x)_3$ with equivalent x values from the work of Ref. [68]. Solid and dashed lines are guides to the eye for the Th-substituted and Pd-substituted data, respectively. Reproduced with the permission of the American Physical Society from Ref. [70].

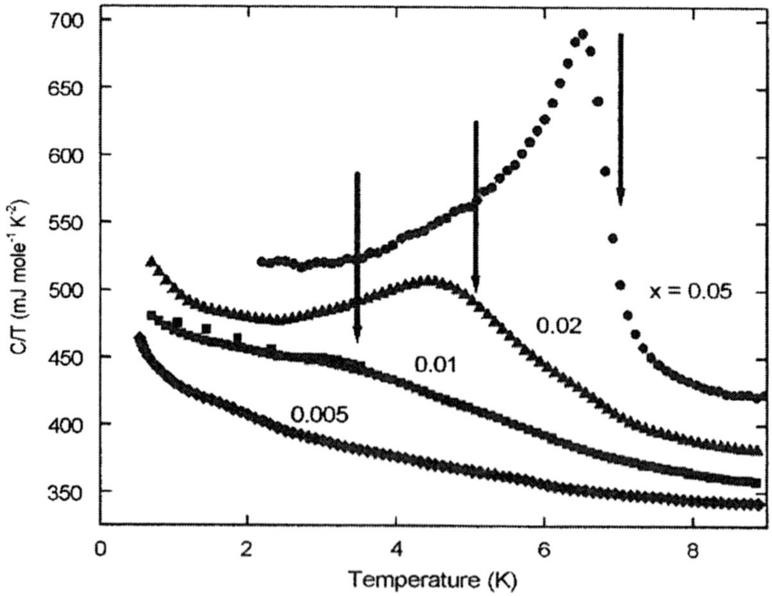

Figure 8.20. Specific heat results for four samples of $U_{1-x}Th_xPt_3$. Arrows mark the Neel temperatures from Ref. [70]. Curves are offset for clarity. Reproduced with the permission of Elsevier from Ref. [63].

It may be noted from Figure 8.19 that all the Pd-substituted samples have narrow transition widths, as does the Th-substituted sample with $x = 0.05$. However, the Th-substituted samples with $x = 0.01$ and 0.02 are quite broad and have a magnetic component up to $T = 7$ K, which is the transition temperature for $x = 0.05$. This indicates that some form of magnetic inhomogeneity is present [70].

Hettinger et al. [63] have presented specific heat measurements of $U_{1-x}Th_xPt_3$ which corroborate the μSR measurements [70] and show signs of an increase near 7 K for $x > 0.01$. Their results are shown in Figure 8.20. This supports the conjecture that Th impurities slow down the fluctuating AF, rendering them observable on the time scale of thermodynamic measurements.

The $x = 0.05$ data show a sharp transition near 7 K, in agreement with μSR measurements [70]. For $x = 0.02$, the size of the anomaly is smaller and broader than for 0.05, as is true for Pd samples. There is also a weak structure near 6.5 K for $x = 0.02$ [63].

8.4. URu_2Si_2

URu_2Si_2 is a moderate heavy fermion compound ($\gamma = 180$ mJ/mol K^2). Palstra et al. [9] made susceptibility, magnetization, and specific heat measurements on single-crystal samples and discovered both a magnetic phase transition at 17.5 K and a superconducting transition at ~1.2 K. In fact, superconductivity coexists with AF order below the superconducting transition temperature T_{sc}. Palstra et al. [9] also found that the magnetic and superconducting properties are highly anisotropic in the sense that the c axis is strongly magnetic while the a axis strongly favors superconductivity. Their results for specific heat measurements are shown in Figure 8.21.

The AF order is characterized by a propagation vector $\mathbf{k} = (001)$ with the ordered dipolar moments pointing along the c axis of the body-centered tetragonal structure (space group **I4**/mmm) [71]. The magnitude of the ordered moment is only (0.03 ± 0.01 μ_B/U atom) [72] at low temperatures. The ordered moment develops at $T_m = 17.5$ K, where clear anomalies are observed in bulk properties. It is indeed difficult to reconcile the small size of the ordered moment with the large jump of $\Delta C/T_m = 0.3$ J/mol K^2 in the specific heat at T_m as seen in Figure 8.21.

Recent experiments of neutron scattering [73, 74] and ^{29}Si NMR [75, 76] performed under hydrostatic pressure have revealed that inhomogeneous AF order with moments of a normal size ($\mu_{ord} \approx 0.25$ μ_B/U atom) is induced by pressure in the range from 0.3 to 0.83 GPa. An extrapolation to ambient pressure indicates that the tiny AF moment is ascribed to a 1% volume fraction of this AF phase. Thus, the 17.5 K transition is due to an unidentified, non-magnetically ordered phase which occupies the rest 99% of the crystal [77]. Luke et al. [78], on the basis of μSR spectra taken at ambient pressure, had pointed out the possibility of the inhomogeneous AF phase in URu_2Si_2. In the hidden-ordered (HO) region, it is widely accepted that weak and isotropic internal fields occur below T_0 [75–84], which are expected to be coupled to the hidden order parameter [85].

To summarize, URu_2Si_2 undergoes a successive phase transition at $T_0 \approx 17.5$ K and $T_{SC} \approx 1.2$ K with lowering temperature, where T_0 and T_{SC} are transition temperatures of a HO and superconductivity (SC), respectively. The HO state is thought to be an SMAF state although it has also been argued that the tiny moment develops below a temperature

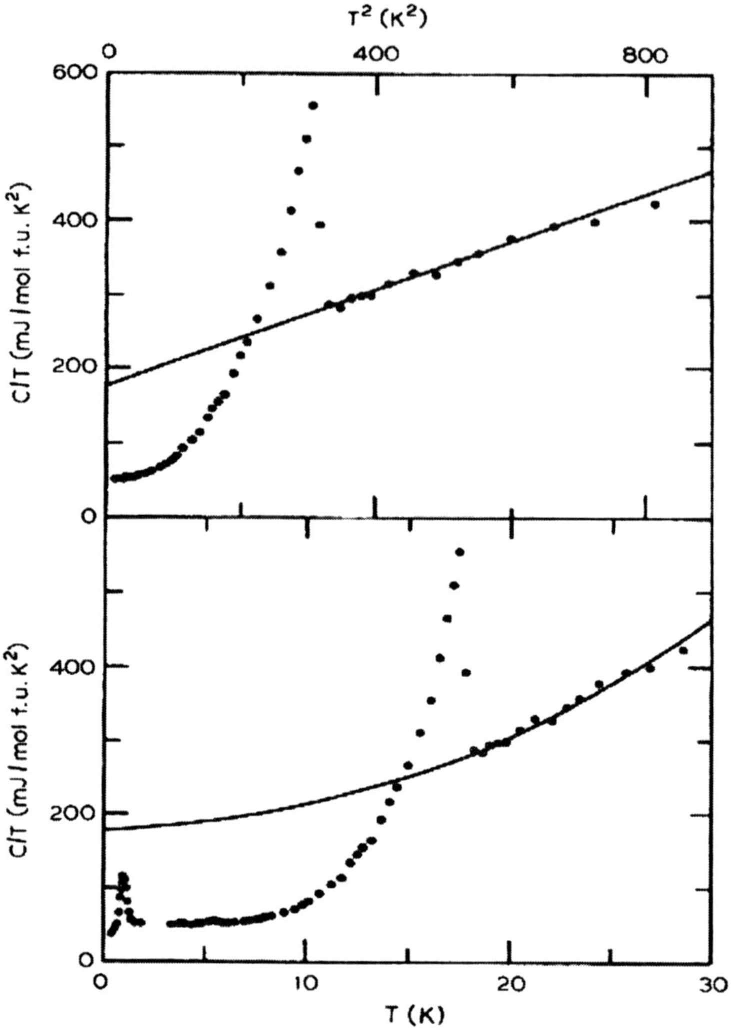

Figure 8.21. Specific heat of URu_2Si_2 plotted as C/T versus T^2 (above) yielding γ and θ_D, and as C/T versus T (below) showing the entropy balance. Reproduced with the permission of the American Physical Society from Ref. [9].

higher than T_0 [86]. In addition to these phase transitions, a weak ferromagnetic anomaly occurs at $T_{FM} \sim 35$ K [71, 87].

Amitsuka et al. [88] have studied the low-temperature phase below 17.5 K in URu_2Si_2 using the zero-field μSR technique under hydrostatic pressures P up to 0.85 GPa. They have confirmed that pressure induces unusual evolution of an inhomogeneous AF phase which spatially competes with some non-magnetic HO. The AF volume fraction develops abruptly around 0.6 GPa, the AF volume evolution is lower than 17.5 K, whereas the nominal Neel temperature is about 20 K. These results, shown in Figure 8.22 suggest that the AF phase exists as a metastable state.

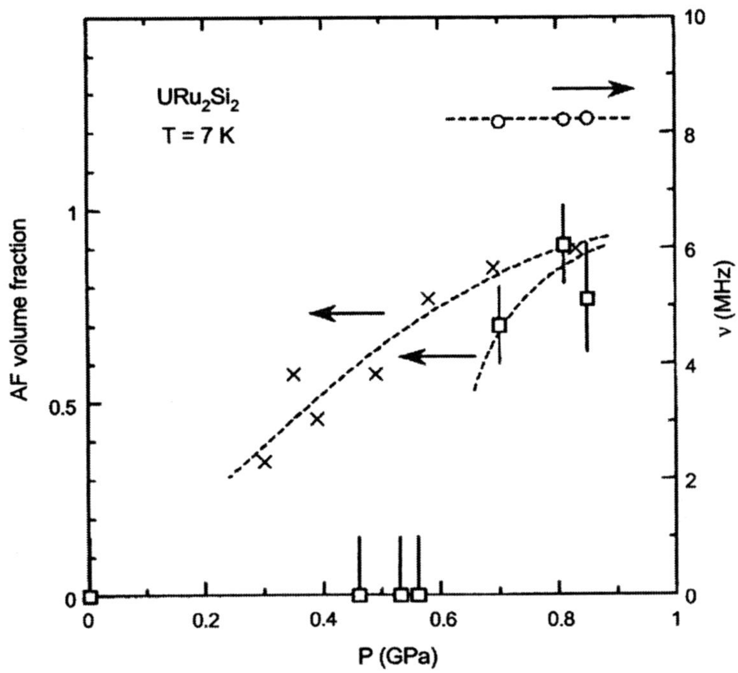

Figure 8.22. Pressure variations of the AF volume fraction estimated by Amitsuka et al. (□) [88] and ^{29}Si-NMR (×) [75, 76]. The frequency ν of the spontaneous muon-spin rotation observed in the AF region (O) is also shown. The broken lines are guide to the eye. Reproduced with the permission of Elsevier from Ref. [88].

Three different groups [89–91] have reported a pressure (P) versus temperature (T) phase diagram. These phase diagrams consist of three transition lines, i.e., $T_0(P)$, $T_M(P)$, and $T_N(P)$. $T_0(P)$ separates HO from a paramagnetic (PM) state; $T_M(P)$, which is a first-order phase transition, separates HO from a pressure-induced LMAFstate, and $T_N(P)$ indicates a direct transition under pressure between PM and LMAF.

In order to probe the origin of the HO, Bourdart et al. [91] have performed specific heat and neutron-scattering measurements on URu_2Si_2 under pressure and doped with 2% Rh. The electronic contribution to the specific heat, C_e, was estimated by subtracting the specific heat from the non-magnetic isomorphic compound $ThRu_2Si_2$. The value of C_e/T of the two compounds as a function of temperature is shown in Figure 8.23.

From Figure 8.23, we note that URu_2Si_2 shows two transitions: a superconducting transition at $T_{SC} \approx 1.5$ K and a lambda-type anomaly at $T_m = 17.5$ K. $U(Ru_{0.98}Rh_{0.02})_2Si_2$ also shows two transitions, a magnetic first-order transition at $T_M = 8.3$ K with an entropy jump of $\Delta S(T_M) = 23$ mJ/K mol and a second-order transition at $T_m = 13.2$ K, which resembles strongly the lambda-type anomaly in the pure compound. However, no superconducting transition is observed in $U(Ru_{0.98}Rh_{0.02})_2Si_2$.

Bourdart et al. [91] have conducted neutron-scattering experiments under pressure on URu_2Si_2 which show strong similarities with the Rh-doped compound. At hydrostatic pressures above 5 kbar, a large-moment phase appears with a transition temperature

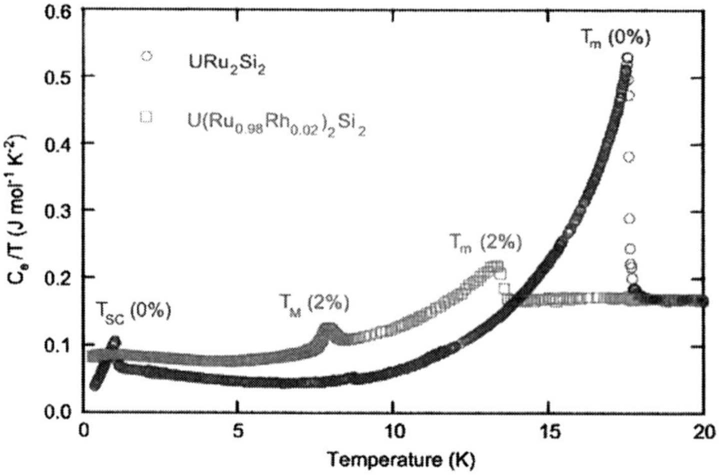

Figure 8.23. Electronic specific heat (after subtraction of the phonon contribution obtained from $ThRu_2Si_2$) divided by temperature for URu_2Si_2 (circles) and for $U(RhU_{0.98}Rh_{0.02})_2Si_2$ (squares). Reproduced with the permission of Elsevier from Ref. [91].

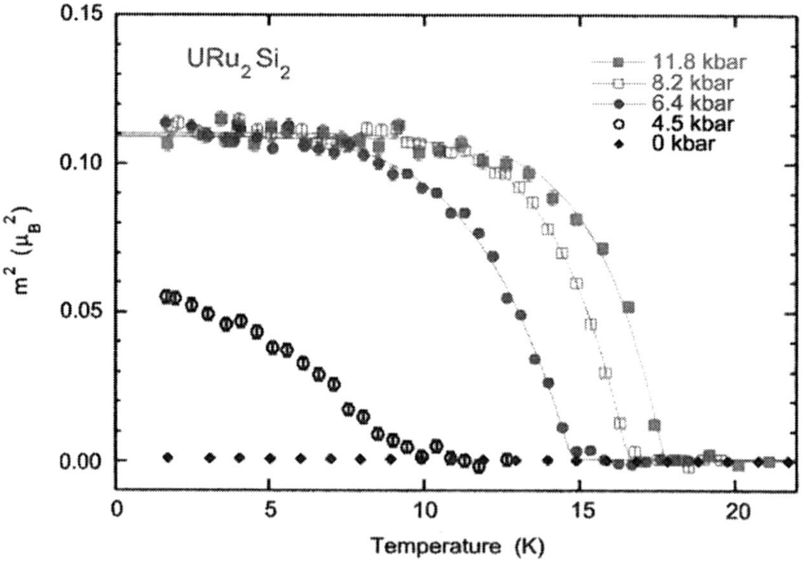

Figure 8.24. Neutron-scattering measurements of the temperature dependence of the [1 0 0] magnetic Bragg peak intensity M^2 in URu_2Si_2 for different applied pressures. The lines are fits to equation (8.2). Reproduced with the permission of Elsevier from Ref. [91].

$T_M(p)$ that is smaller than the small-moment phase, $T_m(p)$. The variation of the ordered moment with temperature is steep at $T_M(p)$, which suggests that the transition is of first order, as in $U(Ru_{0.98}Rh_{0.02})_2Si_2$. The temperature dependence of the magnetic intensity, $M^2(T)$, is shown in Figure 8.24 after background subtraction and normalization to nuclear Bragg peaks.

The temperature variation of the magnetic intensity is much faster near T_M than near T_m. This has been by fitting the expression [91]

$$m^2 = m_0^2\left[1 - \left(\frac{T}{T_N}\right)^{\alpha}\right], \qquad (8.2)$$

to the data, with $N = M$ or m. In the SMAF phase ($N = M$), $\alpha < 2.5$, while in the LMAF phase ($N = M$), $\alpha \sim 8.5$.

We note from Figure 8.24 that the large moment $M = 0.33\ \mu_B$ for $p > 6.4$ kbar is independent of pressure. From the data of Figure 8.24, the transition temperature $T_M(p)$ is determined as the onset of magnetic order as defined in equation (8.2). The p–T phase diagram is shown in Figure 8.25, where the results of resistivity measurements [92–94] have been used to obtain T_m.

The SMAF phase is characterized by a small moment, $\sim 0.03\ \mu_B$ and a finite-correlation length. The transition between the PM phase and the SMAF phase at T_m is of second order while the transition T_M between LMAF and SMAF phases is of first order in character. The LMAF phase is of long-range order and the ordered moment $M = 0.33\ \mu_B$. The first-order transition line between the SMAF and LMAF phases does not join the second-order transition line between the SMAF and the PM phases up to 11.8 kbar. This suggests a linear coupling between the magnetic moment and a hidden order parameter, which breaks time-reversal symmetry.

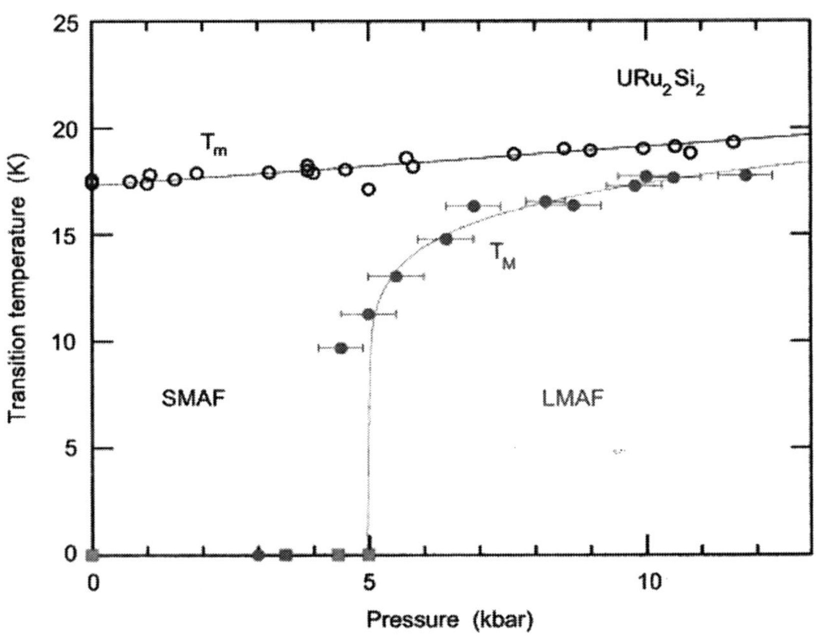

Figure 8.25. Pressure–temperature phase diagram of URu_2Si_2 showing the SMAF and LMAF phases determined from resistivity data (open circles, Refs. [92–94]) and neutron-scattering data (closed circles for pressure clamps, closed squares for hydrostatic helium-cells). The lines are guides for the eye. Reproduced with the permission of Elsevier from Ref. [91].

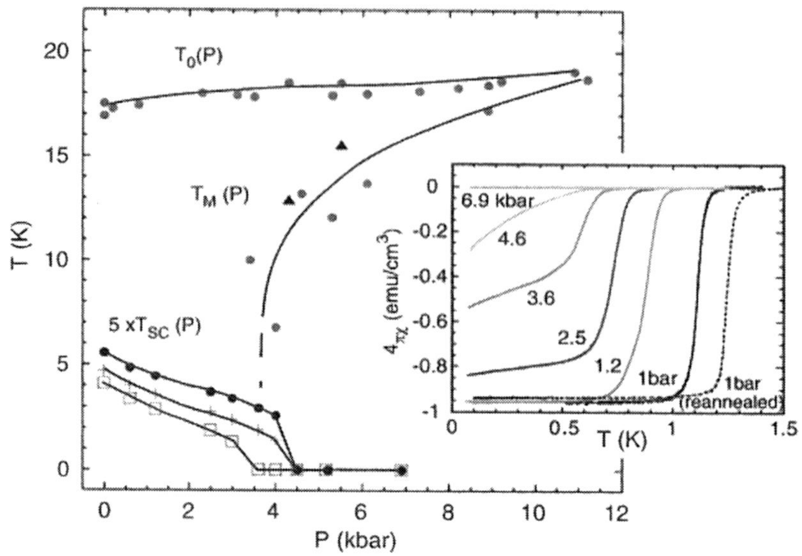

Figure 8.26. *P–T* phase diagram of URu_2Si_2 determined from the ac magnetic susceptibility, dc magnetization, and thermal expansion experiments. $T_0(P)$ and $T_M(P)$ denote the transition between the PM and the HO states and between the HO state and the LMAF states induced under pressure, respectively. Closed circles correspond to T_{SC} at zero external fields, and crosses (squares) are T_{SC} measured under the external field of 6 (10) kOe. Triangles on the $T_M(P)$ line denote the data of the dc magnetization measured at 5000 Oe. Reproduced with the permission of Elsevier from Ref. [95].

Sato et al. [95] have presented a revised pressure–temperature phase diagram of URu_2Si_2 constructed from the ac magnetic susceptibility, dc magnetization, and dilatation measurements under pressure. This phase diagram is shown in Figure 8.26.

The inset shows the T-dependence of the real part of the ac magnetic susceptibility $\chi(T)$ which determines the superconducting transition temperature T_{SC}. At ambient pressure $T_{SC} \sim 1.1$ K. As the pressure increases, T_{SC} decreases monotonically and rapidly decreases to zero at around critical pressure $P_X = 4.0$ kbar. Correspondingly, the magnitude of $\chi(T)$ at low temperatures, which measures the superconducting volume fraction, decreases from unity corresponding to the fully shielding effect at ambient pressure, to zero at around 5 kbar. In fact, P_X almost coincides with a pressure where $T_M(P) = 0$ in the phase diagram. When an external magnetic field is applied, P_X shifts to low pressures thereby implying that it is due to the effect of the magnetic field on the phase boundary between HO and LMAF. In any case, the revised phase diagram shows that the superconducting state competes with the LMAF state, implying the repulsion of superconductivity with the antiferromagnetism.

Barnal et al. [96] have conducted NMR experiments at ambient pressure in single-crystal URu_2Si_2 and have shown that there exists a linewidth enhancement effect below the HO transition temperature T_0. By comparing oriented-powder and single-crystal data, they have observed that the size of the linewidth enhancement below T_0 correlates with the size of the high-T broadening. They have measured a ^{29}Si up-field line shift below T_0 which

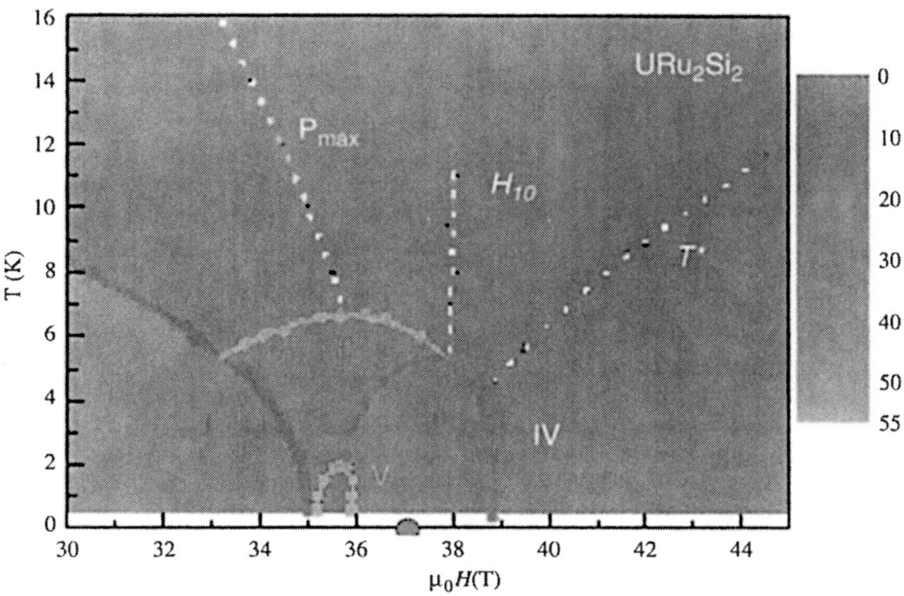

Figure 8.27. Nexus of critical points in URu_2Si_2 determined from a combination of magneto-caloric and resistivity measurements [97]. Color coding is determined by the resistivity. The extrapolation of T^* to zero (red circle) is thought to be a quantum critical end point associated with a metamagnetic transition. Phase (I) is a "hidden order" phase of URu_2Si_2. Phase (IV) is a fully polarized Fermi liquid. Reproduced with the permission of Elsevier from Ref. [98].

indicates the presence of an internal-field average. They have shown that both NMR and NQR measurements suggest a connection between linewidth/disorder effects and the transition to HO. Barnal et al. [96] have confirmed a growth of tiny local fields inside the HO phase that had previously been seen in powdered experiments. These local fields (about 4.8 G) do not change even in a 14.5 T field. This field insensitivity adds support to the idea that these tiny fields derive from staggered orbital currents rather than from spin dipoles.

One of the interesting characteristic of URu_2Si_2 is the "nexus of critical points" which appear to cluster around a field-induced QCP. Figure 8.27 shows such a nexus determined from a combination of magneto-caloric and resistivity measurements [97] which has been presented by Coleman in his brilliant summary of the '05 SCES conference [98].

Recently, Oh et al. [99] have measured the Hall effect of pure and 4% Rh-doped URu_2Si_2 at high magnetic fields and at different temperatures. They have shown low-density (0.03 hole/U atom) high-mobility carriers to be unique to the "hidden order" phase and consistent with an itinerant density-wave OP. The Fermi surface undergoes a series of abrupt changes as the magnetic field is increased. The Hall effect data combined with the de Haas–van Alphen data, shows a strong interplay between the stability of the "hidden order", the degree of polarization of the Fermi liquid, and the Fermi-surface topology. Oh et al. have summarized the results of several experiments [97, 99, 100] in Figure 8.28.

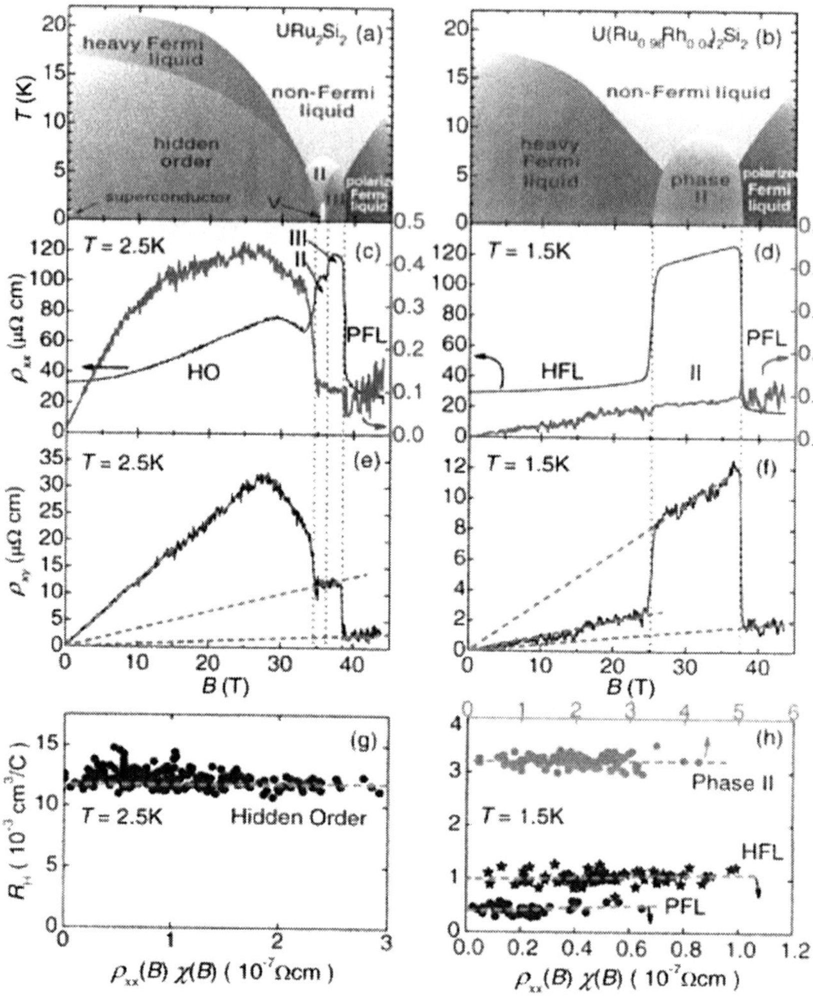

Figure 8.28. Left axes: (a), (b) Phase diagram [97, 100], (c), (d) longitudinal resistivity ρ_{xx}, and (e), (f) Hall resistivity ρ_{xy}. (g), (h) linear extrapolations to estimate the anomalous Hall contribution of URu_2Si_2 at T=2.5 K and $U(Ru_{0.96}Rh_{0.04})_2Si_2$ at T = 1.5 K, respectively. Dotted lines passing through the origin are guides for the eye. The other dotted lines are linear extrapolations. Right axes: (c), (d) ρ_{xy}/ρ_{xx} for both components. Reproduced with the permission of the American Physical Society from Ref. [99].

8.5. UPd_2Al_3

UPd_2Al_3 was discovered as a heavy fermion superconductor in 1991 by Geibel et al. [11]. UPd_2Al_3 crystallizes in the $PrNi_2Al_3$ structure type, a rather simple hexagonal structure, with lattice parameters of a = 5.365 A and c = 4.186 A. UPd_2Al_3 presents a new variant of the interplay between magnetism and superconductivity. It orders antiferromagnetically at $T_N \approx 14.4$ K [11] with a value of AF-ordered moment ~0.85 μ_B [101]. The superconducting transition temperature $T_c \approx 2$ K [11], is quite large for a heavy fermion superconductor. It is also one of the heavy fermion metals which has coexistence of long-range AF order with an LFL state above T_c ($<T_N$) and with HF superconductivity below T_c.

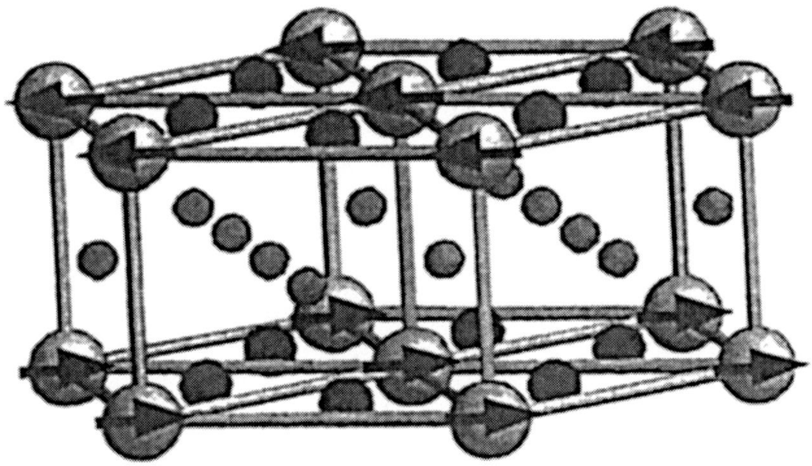

Figure 8.29. The crystal structure of UPd_2Al_3. Reproduced with the permission of Elsevier from Ref. [102].

In Figure 8.29, the crystal structure of UPd_2Al_3 is shown. The large circles represent the positions of uranium ions with the bold arrows marking the relative directions of the magnetic moments. The smaller circles in the same planes represent the positions of the palladium ions, and the smallest circles in the intercalating plane represent the aluminum ions [102].

In Figure 8.30, the experimental results of dc susceptibility of two single crystals of UPd_2Al_3 [103] is shown. The dc susceptibility of single crystals (measured at $B = 10$ mT) is strongly anisotropic, indicating that the basal plane is an easy plane for magnetization. A maximum $T_x \approx 40$ K is observed only in the basal plane but the decrease of χ below T_N is visible in both directions.

In Figure 8.31, the magnetoresistivity, $\Delta\rho_B(T)$, and α/T (α is the thermal expansion coefficient) are plotted as a function of temperature. $\Delta\rho_B(T)$ (Figure 8.31(a)) is positive at low temperature, drops at T_N, changes sign, and assumes a minimum at $T_{min} = 40$ K, which is a phenomenological measure of the Kondo (-lattice) temperature T^*. α/T (Figure 8.31(b)) exhibits a huge negative peak at T_N. Along with the large anisotropy, observed in α at lower temperatures on the single crystals (Figure 8.32(b)), this indicates that the AF state results (upon cooling) in a dominating expansion of the c axis and a relatively small shrinking of the basal plane. The dashed curve in Figure 8.31(b) as well as in Figure 8.32(b) (symbol – p.c.) is for temperature variation of $2/3\ \alpha_{100} + 1/3\ \alpha_{001}$ for polycrystalline crystals.

The specific heat (C/T) versus T of single crystal #1 is plotted for various magnetic fields B ($B \| c$) in Figure 8.32(a). α/T versus T is plotted in Figure 8.32(b) for single crystals #1 and #2 in different orientations. Dashed line in Figure 8.32(b) indicates $2/3\ \alpha_{100} + 1/3\ \alpha_{001}$ and for comparison, low-T data for polycrystalline sample (from Figure 8.31(b)) are also shown.

Neutron-scattering experiments in zero field have revealed a simple AF structure with ordered moments of 0.85 μ_B parallel to the basal plane, coupled ferromagnetically within the basal plane and antiferromagnetically along the c axis. When a field is applied along the basal plane, further anomalies at $T < T_N$ appear in the susceptibility, magnetization, magnetostriction, and magnetoresistance. Figure 8.33 shows the B–T phase diagram

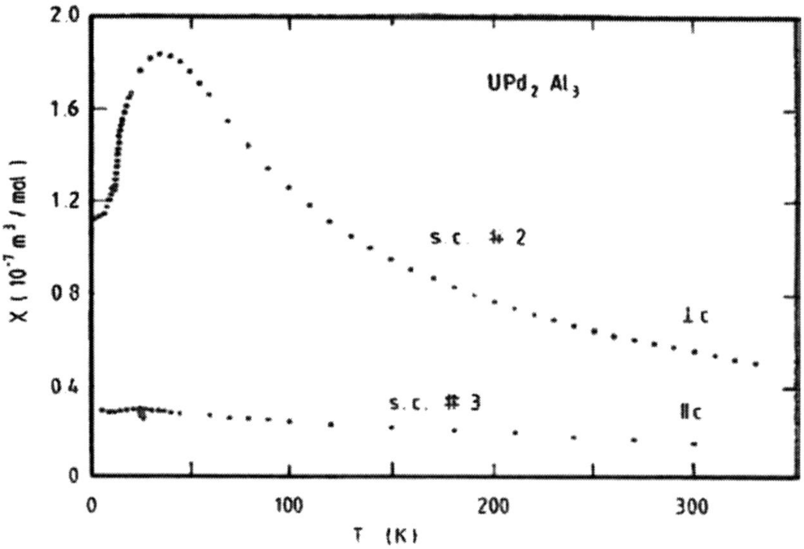

Figure 8.30. dc susceptibility for UPd_2Al_3 single crystal #2 and #3, with B (=10 mT) perpendicular and parallel to c. Reproduced with the permission of Elsevier from Ref. [103].

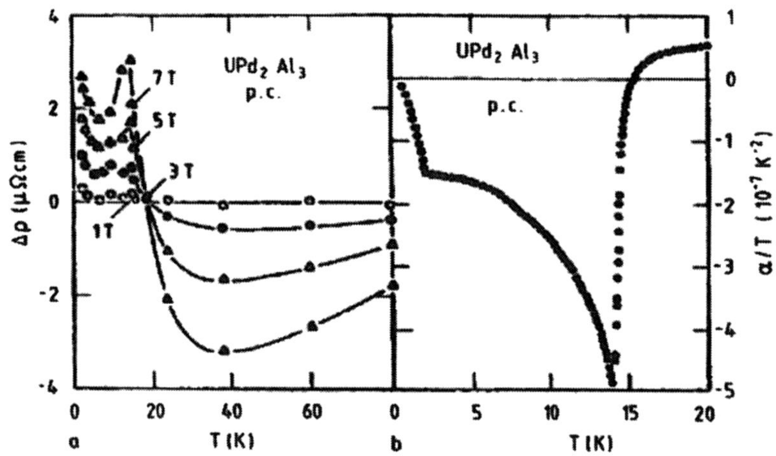

Figure 8.31. Magnetoresistivity for different values of B, $\Delta\rho_B$ (a), and α/T (b), as a function of temperature for polycrystalline UPd_2Al_3. Reproduced with the permission of Elsevier from Ref. [103].

which indicates the complex magnetization process with three AF regimes (I, II, III) [104]. The transition from phase I to phase II (as found from a distinct increase in both $m(B)/B$ and $\Delta l(B)$ measured along [1 1 0]) is of first order (as indicated from the hysteresis found for the anomaly in $M(B)/B$) and persists in the superconducting state. The transition between phases II and III (reflected by a peak in $\partial\chi_\perp/\partial T$ for B > 0.5 T, an inflection point in the magnetization curve, a break of slope in the transverse magnetoresistivity,

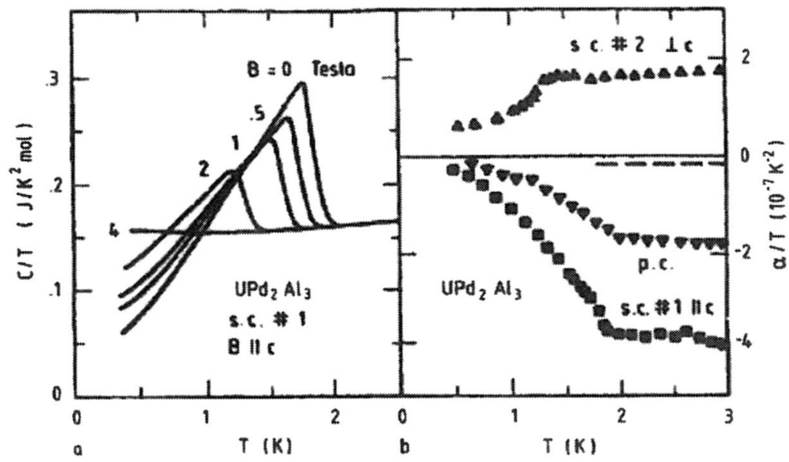

Figure 8.32. (a) C/T versus T and (b) α/T versus T for UPd_2Al_3 single crystals. Reproduced with the permission of Elsevier from Ref. [103].

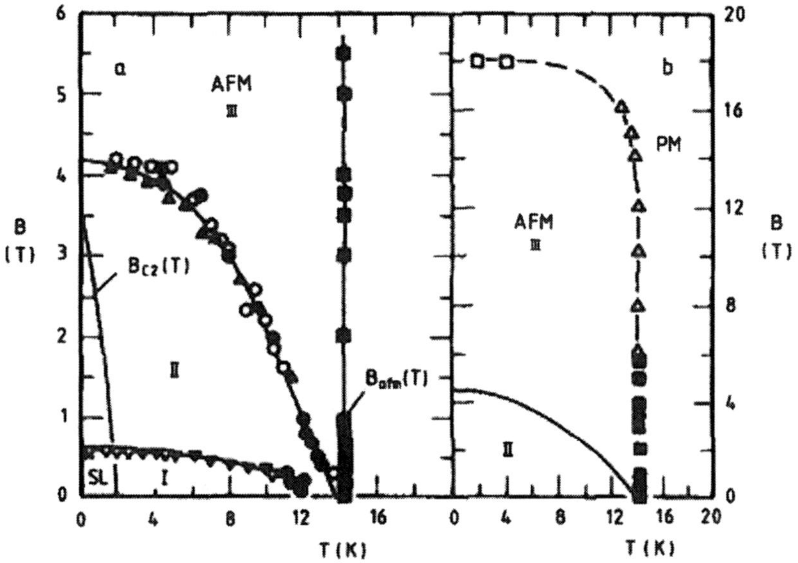

Figure 8.33. Magnetic phase diagram of UPd_2Al_3 for $B \perp c$ as constructed from results of dc susceptibility (●, ■), magnetization (○, ▼, □), magnetostriction (▽, ◆), magnetoresistivity (▲), and ultrasound measurements (Δ). Reproduced with the permission of Elsevier from Ref. [104].

and a very shallow shoulder in the magnetostriction) is of the second (or weakly first) order since no hysteresis effects are resolved. The details of as to how the data were obtained are described in Ref. [104].

In Figure 8.34, the B–T phase diagram of the superconducting state of UPd_2Al_3 is shown. $B_1(T)$ and $T_1(B)$ define a line of first-order phase transition which separates from the $B_{c2}(T)$ line at $\tilde{B} \approx 1$ T and $\tilde{T} = 1.5$ K. It was proposed by Gloos et al. [105] that this

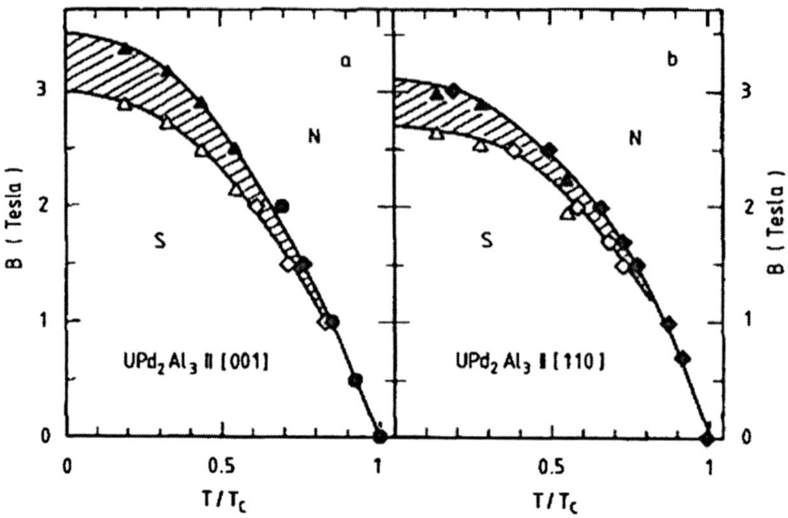

Figure 8.34. B versus T/T_c phase diagram of UPd_2Al_3 for $B\|[0\ 0\ 1]$ (a) and $B\|[1\ 1\ 0]$ (b). Symbols mark positions of jumps in $C(T)$ (●) and in FC $\alpha(T)$ (◆), which both correspond to $T_c(B)$, of the onset (□) of irreversible changes in ZFC $\Delta l(T)$, and of the offset (▲) and onset (Δ) fields of $\delta l(B)$ anomalies. Hatched region marks existence of non-uniform conducting state [105]. Reproduced with the permission of Elsevier from Ref. [104].

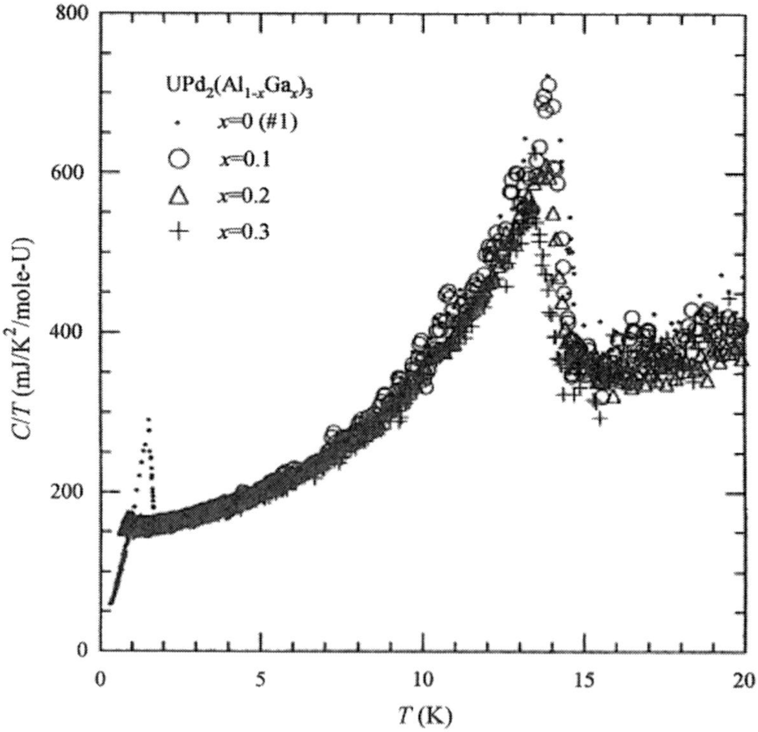

Figure 8.35. Temperature dependence of the specific heat divided by T of $UPd_2(Al_{1-x}Ga_x)_3$. Reproduced with the permission of Elsevier from Ref. [111].

first-order transition indicates the d evelopment of a Fulde–Ferrell–Larkin– Ovchinnikov (FFLO) [106, 107] state.

The ordered moments $\mu_s \approx 0.85\ \mu_B$ for UPd_2Al_3 points to the existence of quasilocalized 5f electrons [108]. On the other hand, an enhanced Sommerfeld coefficient, $\gamma = 140$ mJ/mol K^2, of the electronic specific heat determined slightly above $T_c \approx 2$ K indicates moderate HF behavior due to a strong hybridization of less localized 5f states with the ligand states. The two-component nature of the uranium 5f shell, including both "localized" and "itinerant" 5f states, is evident [108] from the bulk properties, i.e., susceptibility, specific heat, and μSR. It has also been suggested that superconductivity in UPd_2Al_3 is mediated by the spin fluctuations [109, 110]. In order to study the interplay between magnetism and superconductivity, Aso et al. [111] have measured the specific heat and magnetization for single-crystal samples of $UPd_2(Al_{1-x}Ga_x)_3$ for Ga concentrations of $x \leq 0.2$. The superconducting transition temperature T_c indicates a rapid decrease, and the AF transition temperature T_N reduces gradually as the gallium atom is substituted for aluminum. The magnetic entropy released at T_N decreases with the increase of x, suggesting that the increase of spin fluctuations is responsible for the reduction of T_N. The temperature dependence of the specific heat is shown in Figure 8.35.

The temperature dependence of the magnetic susceptibility χ of $UPd_2\,(Al_{1-x}Ga_x)_3$ is plotted in Figure 8.36. At high temperatures, the susceptibility in the external magnetic

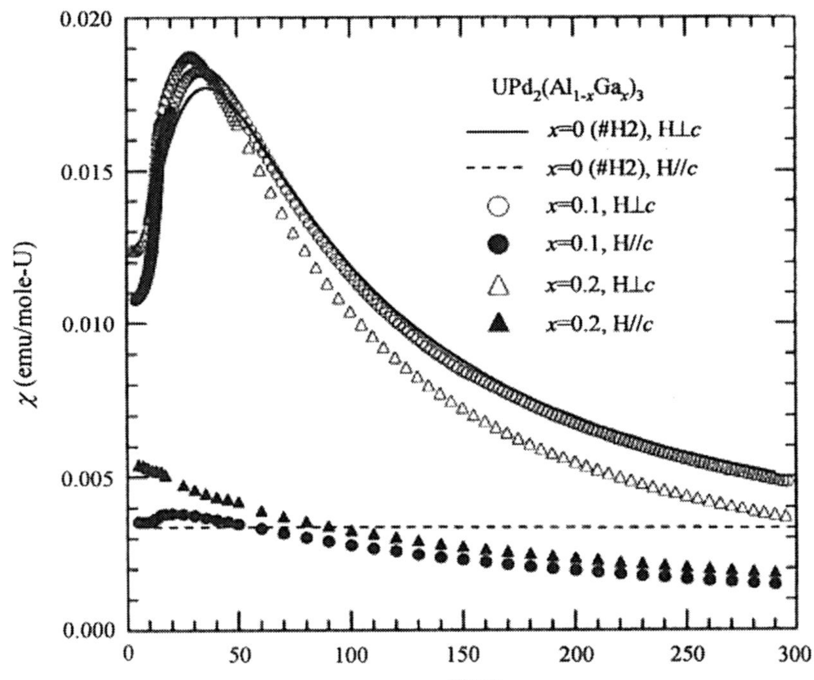

Figure 8.36. Temperature dependence of the magnetic susceptibility χ of $UPd_2(Al_{1-x}Ga_x)_3$ for $x = 0$, 0.1, and 0.2. Reproduced with the permission of Elsevier from Ref. [111].

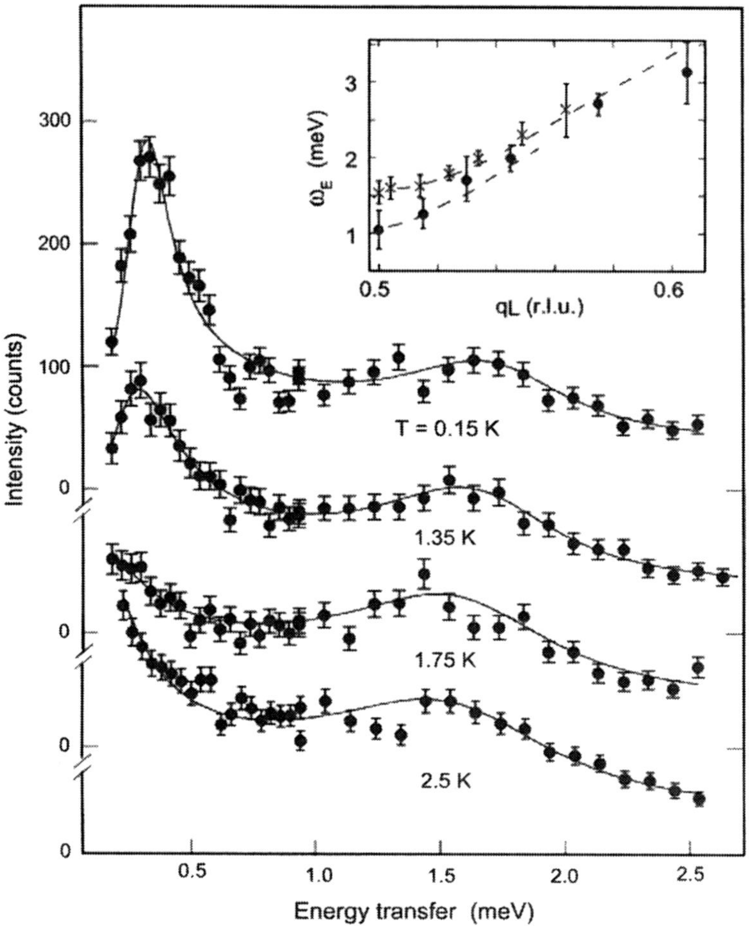

Figure 8.37. Temperature evolution of the INS spectrum for a UPd_2Al_3 spectrum for a UPd_2Al_3 single crystal at $\mathbf{q} = (0,0,q_L) = (0,0,0.5) = \mathbf{Q}_0$ (center of the AF Brillouin zone), through $T_c \approx 1.8$ K. Solid lines represents fits using the microscopic "two-component" model described in Ref. [108]. Inset: The magnetic-excitation energy $\omega_E(q)$ for $T = 2.5$ K (crosses) and 0.15 K (circles). Reproduced with the permission of Elsevier from Ref. [112].

field perpendicular to the c axis follows a Curie–Weiss law $\chi = C/(T + \theta)$ with an effective Bohr magneton $\mu_{eff} \approx 3.6\ \mu_B$ and Weiss temperature $\theta \approx 40$ K for $x = 0$ and 0.1. This value of μ_{eff} is close to a free ion of either U^{3+} or U^{4+}. For $x = 0.2$, Aso et al. [111] obtained $\mu_{eff} \approx 3.0\ \mu_B$ and $\theta \approx 0$.

In fact, tunneling [12] and inelastic neutron-scattering experiments [108] have shown a "strong coupling" between the "localized" and the "itinerant" 5f states of U. In Ref. [12], because of the low-energy structure of the tunneling density of states, UPd_2Al_3 was assumed to be a spin-fluctuation-mediated superconductor in which overdamped AF paramagnons of the Fermi sea provide the effective attractive interaction between the itinerant quasiparticles. However, due to the low energy of ≈1 meV, the role of AF paramagnons in forming Cooper pairs is doubtful.

As analyzed in Ref. [108], the susceptibility results for $T > T_N$ show the existence of two quasilocalized 5f electrons and of another "itinerant" one. The interaction of the localized $5f^2$ configuration with the crystal electric field (CF) separates a singlet ground state from the first excited state (perhaps a doublet), about 7 meV above the ground state. Intersite interactions allow this local CF excitation to progress through the crystal, forming a band of "magnetic excitons" which manifests itself in the inelastic neutron-scattering data, (Figure 8.37 [112]) in a q-dependent broad peak. At the center of the AF Brillouin zone, $\mathbf{Q}_0=(0,0,q_L=1/2)$, the latter occurs for $T = 2.5$ K $\geq T_c$, around $\hbar\omega \approx 1.5$ meV. The large ordered moment $\mu_s = 0.85\ \mu_B$ must be excited by the excited magnetic CF states below $T_N = 14.3$ K, where magnetic order forms due to the same intersite interactions that cause the band magnetic excitations to form above T_N.

The localized 5f states dominate the neutron-scattering cross section due to their large form factor. However, the fact that the low-energy peak has high intensity and follows upon cooling the shift of the quasiparticles to the gap edge of the superconductor (Figure 8.37) confirms the interaction between the localized and itinerant 5f states [108].

UPd_2Al_3 is the first superconductor for which a magnetic-pairing mechanism can be demonstrated experimentally by INS and tunneling results. The CF excitation that is propagating through the lattice of U-ions plays the same role as high-energy phonons do in classical superconductors.

8.6. UNi_2Al_3

UNi_2Al_3 was discovered as a heavy fermion superconductor in 1991 by Geibel et al. [10]. It crystallizes in the hexagonal $PrNi_2Al_3$ structure belonging to the space group P6/mmm. It undergoes AF transition at $T_N \approx 5$ K and superconducting transition at $T_c \approx 1$ K. Neutron-scattering measurements [113] show UNi_2Al_3 which is unique among heavy fermion superconductors displaying long-range magnetic ordering which is incommensurate with its crystal lattice. The magnitude of the ordered moment $\mu_{ord} \approx 0.24\ \mu_B$. Mihalik et al. [114] have measured the heat capacity, ac–susceptibility, and dc susceptibility of UNi_2Al_3 single crystal and their results are summarized below. In Figure 8.38, their specific heat data in the temperature range 0.4–16 K is presented. The electronic specific heat coefficient $\gamma = 130$ mJ/mol K^2 was obtained by extrapolating the specific heat data between 1 and 3 K to $T = 0$ K. The onset of anomaly at $T_c \approx 1$ K is due to transition to superconducting state, and the onset of the second anomaly is at $T_N \approx 4.8$ K due to transition to AF ordered state. The specific heat jumps at T_c, $\Delta C/C_n(T_c) \approx 0.2$. A third anomaly is observed between T_N and 11 K. The magnitude of $C/T \approx 145$ mJ/mol K^2 is almost temperature independent and $C/T(T)$ shows a dull minimum at about 9 K.

Mihalik et al. [114] have measured the ac susceptibility of UNi_2Al_3 in a magnetic field with induction $\mu_0 H \approx 12$ mT applied along a axis. In Figure 8.39, the temperature dependence of real part χ' and imaginary part χ'' of ac susceptibility $\chi_{ac} = (\chi' - i\chi'')$ between 0.1 and 1.2 K is shown.

The diamagnetic response appears below $T_c \approx 1.1$ K. The size of the diamagnetic response is large, indicating that superconductivity is a bulk property of UNi_2Al_3.

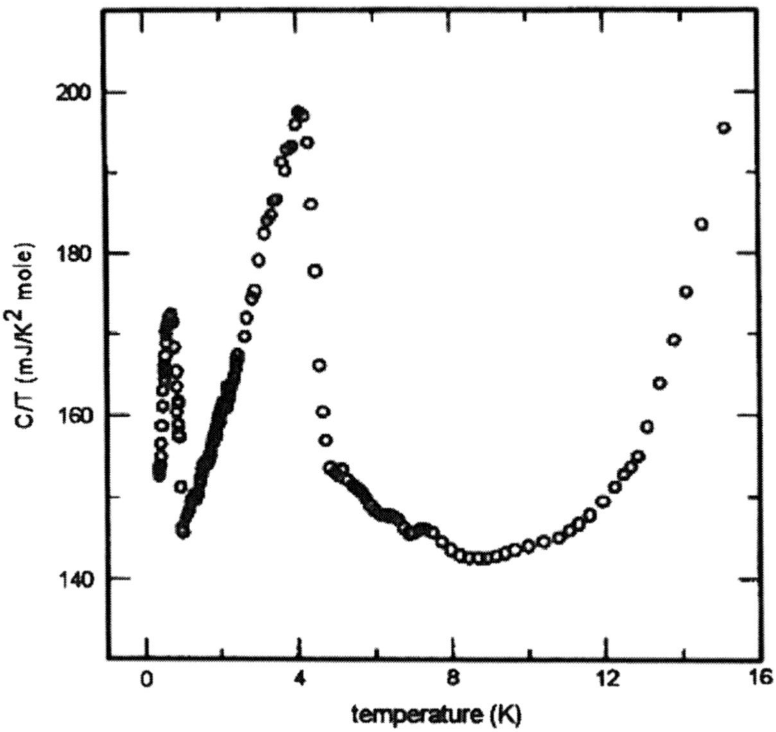

Figure 8.38. Temperature dependence of specific heat C/T of UNi_2Al_3 between 0.4 and 16 K. Reproduced with the permission of Elsevier from Ref. [114].

The transition is broad and no saturation of χ' down to 0.1 K was observed. The "broad transition" is probably due to imperfection in the single crystal which could also account for the lack of any clear indications of the AF transition in the ac susceptibility measurements [114].

The temperature dependence of dc susceptibility between 2 and 350 K, with induction $\mu_0 H = 2$ T applied along and perpendicular to the c axis, is shown in Figure 8.40.

The susceptibility χ is very anisotropic and the maximum of the ratio $\chi_{a\text{-axis}}/\chi_{c\text{-axis}} \approx 3$. There are two pronounced maxima in $\chi_{a\text{-axis}}(T)$ data. The maximum at $T_{max} \approx 100$ K corresponds to the flattening of the resistivity versus temperature data [10, 115] and indicates Kondo-lattice behavior. A cusp in $\chi_{c\text{-axis}}(T)$ data at $T_N \approx 4.8$ K corresponds to the AF transition. The $1/\chi_{a\text{-axis}}(T)$ are linear above 200 K but $\mu_{eff} = 3.95\ \mu_B$ calculated from the slope $1/\chi_{a\text{-axis}}(T)$ is larger than the free ion value. The $\chi_{c\text{-axis}}(T)$ data linearly increases with decreasing temperature but the sharp increase below 30 K is probably due to magnetic impurities [114].

Ishida et al. [116] have reported ^{27}Al Knight shift (^{27}K) measurements on a high-quality single-crystal UNi_2Al_3 which exhibits a sharp superconducting transition at $T_c \approx 0.9$ K. Their experiments reveal a coexistence of superconductivity and a spin-density-wave (SDW) type of magnetic ordering ($T_{SDW} = 4.5$ K). In general, the shift in the NMR spectrum is expressed as $K(T) = K_s(T) + K_{orb} + K_Q(H)$, which are the spin, orbital, and quadrupolar

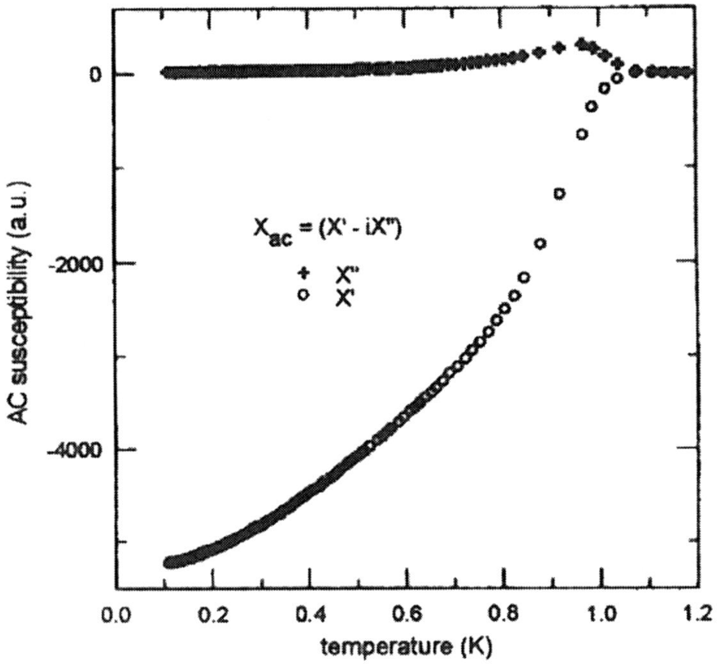

Figure 8.39. Temperature dependence of ac susceptibility of UNi_2Al_3 between 0.1 and 1.2 K. Reproduced with the permission of Elsevier from Ref. [114].

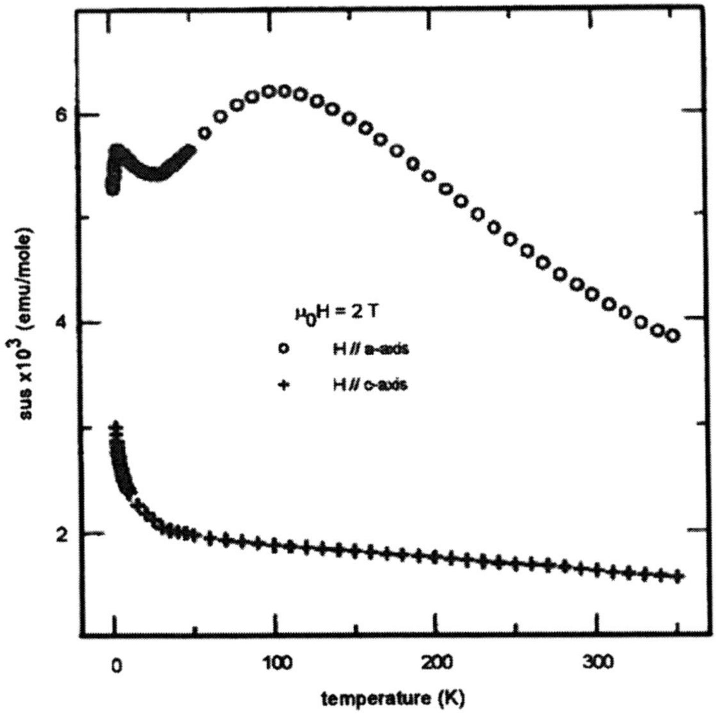

Figure 8.40. Temperature dependence of dc susceptibility between 2 and 350 K. Reproduced with the permission of Elsevier from Ref. [114].

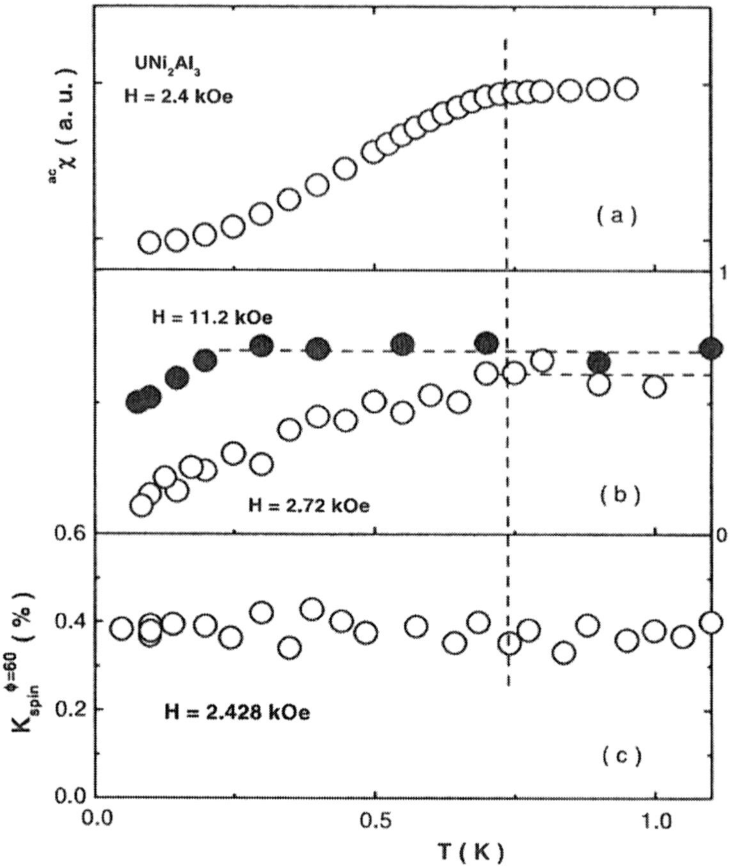

Figure 8.41. T dependence of (a) ac susceptibility, (b) $1/T_1T$ for a$\|\mathbf{H}_0$ = 2.72 kOe (open circles), and 11.2 kOe (solid circles) and Knight shift at the $\phi = 60°$ peak. The vertical dotted line indicates T_c (2.4 kOe). Reproduced with the permission of Elsevier from Ref. [117].

components of the Knight shift. The quadrupolar shift originates from the second-order electric quadrupole interaction and is inversely proportional to H^2.

The results of T dependence of (a) ac susceptibility, (b) $1/T_1T$ for **a**$\|\mathbf{H}_0 = 27.2$ kOe (open circles), and 11.2 kOe (solid circles), and (c) Knight shift at $\phi = 60°$ peak are shown in Figure 8.41 [117]. The vertical dotted line indicates T_c (2.4 kOe).

For $\mathbf{H}_0\|\mathbf{a}$, ($\mathbf{H}_\mathbf{a}$), there exist two inequivalent Al sites. For one-third of the Al sites $\mathbf{H}_0$ is parallel to **a** (denoted as the $\phi = 0°$ site) and for two-thirds of the Al sites, the angle between $\mathbf{H}_0$ and **a** is ±60° (denoted as the $\phi = 60°$ site). In order to deduce K_s, $K(T)$ was measured under a strong field of $\mathbf{H}_0$ so that K_Q could be neglected. Normally, only K_s is T-dependent while K_{orb} is independent of T. $^{27}K_s$ was measured at $\phi = 60°$. However, in case of UNi_2Al_3, $^{27}K_s$ does not change down to 50 mK across the superconducting transition temperature $T_c \approx 0.90$ K as evident from Figure 8.41. The invariance of ^{27}K suggests strongly that a spin-triplet pairing state is realized in UNi_2Al_3, with parallel spin in the *ab* plane. In addition, the T dependence of $1/T_1$ in UNi_2Al_3 was consistent with the itinerant SDW type [118]. The 5f electrons in UNi_2Al_3 possess an itinerant character and are in an AF spin-fluctuation regime in the PM state as evidenced by the T_1 measurements.

8.7. UGe_2

The discovery of superconductivity in UGe_2 [119, 120] in single crystals of UGe_2 under pressure below $P_c \sim 16$ kbar was very surprising. The sensational part of this discovery is that the pressure $P \sim 12$ kbar, where the superconducting temperature $T_S = 0.75$ K is strongest and the Curie temperature $T_C \sim 35$ K is two orders of magnitude higher than T_S: superconductivity occurs in a very highly polarized state ($\mu(T \to 0\text{ K}) \sim \mu_B$).

The superconductivity in UGe_2 disappears above a pressure $P_c \approx 16$ kbar that coincides with the pressure at which the ferromagnetism is suppressed. The pressure–temperature phase diagram of UGe_2 is shown in Figure 8.42 [121].

The existence of an additional phase transition, not shown in this figure, has been previously suggested by Oomi et al. [122]. We shall show this transition in Figure 8.45. The striking fact in the p–T phase diagram of UGe_2 is the disappearance of superconductivity, just at P_c, as if superconductivity and ferromagnetism are cooperative. The field reentrant "metamagnetic behavior" above P_c, observed earlier in UGe_2 [123], indicates that the magnetic transition becomes first order just below P_c. This suggests that one does not expect low-energy spin fluctuations and another soft mode, directly related to the lattice structure, exists in UGe_2.

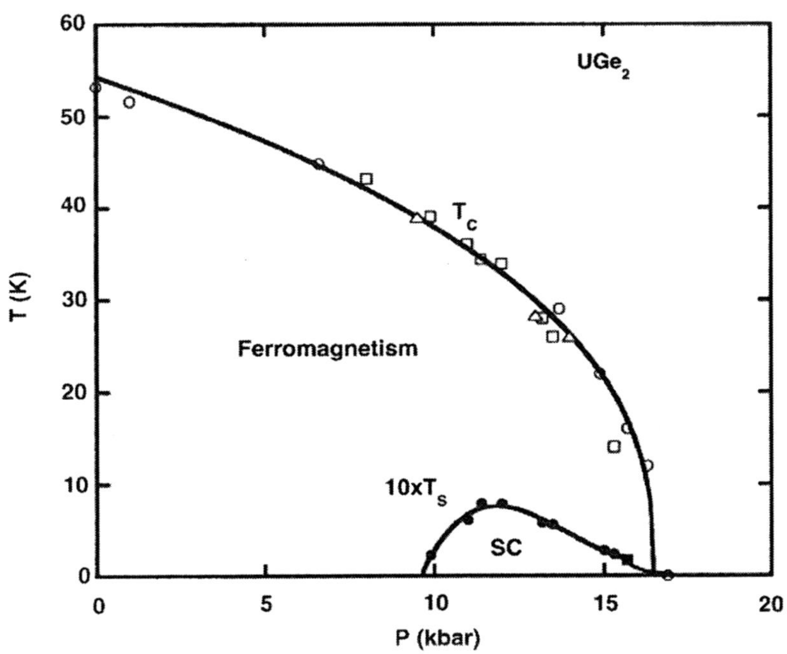

Figure 8.42. The pressure–temperature phase diagram of UGe_2. Reproduced with the permission of Elsevier from Ref. [121].

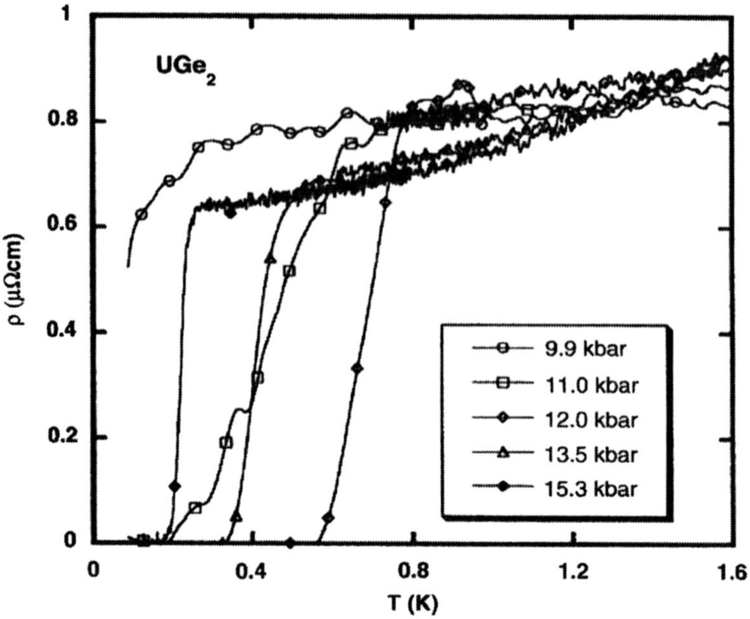

Figure 8.43. Resistivity of UGe_2 as a function of temperature. Reproduced with the permission of Elsevier from Ref. [121].

The resistivity is plotted as a function of temperature and pressure in Figure 8.43 [121]. One can notice that superconductivity occurs in a highly polarized state.

A clear indication of another transition at T_x ($T_S < T_X < T_C$), which is detected between 9 and 12 kbar, comes from the observation of supplementary marked drop of ρ as shown in Figure 8.44. This transition, T_X, seems to be a low-temperature continuation of a broad maximum in $\partial\rho/\partial T$ already observed at $P = 0$ by Oomi et al. [122].

From Figure 8.44, one notes the remarkable phenomenon that T_S reaches its maximum when T_X seems to cross the superconducting boundary. Huxley et al. [120] have carefully analyzed the effect of the magnetic field and suggest that the magnetic field shifts the line $T_X(P)$ measured in zero field to higher pressure. This supports the view that T_X might correspond to a phase transition facilitated by a special geometry of the Fermi surfaces. The Fermi surface for the majority and the minority spin directions are expected to expand and contract as the spin polarization is increased. When the splitting approaches a particular value, the nesting of the spin majority Fermi surface might be fulfilled which could give rise to a phase line in the ferromagnetic state below $T_{Curie}(P)$ corresponding to a constant value of the ordered ferromagnetic moment.

Neutron-scattering experiments have shown that the ferromagnetic component detected on the (0,0,1) peak persists even in the superconducting state of UGe_2. Figure 8.45 shows the temperature dependence of the magnetic intensity, at different pressures, measured on the (0,0,1) reflection in UGe_2.

Harada et al. [123] have conducted ^{73}Ge-NMR/NQR measurements in UGe_2 under pressure. The NQR spectrum reveals that the ferromagnetic phases are separated into

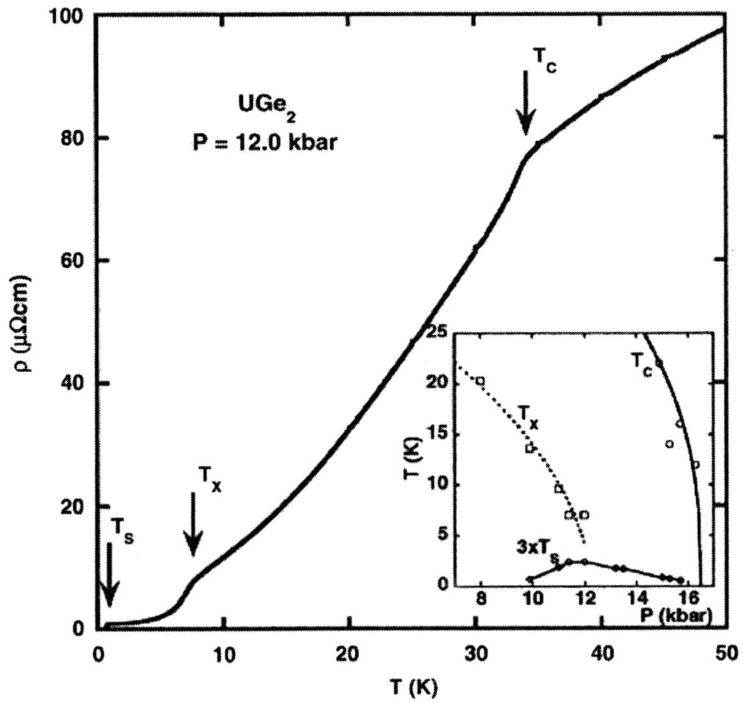

Figure 8.44. A new marked resistivity anomaly appears at T_X at 12 kbar. The inset shows the respective pressure variation of T_C, T_X, and $3T_S$. Reproduced with the permission of Elsevier from Ref. [121].

weakly and strongly polarized phases around a critical value of $P_c^* \sim 12$ kbar, indicating a first-order transition around P_c^*. They have also shown that there is a phase separation into ferromagnetic and PM phases around a critical pressure $P_c \sim 16$ kbar. The measurements of spin-lattice relaxation rate $1/T_1$ have probed that superconductivity sets only in the ferromagnetic phase at $T_{sc} \sim 0.2$ K, but not in the PM phase. Their results for temperature dependence of $1/T_1$ are shown in Figure 8.46.

To summarize, UGe_2 is a ferromagnetic spin-triplet superconductor with unusual properties. The fact that superconductivity is confined to the ferromagnetic phase makes the theoretical study of UGe_2 more interesting.

8.8. URhGe

URhGe orders ferromagnetically below 9.5 K in an orthorhombic collinear magnetic structure, with a tiny ordered moment of $\approx 0.4\ \mu_B$, oriented along the $\vec{c}$-easy axis [124]. The linear term of the specific heat is $\gamma = 160$ mJ/mol K^2. At lower temperature, URhGe exhibits a pronounced jump in the heat capacity around 0.25 K which coincides with the onset of zero resistivity establishing the fact that the compound is a bulk superconductor [124]. Superconductivity with a non-negligible band polarization due to ferromagnetism can be explained by spin-triplet OP.

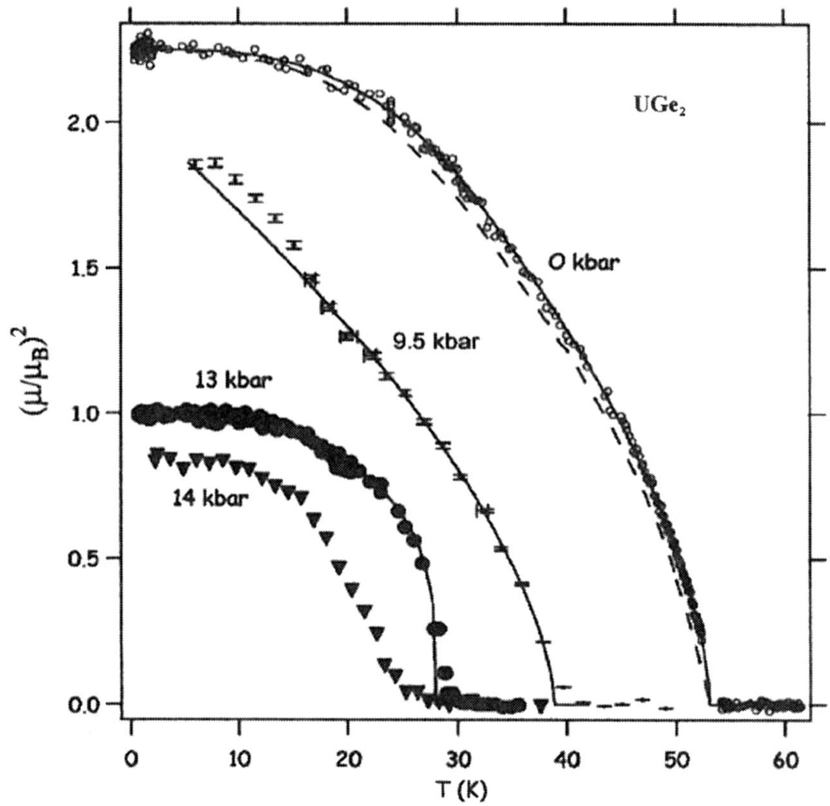

Figure 8.45. Temperature dependence of the magnetic intensity measured on the (0,0,1) reflection in UGe_2. Reproduced with the permission of Elsevier from Ref. [121].

At ambient pressure, the magnetic and superconducting properties of URhGe are of the same order as that of UGe_2 around 15 kbar. The Curie temperature of UGe_2 decreases monotonically from 53 K at ambient pressure to zero at $P \approx 16$ kbar where superconductivity also collapses [120]. Thus, one can reasonably expect that URhGe is close to such a critical pressure. Hardy et al. [125] have measured the temperature dependence of C/T for different pressures which is shown in Figure 8.47.

In Figure 8.47, the observed jump corresponds to the ferromagnetic anomaly at the Curie point and the Curie temperature is defined as the top of the anomaly. On increasing the pressure, the transition becomes broader while the height of the anomaly decreases. This decrease can be partly related to a slowly decreasing ordered moment observed by magnetization measurements.

In Figure 8.48, the temperature dependence of the resistivity of URhGe at low temperatures for different values of the applied pressure is shown [125].

The pressure–temperature phase diagram of URhGe [125] is shown in Figure 8.49.

The Curie temperature is found to increase in URhGe while ferromagnetism disappears around 17 kbar for UGe_2. In fact, Sheng and Cooper [126] had predicted that

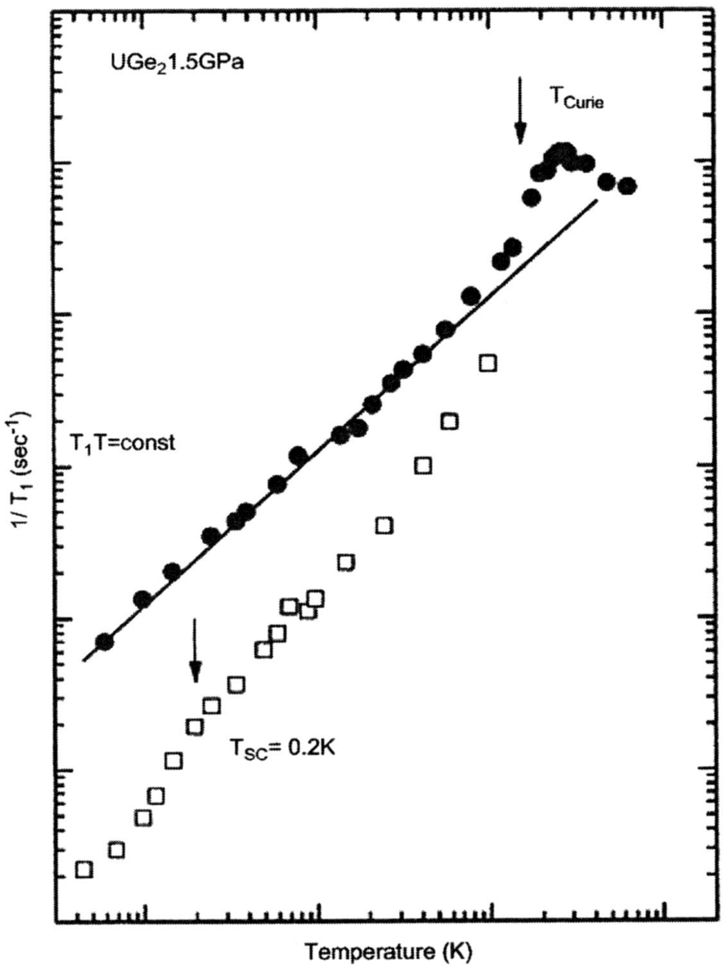

Figure 8.46. The T dependence of $1/T_1$ for the PM and ferromagnetic phases. The long component in $1/T_1$ for the ferromagnetic phase indicates that Sc sets at $T_{sc} \sim 0.2$ K, but the short one does not for the PM phase. Reproduced with the permission of Elsevier from Ref. [123].

pressure increases the hybridization of 5f electrons with band states leading to a suppression of ordered moment. However, this hybridization also enhances the exchange coupling, thus increasing the Curie temperature. The latter mechanism prevails up to 130 kbar for URhGe. However, moment reduction is expected to occur at higher pressures as observed in $U(In_{0.6}Sn_{0.4})_3$ [127].

Roussev et al. [128] have suggested that for triplet superconductivity in an Ising ferromagnet, T_c can be affected by the distance to a quantum phase transition. Mineev [129] has suggested that in such a case, T_c can also be affected by the value of the ordered moment. The P–T phase diagram of Hardy et al. [125], as shown in Figure 8.49, supports the former model but contradicts the latter.

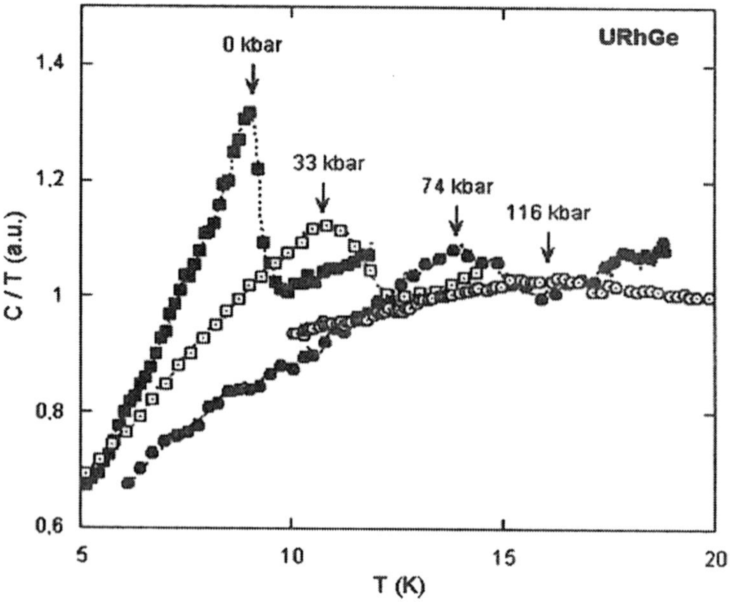

Figure 8.47. Temperature dependence of the specific heat divided by temperature (C/T) of URhGe for different pressures. Reproduced with the permission of Elsevier from Ref. [125].

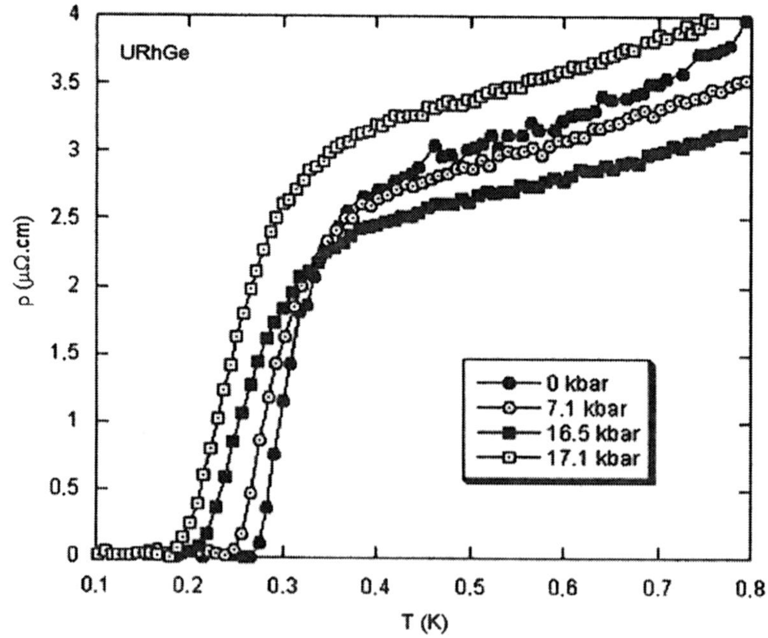

Figure 8.48. Temperature dependence of the resistivity of URhGe for different pressures at low temperatures. The critical temperature is defined by the zero-resistivity point. Reproduced with the permission of Elsevier from Ref. [125].

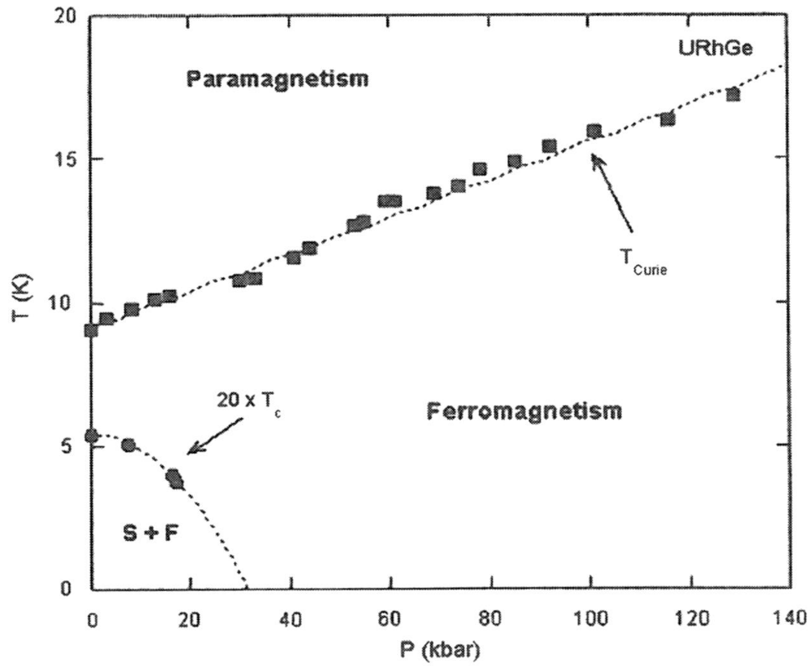

Figure 8.49. Pressure–temperature phase diagram of URhGe from ac calorimetry (■) and resistivity measurements (●). Lines are guide to the eye. Reproduced with the permission of Elsevier from Ref. [125].

8.9. UIr

The pressure-induced superconductivity in UIr was discovered recently by Akazawa et al. [130]. UIr is an itinerant ferromagnet with a lack of inversion symmetry in its crystal structure. The superconductivity in ferromagnets UGe_2 and URhGe coexists with the ferromagnetic order. The itinerant 5f electrons which produce the band ferromagnetism are considered to form Cooper pairs in the superconducting state. The existence of crystal symmetry in their crystal structure is supposed to favor superconductivity since it guarantees the degeneracy of equal-spin states of the spin-triplet pairing with opposite momentum. In contrast, for a system without inversion symmetry such as $CePt_3Si$, the Cooper pair is a mixture of spin-singlet and spin-triplet states since the spatial part of the wave function cannot be taken to be symmetric or antisymmetric under electron exchange [131].

UIr is an Ising-like ferromagnet with the easy axis parallel to $[1\ 0\ \bar{1}]$ direction. The Curie temperature is $T_C = 46$ K at ambient pressure [132]. The magnetic susceptibility follows the Curie–Weiss law with the effective moment $\mu_{eff} = 2.4\ \mu_B$/U atom while the saturation moment in the ordered state is 0.5 μ_B/U atom. These results indicate an itinerant nature of 5f electrons [133]. Superconductivity is observed in the narrow pressure range near the ferromagnetic QCP, where the temperature dependence of resistivity follows the NFL form of $T^{1.6}$. The superconductivity is supposed to be associated with critical fluctuations with disappearance of the ferromagnetic phase.

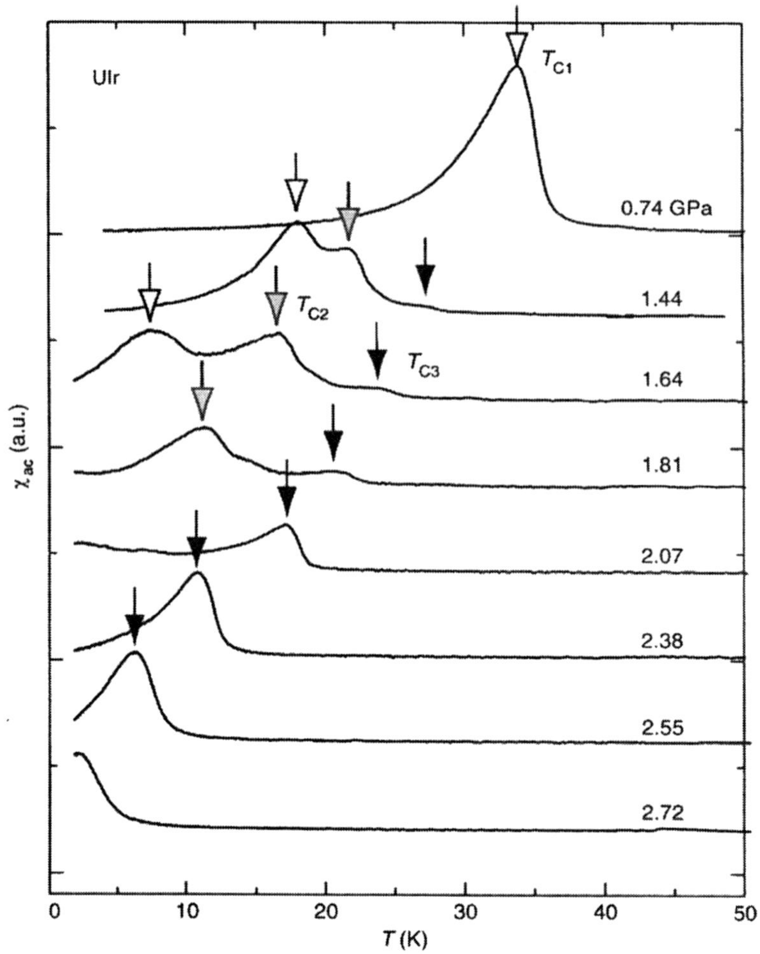

Figure 8.50. Temperature dependence of ac susceptibility along the [1 0 $\bar{1}$] direction under high pressure. Reproduced with the permission of Elsevier from Ref. [134].

The temperature dependence of ac susceptibility (AC-χ) along the [1 0 $\bar{1}$] direction under high pressure is shown in Figure 8.50.

The three series of peaks indicate the successive ferromagnetic transitions with transition temperatures $T_{C1}-T_{C3}$. The P–T phase diagram is shown in Figure 8.51.

Kobayashi et al. [134] have also investigated the possibility of the canted spin structure (expected in a system without inversion symmetry due to Dzialoshinsky–Moriya interaction) by analyzing the anisotropic magnetization process in the FM3 phase as well as the temperature dependence of magnetization. Their data show that there is no evidence of canted ferromagnet.

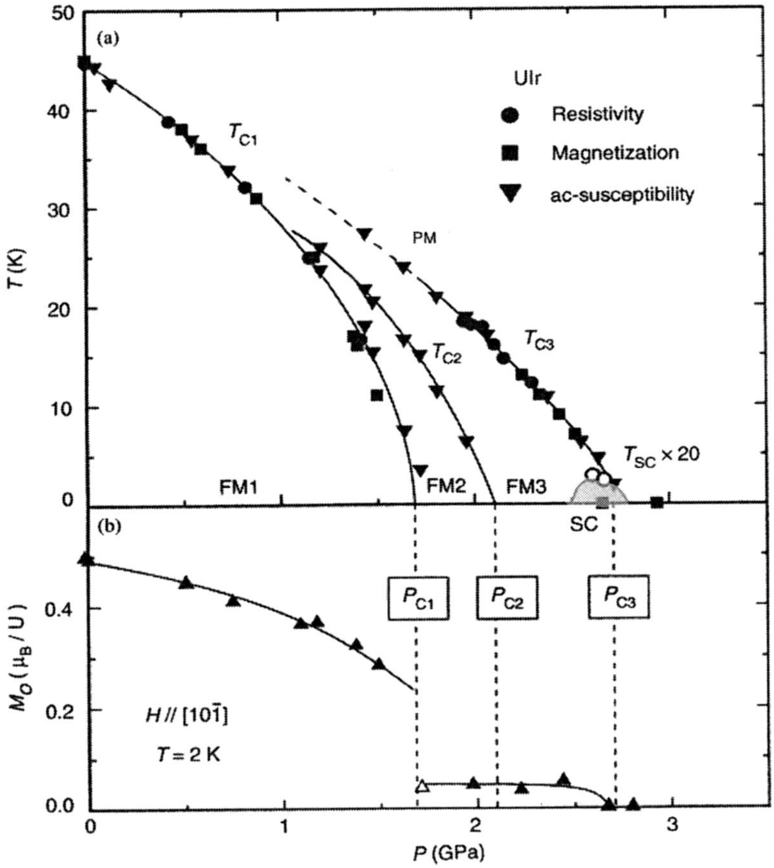

Figure 8.51. (a) Pressure–temperature phase diagram of UIr. The phase boundaries determined by the resistivity, ac–susceptibility, and magnetization measurements have been plotted. The phase diagram consists of three ferromagnetic phases FM1–3 and a superconducting phase in the vicinity of the QCP of FM3. (b) Pressure dependence of the ordered moment along the [1 0 $\bar{1}$] direction. In the FM2 and FM3 phases, the ordered moment is less than 0.05 μ_B/U atom. Reproduced with the permission of Elsevier from Ref. [134].

The detailed *P–T* phase diagram determined by the AC-χ measurement is shown in Figure 8.52. The Meissner effect is observed successfully as shown in the inset of Figure 8.52.

The onset temperature of Meissner effect (inset of Figure 8.52) is $T_{SC} \sim 0.09$ K at 2.67 GPa which is close to $T_{SC} \sim 0.14$ K at 2.61 GPa defined as the midpoint of resistance drop [130]. The Meissner effect is observed in 2.64 GPa $< P <$ 2.74 GPa where the obvious peak corresponding to the FM3 transition exists and vanishes with the FM3 phase. Thus, the superconducting phase is embedded in the FM3 phase. This indicates that the time-reversal symmetry is also broken similar to the cases of UGe_2 and URhGe.

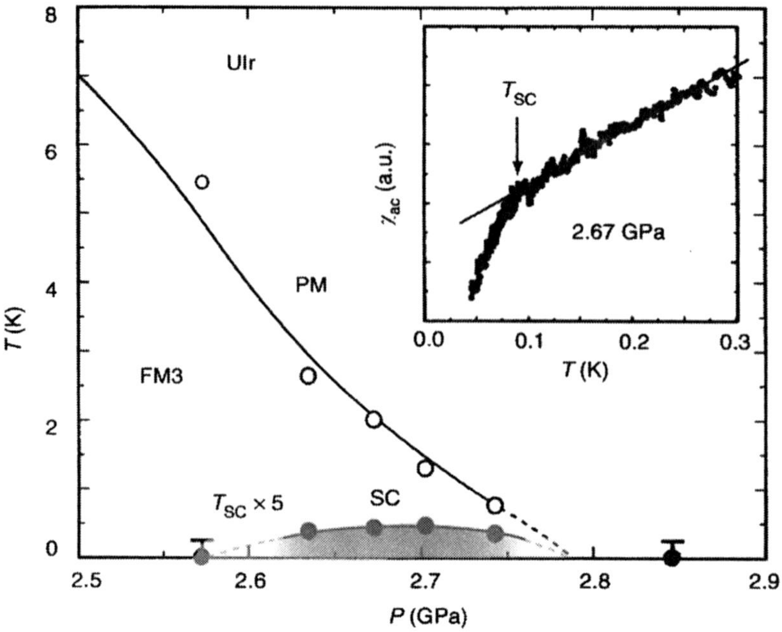

Figure 8.52. Detailed P–T phase diagram in the vicinity of superconducting phase. Inset shows the Meissner effect in ac susceptibility measurement. The superconducting phase is embedded in the FM3 phase. Reproduced with the permission of Elsevier from Ref. [134].

To summarize, because of the lack of inversion symmetry in the crystal structure of UIr and the appearance of superconductivity adjacent to the ferromagnetic QCP in the ferromagnetic phase FM3, the Cooper pair is the mixture of spin-singlet and spin-triplet states and also lacks time-reversal symmetry. It is possible that the ferromagnetic order enhances the superconductivity mediated by spin fluctuations due to the longitudinal magnetic susceptibility [135].

References

[1] H.R. Ott, H. Rudigier, P. Delsing, and Z. Fisk, Phys. Rev. Lett. **50**, 1595 (1983).
[2] G.R. Stewart, Z. Fisk, J. Willis, and J.L. Smith, Phys. Rev. Lett. **52**, 679 (1984).
[3] G.R. Stewart, Rev. Mod. Phys., **56**, 755 (1984).
[4] R.A. Fischer, S. Kim, B.F. Woodfield, N.E. Phillips, L.Taillefer, K. Hasselbach, J. Flouquet, Al. Giorgi, and J.L. Smith, Phys. Rev. Lett., **62**, 1411 (1989).
[5] K. Hasselbach, L. Taillefer, and J. Flouquet, Phys. Rev. Lett. **63**, 93 (1989).
[6] M. Sigrist and K. Ueda, Rev. Mod. Phys. **63**, 239 (1991).
[7] H.R. Ott, H. Rudigier, E. Felder, Z. Fisk, and J.L. Fisk, Phys. Rev. B **33**, 126 (1986).
[8] Y. Kuramato and Y. Kitaoka, in *Dynamics of Heavy Electrons*, eds. J. Birman, S.F. Edwards, R. Friend, C.H. Llewellyn Smith, M. Rees, D. Sherrington, and G. Veneziano (Oxford, New York, 2000).

[9] T.T.M. Palstra, A.M. Menovsky, J. van den Berg, A.J. Dirkmaat, P.H. Kes, J. Nieuwenhuys, and J.A. Mydosh, Phys. Rev. Lett. **55**, 2727 (1985).
[10] C. Geibel, S. Thies, D. Kaczorowski, A. Mehner, A. Grauel, B. Seidel, U. Ahlheim, R. Helfrich, K. Peterson, C.D. Bredl, and F.Steglich, Z. Phys. B **83**, 305 (1991).
[11] C. Geibel, C. Schank, S. Thies, H. Kitazawa, C.D. Bredl, A. Bohm, M. Rau, A. Grauel, R. Caspary, R. Helfrich, U. Ahlheim, G. Weber, and F. Steglich, Z. Phys. B **84**, 1 (1991).
[12] M. Jourdan, M. Huth, and H. Adrian, Nature **398**, 47 (1999).
[13] N.K. Sato, N. Aso, K. Miyake, R. Shiina, P. Thalmeier, G. Varelogiannis, C. Geibel, F. Steglich, P. Fulde, and T. Komatsubara, Nature **410**, 340 (2001).
[14] D.L. Cox, Phys. Rev. Lett. **59**, 1240 (1987).
[15] H.R. Ott, H. Rudigier, Z. Fisk, and J.L. Smith, Phys. Rev. B **31**, 1651 (1985).
[16] S.E. Lambert, Y. Dalichaouch, M.B. Maple, J.L. Smith, and Z. Fisk, Phys. Rev. Lett. **57**, 1619 (1986).
[17] H.R. Ott, Physica B **378–380**, 1 (2006).
[18] P.W. Anderson, Phys. Rev. B **30**, 1549 (1984).
[19] H.R. Ott, H. Rudigier, T.M. Rice, K. Ueda, Z. Fisk, and J.L. Smith, Phys. Rev. Lett. **52**, 1915 (1984).
[20] D.E. MacLaughlin, C. Tien, G.W. Clark, M.D. Lan, Z. Fisk, and J.L. Smith, Phys. Rev. Lett. **53**, 1833 (1984).
[21] B. Golding, D.J. Bishop, B. Batlogg, W.H. Haemmerle, Z. Fisk, J.L. Smith, and H.R. Ott, Phys. Rev. Lett. **55**, 2479 (1985).
[22] M.B. Maple, J.W. Chen, S.E. Lambert, Z. Fisk, H.R. Ott, J.S. Brooks, and M.J. Naughton, Phys. Rev. Lett. **54**, 477 (1984).
[23] F. Gross, B.S. Chandrasekhar, D. Einzel, K. Andres, P.J. Hirschfeld, H.R. Ott, J. Beuers, Z. Fisk, and J.L. Smith, Z. Phys. B **64**, 175 (1986).
[24] J.L. Smith, Z. Fisk, J.O. Willis, A.L. Giorgi, R.B. Roof, and H.R. Ott, Physica B **135**, 3 (1985).
[25] M. Sigrist and T.M. Rice, Phys. Rev. B **39**, 2200 (1989).
[26] R.H. Heffner, W.P. Beyermann, M.F. Hundley, J.D. Thompson, J.L. Smith, Z. Fisk, K. Bedell, P. Birrer, C. Baines, F.N. Gygax, B. Hitti, E. Lippelt, H.R. Ott, C. Schank, and D.E. MacLaughlin, J. Appl. Phys. **69**, 5481 (1991).
[27] E. Dumont and A.C. Mota, Phys. Rev. B **65**, 144519 (2002).
[28] H.R. Ott, H. Rudigier, E. Felder, Z. Fisk, and J.l. Smith, Phys. Rev. B **33**, 126 (1986).
[29] Z. Fisk, H. Borges, M. McElfresh, J.L. Smith, J.D. Thompson, H.R. Ott, G. Aeppli, E. Bucher, S.E. Lambert, M.B. Maple, C. Broholm, and J.K. Kjems, Physics C **153–155**, 1728 (1998).
[30] H.R. Ott, E. Felder, Z. Fisk, R.H. Heffner, and J.L. Smith, Phys. Rev. B **44**, 7081 (1991).
[31] P. Gegenwart, C. Langhammer, R. Helfrich, N. Oeschler, M. Lang, J.S. Kim, G.R. Stewart, and F. Steglich, Physica C **408–410**, 157 (2004).
[32] F. Kromer, R. Helfrich, M. Lang, F. Steglich, C. Langhammer, A. Bach, T. Michels, J.S. Kim, and G.R. Stewart, Phys. Rev. Lett. **81**, 4476 (1998).
[33] C.J. Pethick, D. Pines, K.E. Quader, K.S. Bedell, and G.E. Brown, Phys. Rev. Lett. **57**, 1955 (1986).

[34] L. Taillefer, J. Flouquet, and G.G. Lonzarich, Physica B **169**, 257 (1991).
[35] R. Joynt, J. Magn. Magn. Mater. **108**, 31 (1992).
[36] L. Taillefer, Hyperfine Interact. **85**, 379 (1994).
[37] H. Lohneysen, Physica B **197**, 551 (1994).
[38] R.H. Heffner and M.R. Norman, Comments Condens. Matter Phys. **17**, 361 (1996).
[39] R. Joynt and L. Taillefer, Rev. Mod. Phys. **74**, 235 (2002).
[40] B. Lussier, B. Ellman, and L. Taillefer, Phys. Rev. Lett. **73**, 3294 (1994).
[41] U. Rauchschwalbe, U. Ahlheim, F. Steglich, D. Rainer, and J.J.M. Franse, Z. Phys. B **60**, 379 (1985).
[42] V. Muller, Ch. Roth, D. Maurer, E.W. Scheidt, K. Luders, E. Bucher, and H.E. Bommel, Phys. Rev. Lett. **58**, 1224 (1987).
[43] A. Schenstrom, M.-F. Xu, Y. Hong, D. Bein, M. Levy, B.K. Sarma, S. Adenwalla, Z. Zhao, T. Tokuyasu, D.W. Hess, J.B. Ketterson, J.A. Sauls, and D.G. Hinks, Phys. Rev. Lett. **62** (1989) 1411.
[44] K. Hasselbach, A. Lacerda, K. Behnia, L. Taillefer, J. Flouquet, and A. de Visser, J. Low Temp. Phys. **81**, 299 (1990).
[45] S. Adenwalla, S.W. Lin, Z. Zhao, Q.Z. Ran, J.B. Ketterson, J.A. Sauls, L. Taillefer, D.G. Hinks, M. Levy, and B.K. Sarma, Phys. Rev. Lett. **65**, 2298 (1990).
[46] B. Bogenberger, H.v. Lohneysen, T. Trappmann, and L. Taillefer, Physica B **186–188**, 248 (1993).
[47] S. Wuchner, N. Keller, J.L. Tholence, and J. Flouquet, Solid State Commun. **85**, 355 (1993).
[48] F. Vincent, J. Hammann, L. Taillefer, K. Behnia, N. Keller, and J. Flouquet, J. Phys. Condens. Matter **3**, 3517 (1991).
[49] K. Behnia, J. Taillefer, and J. Flouquet, J. Appl. Phys. **67**, 5200 (1990).
[50] T. Trappmann, H.v. Lohneysen, and L. Taillefer, Phys. Rev. B. **43**, 13714 (1991).
[51] H.v. Lohneysen, T. Trappmann, and L. Taillefer, J. Magn. Magn. Mater. **108**, 49 (1992).
[52] L. Taillefer, K. Behnia, K. Hasselbach, J. Flouquet, S.M. Hayden, and C. Vettier, J. Magn. Magn. Mater. **90–91**, 623 (1990).
[53] S.M. Hayden, L. Taillefer, C. Vettier, and J. Flouquet, Phys. Rev. B **46**, 8675 (1992).
[54] G. Aeppli, E. Bucher, C. Broholm, J.K. Kjems. J. Baumann, and J. Hufnagl, Phys. Rev. Lett. **60**, 615 (1988).
[55] E.D. Isaacs, P. Zschack, C.L. Broholm, C. Burns, G. Aeppli, A.P. Ramirez, T.T.M. Palstra, R.W. Erwin, N. Stucheli, and E. Bucher, Phys. Rev. Lett. **75**, 1178 (1995).
[56] N.H. van Dijk, P. Rodiere, B. Fak, A. Huxley, and J. Flouquet, Physica B **319**, 220 (2002).
[57] J. Sauls, Adv. Phys. **43**, 113 (1994).
[58] K. Machida and M. Ozaki, Phys. Rev. Lett. **66**, 3293 (1991).
[59] K. Machida, T. Ohmi, and M. Ozaki, J. Phys. Soc. Jpn. **62**, 3216 (1993).
[60] D.-C. Chen and A. Garg, Phys. Rev. Lett. **70**, 1689 (1993).
[61] A. Garg and D.-C. Chen, Phys. Rev. B **49**, 479 (1994).
[62] A. de Visser, Physica B **319**, 233 (2002).

[63] J.D. Hettinger, J.C. Cooley, R.E. Hackenberg, E.J. Peterson, A.M. Kelley, P.A. Papin, J.L. Smith, A. de Visser, and M.J. Graf, Physica B **359–361**, 1066 (2005).
[64] R.J. Keizer, A. de Visser, A.A. Menovsky, J.J.M. Franse, B. Fak, and J.-M. Mignot, Phys. Rev. B **60**, 6668 (1999).
[65] R.J. Keizer, A. de Visser, M.J. Graf, A.A. Menovsky, and J.J.M. Franse, Phys. Rev. B **60**, 10527 (1999).
[66] M.J. Graf, R.J. Keizer, A. de Visser, and J.J.M. Franse, Physica B **259–261**, 666 (1999); M.J. Graf, R.J. Keizer, A. de Visser, A.A. Menonosky, and J.J.M. Franse, Phys. Rev. B **60**, 3065 (1999).
[67] A. de Visser, J.C.P. Klasse, M. van Sprang, J.J.M. Franse, A. Menovsky, T.T.M. Palstra, and A.J. Dirkmaat, Phys. Lett. A **113**, 489 (1986).
[68] R.J. Keizer, A. de Visser, A.A. Menovsky, J.J.M. Franse, A. Amato, F.N. Gygax, M. Pinkpank, and A. Schenck, J. Phys. Condens. Matter **11**, 8591 (1999).
[69] A. de Visser, M.J. Graf, P. Estrela, A. Amato, C. Baines, D. Andreica, F.N. Gygax, and A. Schenk, Phys. Rev. Lett. **85**, 3005 (2000).
[70] M.J. Graf, A. de Visser, C.P. Opeil, J.C. Cooley, J.L. Smith, A. Amato, C. Baines, F. Gygax, and A. Dchenk, Phys. Rev. B **68**, 224421 (2003).
[71] C. Broholm, H. Lin, P.T. Matthews, T.E. Masson, W.J.L. Buyers, M.F. Collins, A.A. Menovsky, J.A. Mydosh, and J.K. Kjerns, Phys. Rev. B **43**, 12809 (1991).
[72] C. Broholm, T.J. Kjerns, W.J.L. Buyers, P. Matthews, T.T.M. Palstra, A.A. Menovsky, and J.A. Mydosh, Phys. Rev. Lett. **58**, 1467 (1987).
[73] H. Amitsuka, M. Sato, N. Metoki, M. Yokoyama, K. Kuwahara, T. Sakakibara, H. Morimoto, S. Kawarzaki, Y. Miyako, and J. Mydosh, Phys. Rev. Lett. **83**, 5114 (1999).
[74] H. Amitsuka et al., J. Phys. Soc. Jpn. **69** (Suppl. A), 5 (2000).
[75] K. Matsuda, Y. Kohori, T. Kohara, K. Kuwahara, and H. Amitsuka, Phys. Rev. Lett. **87**, 87203 (2001).
[76] K. Matsuda, Y. Kohri, T. Kohara, H. Amitsuka, and T. Matsumoto, Physica B **312–313**, 504 (2002).
[77] H. Amitsuka, M. Yokoyama, S. Niyazaki, K. Tonya, T. Sakakibara, W.H. Higemoto, K. Nogamine, K. Matsuda, Y. Kohori, and T. Kohora, Physica B **312–313**, 390 (2002).
[78] G.M. Luke, A. Keren, L.P. Le, Y.J. Uemara, W.D. Wu, D. Bonn, L. Taillefer, J.D. Garrett, and Y. Onuki, Hyperfine Interactions **85**, 397 (1994).
[79] D.E. MacLaughlin, D.M. Cooke, R.H. Heffner, R.L. Huston, W.W. McElfresh, M.E. Schillaci, H.D. Rempp, J.L. Smith, J.O. Willis, E. Zimgiebt, C. Boekema, R.L. Lichti, and J. Oostens, Phys. Rev. B **37**, 3153 (1988).
[80] E.A. Knetsch, A.A. Menovsky, G.J. Nieuwenhuys, J.A. Mydosh, A. Amato, R. Feyerherm, F.N. Gygax, A. Schenk, R.H. Heffner, and D.E. MacLaughlin, Physica B **186–188**, 300 (1993).
[81] A. Yaouanc, P. Dalmas De Reotier, A.D. Huxley, P. Bonville, P.C.M. Gubbens, A.M. Mulders, P. Lejay, and S. Kunii, Physica B **230–232**, 269 (1997).
[82] O.O. Bernal, B. Becker, J.A. Mydosh, G.J. Nieuwenhys, A.A. Menovsky, P.M. Paulus, H.B. Brom, D.E. MacLaughlin, and H.G. Lukefahr, Physica B **281–282**, 236 (2000).

[83] O.O. Bernal, C. Rodriguez, A. Martinez, H.G. Lukefahr, D.E. MacLaughlin, A.A. Menovsky, and J.A. Mydosh, Phys. Rev. Lett. **87**, 196402 (2001).
[84] O.O. Bernal, J.A. Mydosh, D.E. MacLaughlin, H. Lukefahr, and A.A. Menovsky, Physica B **312–313**, 501 (2002).
[85] P. Chandra, P. Coleman, and J.A. Mydosh, Physica B **312–313**, 397 (2002).
[86] S. Uemura, G. Motoyama, Y. Oda, T. Nishioka, and N.K. Sato, J. Phys. Soc. Jpn. **74**, 2667 (2005).
[87] A.P. Ramirez, T. Siegrist, T.T.M. Palstra, J.D. Garrett, E. Bruck, A.A. Menovsky, and J.A. Mydosh, Phys. Rev. B **44**, 5392 (1991).
[88] H. Amitsuka, K. Tenya, M. Yokoyama, A. Schenk, D. Andrecia, F.N. Gygax, A. Amato, Y. Miyako, Y.K. Huang, and J.A. Mydosh, Physica B **326**, 418 (2003).
[89] G. Motoyama, T. Nischoka, and N.K. Sato, Phys. Rev. Lett. **90**, 166402 (2003).
[90] A. Amato, M.J. Graf, A. de Visser, H. Amitsuka, D. Andreica, and A. Schenk, J. Phys. Condens. Matter **16**, SS4403 (2004).
[91] F. Bourdart, A. Bombardi, P. Burlet, M. Endrele, P. Lejay, N. Kernavonis, V.P. Mineev, L. Paolasini, M.E. Zhitomirsky, and B. Fak, Physica B **359–361**, 986 (2005).
[92] E. Louis, A. de Visser, A. Menovsky, and J.J.M. Franse, Physica B+C **144**, 48 (1986).
[93] M.M. McElfresh, J.D. Thomson, J.O. Willis, M.B. Maple, T. Kohara, and M.S. Torikachville, Phys. Rev. B **35**, 43 (1987).
[94] G. Oomi, T. Kagayama, Y. Onuki, and T. Komatsubara, Physica B **199–200**, 148 (1994).
[95] N.K.Sato, S. Uemura, G. Motoyama, and T. Nisioka, Physica B **378–380**, 576 (2006).
[96] O.O. Barnal, M.E. Moroz, K. Ishida, H. Murakawa, A.P. Reyes, P.L. Kuhns, D.E. MacLaughlin, J.A. Mydosh, and T.J. Gortenmulder, Physica B **378–380**, 574 (2006).
[97] H. Kim, N. Harrison, M. Jaime, G.S. Boebinger, and J.A. Mydosh, Phys. Rev. Lett. **91**, 256401 (2003).
[98] P. Coleman, Physica B **378–380**, 1160 (2006).
[99] Y.S. Oh, K.H. Kim, P.A. Sharma, N. Harrison, H. Amitsuka, and J.A. Mydosh, Phys. Rev. Lett. **98**, 016401 (2007).
[100] K.H. Kim, N. Harrison, H. Amitsuka, G.A. Jorge, M. Jaime, and J.A. Mydosh, Phys. Rev. Lett. **93**, 206402 (2004).
[101] A. Krimmel, P. Fischer, B. Roessli, H. Maletta, C. Geibel, C. Schank, A. Grauel, A. Loidl, and F. Steglich, Z. Physica B **86**, 161 (1992).
[102] N. Bernhoeft, B. Roessli, N. Sato, N. Aso, A. Hiess, G.H. Lander, Y. Endoh, and T. Kpmatsubara, Physica B **259–261**, 614 (1999).
[103] C. Geibel, U. Alheim, C.D. Bredl, J. Diehl, A. Grauel, R. Helfrich, H. Kitazawa, R. Kohler, R. Modler, M. Lang, C. Schank, S. Thies, F. Steglich, N. Sato, and T. Komatsubara, Physica C **185–189**, 2651 (1991).
[104] C. Geibel, A. Bohm, R. Caspary, K. Gloos, A. Grauel, P. Hellmann, R. Modler, C. Schank, G. Weber, and F. Steglich, Physica B **186–188**, 188 (1993).
[105] K. Gloos, R. Modler, H. Schimanski, C.D. Bredl, C. Geibel, F. Steglich, A.I. Buzdin, N. Sato, and N. Komatsubara, Phys. Rev. Lett. **70**, 501 (1993).

[106] P. Fulde and R.A. Ferrell, Phys. Rev. A **135**, 550 (1964).
[107] A.I. Larkin and Y.N. Ovchinnikov, Zh. Eksp. Teor. Fiz. **47** (1964) 1136; Sov. Phys. JETP **20** (1965).
[108] N.K. Sato, N. Aso, K. Miyake, R. Shiina, P. Thalmeier, G. Varelogiannis, C. Geibel, F. Steglich, and P. Fulde, Nature **410**, 340 (2001).
[109] N. Sato, H. Matsuda, T. Fujiina, K. Yameda, and M. Yakeshima, J. Alloy Comp. **271–273**, 433 (1998).
[110] N.K. Sato, N. Aso, G.H. Lander, B. Roessli, T. Komatsubara, and Y. Endoh, J. Phys. Soc. Jpn. **66**, 2981 (1997).
[111] N. Aso, N. Sato, and T. Komatsubara, Physica B **259–261**, 636 (1999).
[112] F. Steglich, N.K. Sato, N. Aso, P. Gegenwart, J. Custers, K. Neumaier, H. Wilhelm, C. Geibel, and O. Tovarelli, Physica B **329–333**, 443 (2003).
[113] A. Schroder, J.G. Lussier, B.D. Gaulin, J.D. Garret, W.J. Buyers, L. Rebelsky, and M. Shapiro, Phys. Rev. Lett. **72**, 136 (1994).
[114] M. Mihalik, F.E. Kayzel, T. Yoshida, K. Kuwahara, H. Amitsuka, T. Sakakibara, A.A. Menovsky, J.A. Mydosh, and J.J.M. Franse, Physica B **230–232**, 364 (1997).
[115] N. Sato, N. Koga, and T. Komatsubara, J. Phys. Soc. Jpn. **65**, 1555 (1996).
[116] K. Ishida, D. Ozaki, T. Kamatsuka, H. Tou, M. Kyogaku, Y. Kitaoka, N. Tateiwa, N.K. Sato, N. Aso, C. Geibel, and F. Steglich, Phys. Rev. Lett. **89**, 037002 (2002).
[117] K. Ishida, H. Tou, N. Tateiwa, Y. Kitaoka, K.N. Sato, N. Aso, C. Geibel, and F. Steglich, Physica B **388–389**, 535 (2003).
[118] M. Kyogaku, Y. Kitaoka, K. Asayama, C. Geibel, C. Schank, and F. Steglich, J. Phys. Soc. Jpn. **62**, 4016 (1993).
[119] S.S. Saxena, P. Agarwal, K. Ahilan, R.K.W. Hasselwimmer, M.J. Steiner, E. Pugh, I.R. Walker, S.R. Julian, P. Monthoux, G.G. Lonzarich, A. Huxley, I. Sheikin, D. Braithwaite, and J. Flouquet, Nature (London) **406**, 587 (2000).
[120] A. Huxley, I. Sheikin, E. Ressouche, N. Kernavanois, D. Braithwaite, R. Calemeczuk, and J. Flouquet, Phys. Rev. B **63**, 144519 (2001).
[121] A. Demuer, I. Sheikin, D. Braithwaite, B. Fak, A. Huxley, S. Raymond, and J. Flouquet, J. Magn. Magn. Mater. **226–230** (2001).
[122] G. Oomi, T. Kagayama, and Y. Onuki, J. Alloys Comp. **271–273**, 482 (1998).
[123] A. Harada, H. Kotegawa, Y. Kawasaki, G.-Q. Zheng, Y. Kitaoka, E. Yamamato, Y. Haga, Y. Onuki, K. Itoh, and E.E. Haller, Physica B **359–361**, 1057 (2005).
[124] D. Aoki, A. Huxley, E. Ressouche, D. Braithwaite, J. Flouquet, J.-P. Brison, E. Lhotel, and J.C. Paulsen, Nature **413**, 613 (2001).
[125] F. Hardy, A. Huxley, J. Flouquet, B. Salce, G. Knebel, D. Braithwaite, D. Aoki, M. Uhlraz, and C. Pfleiderer, Physica B **359–361**, 1111 (2005).
[126] Q.G. Sheng and B.R. Cooper, J. Appl. Phys. **75**, 7035 (1985).
[127] A. Barla, J.P. Sanchez, B. Ni, B.P. Doyle, P. Vulliet, O. Leupold, R. Ruffer, D. Kaczouski, J. Plessel, and M. M. Abdflemquid, Phys. Rev. B **66**, 94425 (2002).
[128] R. Roussev and A.J. Millis, Phys. Rev. B **63**, 140504 (2001).
[129] V.P. Mineev, Phys. Rev. B **66**, 134504 (2002).
[130] T. Akazawa, H. Hidaka, H. Kotegawa, T. Fujiwara, T.C. Kobayashi, E. Yamamoto, Y. Haga, R. Settai, and Y. Onuki, J. Phys. Soc. Jpn. **73**, 3129 (2004).
[131] S. Saxena, and P. Monthoux, Nature **427**, 799 (2004).

[132] E. Yamamoto, Y. Haga, T.D. Matsuda, A. Nakamura, R. serrai, Y. inada, H. Sugawara, H. Sato, and Y. Onuki, J. Nucl. Sci. Technol. (Suppl. 3), 87 (2002).

[133] T. Akazawa, H. Hidaka, H. Kotegawa, T.C. Kobayashi, T. Fujiwara, E. Yamamoto, Y. Haga, R. Settai, and Y. Onuki, Physica B **359–361**, 1138 (2005).

[134] T.C. Kobayashi, S. Fukushima, H. Hidaka, H. Kotegawa, T. Akazawa, E. Yamamoto, Y. Haga, R. Settai, and Y. Onuki, Physica B **378–380**, 355 (2006).

[135] T.R. Kirkpatrick, D. Belitz, T. Vojta, and R. Narayanan, Phys. Rev. Lett. **87**, 127003 (2001).

Chapter 9

Filled Skutterdites and Transuranium Heavy Fermion Superconductors

9.1. Filled skutterdites

Jeitschko and Braun [1] first synthesized the filled skutterdites with a chemical formula RT_4X_{12}, where R is rare earth or U; T is Fe, Ru, or Os; and X is P, As, or Sb. The compounds with this crystal structure have been extensively investigated. One can obtain widely different features such as heavy fermion superconductivity, metal-insulator transition, quadrupolar ordering, or field-induced heavy fermion state by replacing the constituting elements in the same crystal structure [2–4]. The crystal structure of filled skutterdite compounds is shown in Figure 9.1.

The RE-ions form a bcc structure and each one is surrounded by 12 pnictogen ions forming slightly distorted icosahedron and eight transition elements forming a cube. The pnictogens are not shared with neighboring R which results in strong isotropic c–f hybridization. In addition, R in the pnictogen cage feels isotropic field compared to that in the low dimensional systems where the RE-ions feel uniaxial CEF. The difference in local symmetry T_h of rare earth sites compared to O_h brings about qualitative difference in the CEF levels [6]. The two pure triplets Γ_4 and Γ_5 in O_h symmetry are mixed together into $\Gamma_4^{(1)}$ and $\Gamma_4^{(2)}$ in T_h symmetry, resulting in different response to dipoles and quadrupoles. In addition, the tunneling of an R-ion between off-center positions in a pnictogen cage could be a key mechanism in the variety of unusual properties of filled skutterdites [7].

9.2. $PrOs_4Sb_{12}$

Bauer et al. [8] discovered superconductivity in $PrOs_4Sb_{12}$ at $T_c = 1.85$ K which appears to involve heavy fermion quasiparticles with the effective mass $m^* \sim 50m_e$. This was inferred from the jump in specific heat at T_c, the slope of the upper critical field at T_c, and the electronic specific heat coefficient γ. The ρ versus T curve displayed a metallic behavior and a distinct decrease below ~7 K before abruptly dropping to zero at $T = T_c$. Bauer et al. [8] argued that since the ground state of Pr^{3+} ions in the cubic CEF appears to be the Γ_3 non-magnetic doublet, the heavy fermion behavior possibly involved the interaction of the Pr^{3+} Γ_3 quadrupole moments and the charges of the conduction electrons. In such a case, the quadrupolar fluctuations would play a role in the heavy fermion superconductivity of $PrOs_4Sb_{12}$.

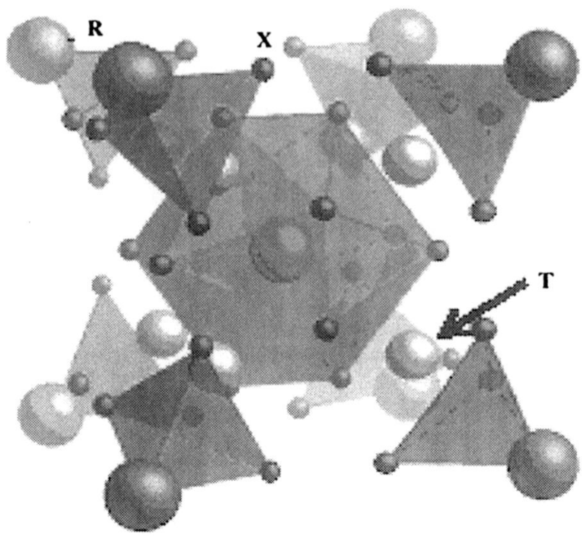

Figure 9.1. Filled skutterdite structure: R = rare earth; T = Fe, Ru, Os; X = P, As, Sb. Reproduced with the permission of Elsevier from Ref. [5].

Kotegawa et al. [9] have reported the observation of unconventional superconducting property probed by the nuclear-spin-lattice-relaxation time T_1, in $PrOs_4Sb_{12}$ through $^{121,123}Sb$ nuclear quadrupolar resonance (NQR) experiments at zero field. The summary of their experimental results of the T dependence of $1/T_1T$ revealed that the 4f-derived moments behave as if localized at the higher T than $T_0 \sim 10$ K. The NQR frequency at the Sb site increased considerably while $1/T_1$ decreased below T_0, which indicated a low-lying crystal-electric-field splitting, associated with the $Pr^{3+}(4f^2)$-derived ground state. For $T < 4$ K, the relaxation process is well accounted for by incorporating both the CEF contributions arising from the first excited state and the $(T_1/T)_{qp}$ = constant contribution from the heavy-quasiparticle state. In the SC state, $PrOs_4Sb_{12}$ has a large and isotropic energy gap $2\Delta/k_BT_c \sim 5.2$ indicating a new type of strong-coupling regime. This would support the idea of Cooper pairing via quadrupolar fluctuations.

Vollmer et al. [10] have measured the magnetic field dependence of the specific heat (C) of single crystals of $PrOs_4Sb_{12}$. The variation of C at low temperature and the magnetic phase diagram inferred from C, the resistivity, and magnetization showed that there was a doublet ground state. The two distinct superconducting anomalies in C provided evidence of two superconducting critical temperatures at $T_{c1} = 1.75$ K and $T_{c2} = 1.85$ K. This could arise from a weak lifting form of the ground-state degeneracy which supports the theory of quadrupolar pairing, i.e., superconductivity in $PrOs_4Sb_{12}$ is neither of electron-phonon nor of magnetically mediated origin.

The temperature-dependence of specific heat (C_p/T) at zero field as well as the real part χ'_{ac} of the ac susceptibility measured with an ac magnetic field of 0.278 Oe at 2.11 Hz, of $PrOs_4Sb_{12}$, was measured by Measson et al. [11] and is shown in Figure 9.2. They found two sharp distinct anomalies at $T_{c1} = 1.887$ K and $T_{c2} = 1.716$ K. The high absolute value of the specific heat ($C_p/T = 3.65$ J/K^2 mol at $T = 1.7$ K) and the large

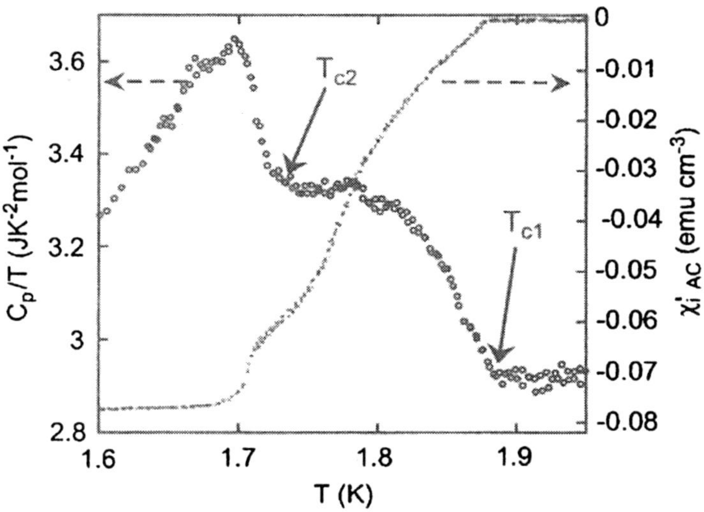

Figure 9.2. Specific heat of $PrOs_4Sb_{12}$ as C_p/T versus T at zero field measured with a quasiadiabatic method. Real part χ' of the ac susceptibility measured with an ac magnetic field of 0.287 Oe at 2.11 Hz. Reproduced with the permission of Elsevier from Ref. [11].

height of the two superconducting jumps and the sharpness of the two transitions (16 and 58 K) provide evidence for the high quality of the sample. However, Figure 9.2 also shows that the transition at T_{c1} is not bulk-homogeneous in which case the susceptibility would have shown perfect diamagnetism far before T_{c2} is reached.

Izawa et al. [12] have investigated the superconducting gap structure of $PrOs_4Sb_{12}$ using thermal transport measurements in magnetic field rotated related to the crystal axes. They have demonstrated that a novel change in the symmetry of the superconducting gap function occurs deep inside the superconducting state. This clearly indicates the presence of two superconducting phases with twofold (B phase) and fourfold (A phase) symmetries. They infer that the gap function in both phases have a point node singularity, in contrast to the line node singularity observed in most unconventional superconductors. The H–T phase diagram obtained by Izawa et al. [12] is shown in Figure 9.3.

Measson et al. [11] have also drawn the H–T superconducting phase diagram of $PrOs_4Sb_{12}$ from ac specific heat measurements which is shown in Figure 9.4. It has the same features as reported by Tayama et al. [13] from magnetization measurements.

Comparing Figures 9.3 and 9.4, we note that $H'(T)$ does not match the line $H^*(T)$ drawn by the angular dependence of thermal conductivity under magnetic field [14]. One possible explanation could be the strong sample dependence of these lines. We also note that the paramagnetic limit plays an important role in H_{c2}, since a different fit is obtained without the paramagnetic limit. However, $H'(T)$ can be reproduced with the same model with the same parameters (except for a scaling of T_c). This means that T_{c2} has the same field dependence as T_{c1}, and the second transition cannot be identified as a symmetry change of the order parameter in contrast to the case of UPt_3.

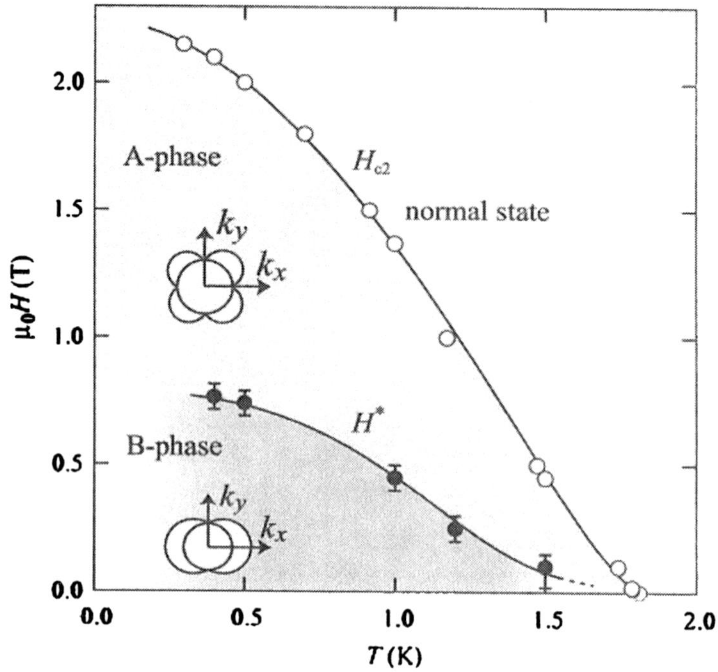

Figure 9.3. The H–T phase diagram of the superconducting gap symmetry of $PrOs_4Sb_{12}$ from thermal conductivity measurements. The filled circles represent the magnetic field H^* at which the transition from fourfold to twofold symmetry takes place. The open circles represent H_{c2}. The area of the gap function with fourfold symmetry is shown by A phase and the area of the gap function with twofold symmetry is shown by blue B phase. Reproduced with the permission of Elsevier from Ref. [14].

A third B–T phase diagram derived from the temperature and magnetic field dependence of the ac susceptibility by Drobnik et al. [15] is shown in Figure 9.5. They have obtained the phase diagram from the real part χ' and the imaginary part χ'' of the ac susceptibility. As a function of temperature at zero field, χ' displays a drop at the superconducting transition, while χ'' displays a broad maximum. They have defined two superconducting transitions, $T_{c1} = 1.85$ K at a change of slope and $T_{c2} = 1.74$ K at a shoulder of χ', respectively, consistent with the two transitions in the specific heat (see Figure 9.2). B_{c2} and B^*_{c2} are the upper critical field corresponding to T_{c1} and T_{c2}, respectively. In addition, they have observed a well developed minimum in $\Delta\chi' = \chi(T) - \chi(2\text{ K})$, and have defined a field value B_{pk} at the low-field side of the minimum.

Figure 9.5 indicates that the first observation of superconductivity at T_{c1}/B_{c2} decreases at $B_{c2} = 2.3$ T. The signature of the second transition at T_{c2}/B^*_{c2} tracks T_{c1}/B_{c2} and disappears at a slightly lower field $B \approx 2.1$ T consistent with specific heat studies [11]. The evidence for a peak effect is confined to the phase bounded by T_{c2}/B^*_{c2} , where B_{pk} drops to zero at $B \approx 1.6$ T. This places the line $B_{pk}(T)$ much lower than B^*_{c2} (T). According to Drobnik et al. [15], in the phase diagram obtained from thermal conductivity (κ) results (Figure 9.3) which suggests lower values of B^*_{c2} (T), κ essentially captures the same property as the peak effect $B_{pk}(T)$ in the ac susceptbilty.

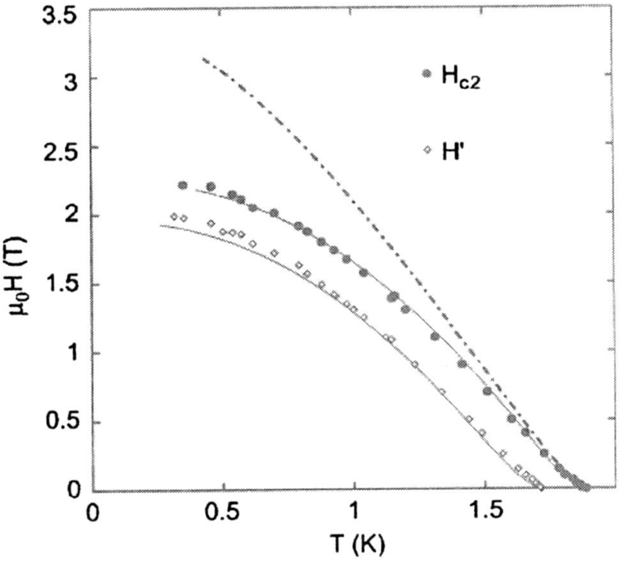

Figure 9.4. *H–T* superconducting phase diagram of $PrOs_4Sb_{12}$ determined by specific heat measurements. The field dependence of T_{c1} and T_{c2} are identical. The lines are fits by a two-band model of the upper critical field, including the paramagnetic limitation. The dash-dotted line is the same fit with the same parameters but without paramagnetic limitation. Reproduced with the permission of Elsevier from Ref. [11].

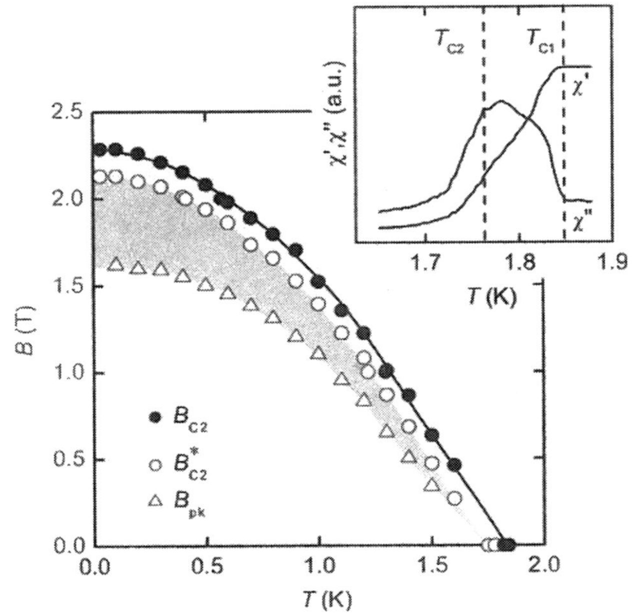

Figure 9.5. Phase diagram derived from the temperature and magnetic field dependence of the ac susceptibility. The inset displays the ac susceptibility as a function of temperature in zero dc magnetic field, where the arrows mark T_{c1} and T_{c2}, defined in correspondence to the specific heat. Reproduced with the permission of Elsevier from Ref. [15].

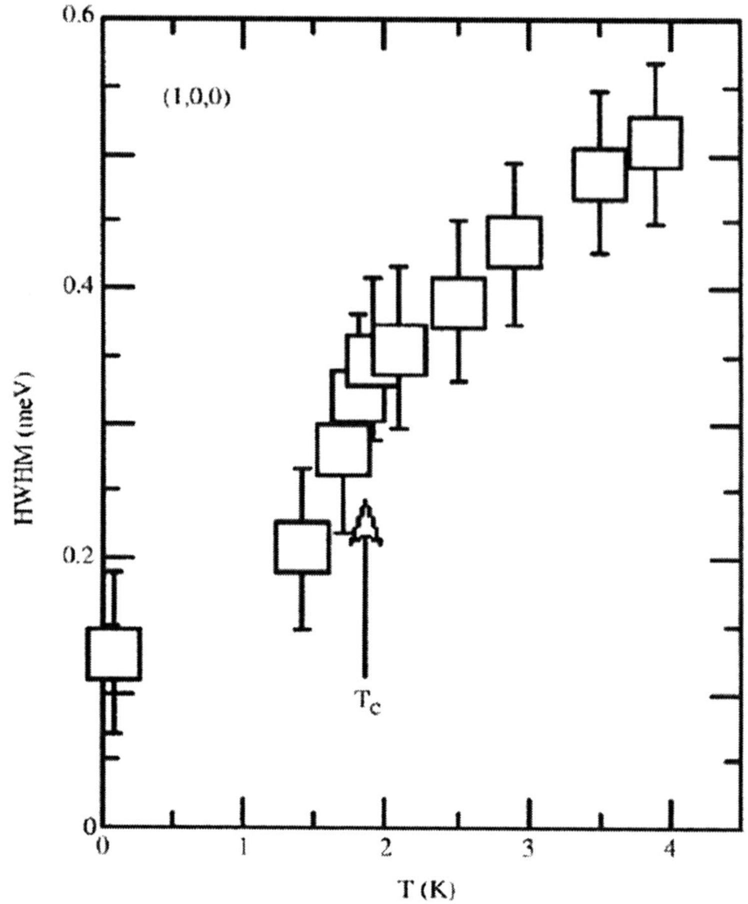

Figure 9.6. Temperature dependence of the linewidth of the peak in the dynamic structure factor $Q = (1, 0, 0)$ in $PrOs_4Sb_{12}$. Reproduced with the permission of Elsevier from Ref. [5].

In order to study the origin of the attractive interaction between superconducting electrons in $PrOs_4Sb_{12}$, inelastic neutron scattering experiments have been performed by Kuwahara et al. [16] on single crystals. The observed excitation softens at a wave vector $\mathbf{Q} = (1, 0, 0)$ which is the same as the modulation vector of the field-induced anti-quadrupolar ordering. Taking into account the fact that the intensity at $\mathbf{Q} = (1, 0, 0)$ is reduced compared to that around the zone center, this excitonic behavior is derived mainly from non-magnetic quadrupolar interactions. Furthermore, the narrowing of the linewidths (full widths at half maximum) of the excitations, as shown in Figure 9.6, indicates the close connection between the superconductivity and excitons.

Nishiyama et al. [17, 18] report an NQR study of $PrOs_4Sb_{12}$ using Ru impurity as a probe for the gap function. They have replaced Os with Ru and studied its effect on the superconducting density of states (DOS). Figure 9.7 shows the T dependence of $1/T_1$ in the Ru-doped samples.

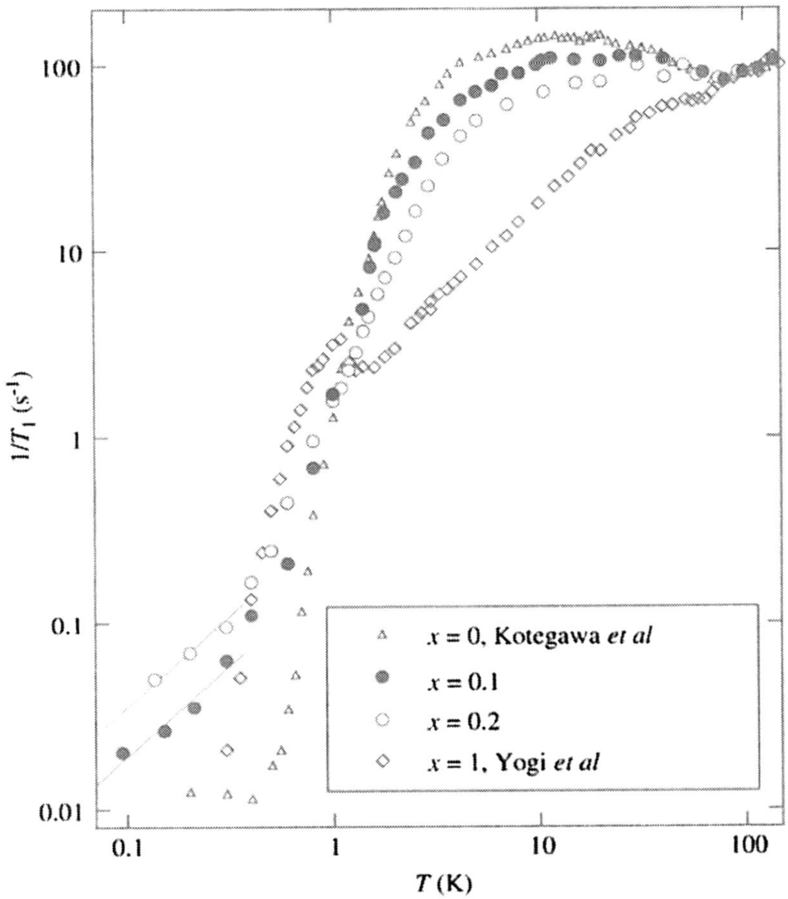

Figure 9.7. Temperature dependence of $1/T_1$ in $Pr(Os_{1-x}Ru_x)_4Sb_{12}$ (x = 0.1, 0.2). The straight lines indicate the T_1T = const relations. Reproduced with the permission of Elsevier from Ref. [18].

From Figure 9.7, we note that above $T \sim 100$ K, the data for all x merge to a same line indicating that at high T, the relaxation is not governed by the electronic state. Below $T \sim 100$ K, $1/T_1$ is reduced as x increases. Below $T \sim 1$ K, $1/T_1$ is proportional to T and the magnitude of $1/T_1T$ increases with increasing x. The fact that $1/T_1T$ is constant indicates that a finite DOS is induced by the impurity which provides strong evidence for the existence of nodes in the gap function [19]. In addition, the relatively weaker suppression of T_c by the impurity with respect to x (Figure 2 of Ref. [18]) suggests that point nodes in the gap function are more plausible [20].

In order to investigate the nature of the superconducting pairing state in $PrOs_4Sb_{12}$, Ozcan et al. [21] have measured the temperature dependence of the in-plane magnetic penetration depth, at temperatures as low as 20 mK. The measurements were made using a superconducting resonant coil with the RF magnetic field pointing along the coil axis

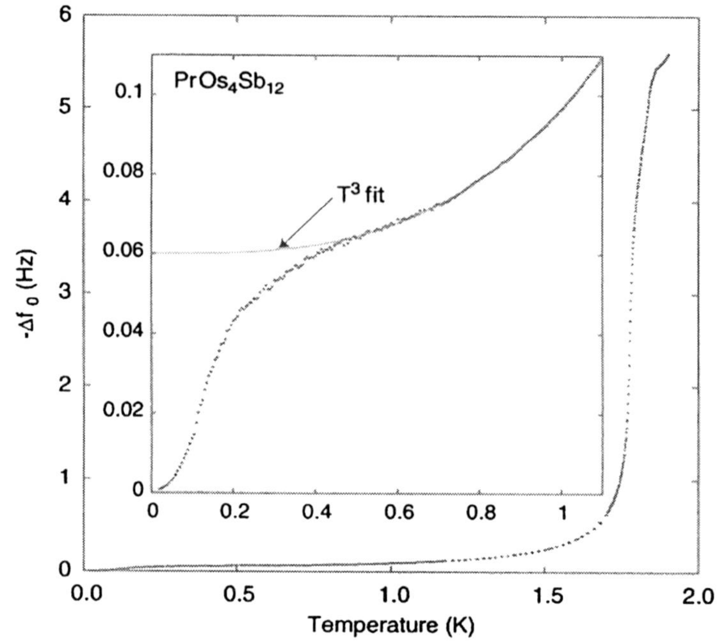

Figure 9.8. Temperature variation of the frequency shift $-\Delta f_0(T)$ due to the expulsion of magnetic flux from the interior of $PrOs_4Sb_{12}$. The inset shows a fit of $-\Delta f_0(T) = a + bT^3$ for 450 mK $\leq T \leq$ 1100 mK. At $T \approx 450$ mK a superconducting-transition-like drop in the magnetic penetration depth sets in. Reproduced with the permission of Elsevier from Ref. [21].

which was parallel to the *c* axis of the crystal. When $T < T_c$, the expulsion of magnetic flux from the interior of the sample causes a temperature-dependent shift in the resonant frequency f_0. The shift in the oscillator frequency $-\Delta f_0$ is related to the penetration depth $\Delta\lambda$ (for small changes in the inductance) by the relation $\Delta\lambda(T) = -\Gamma\Delta f_0(T)$, where Γ is a constant scaling factor. The temperature variation of $-\Delta f_0(T)$ is plotted in Figure 9.8.

The most interesting results of this experiment are the existence of gapless excitations and the onset of a superconducting-transition-like drop in the magnetic penetration depth at about 450 mK. This additional superconducting transition in $PrOs_4Sb_{12}$ has not been reported in earlier experiments.

Custers et al. [22] have investigated the superconducting gap of $PrOs_4Sb_{12}$ by field-angle-dependent specific heat $C(H, \varphi)$ measurements. H was varied at fixed temperatures and rotated in the basal planes. Their results are shown in Figure 9.9.

Here, two samples of different quality have been measured. Sample #1 shows a broad single jump in $C(T)$ at $T_{c1}^{\#1} = 1.81\mathrm{K}(\mu_0 H_{c2}^{\#1} = 2.1\mathrm{T})$ while sample #2 demonstrated two SC jumps at $T_{c1}^{\#2} = 1.85\mathrm{K}$ and $T_{c2}^{\#2} = 1.67\mathrm{K}(\mu_0 H_{c2}^{\#2} = 2.35\mathrm{T})$, respectively. No evidence was found for a symmetry transition which would have resulted in a change of a four to twofold oscillation of $C(H, \varphi)$ as found in $\kappa(H, \varphi)$ [12] from thermal conductivity experiments.

As shown in Figure 9.9, a fourfold oscillation of $C(H, \varphi)$ was observed in the H–T SC phase diagram. The minima are located along the [1 0 0] directions suggesting the existence

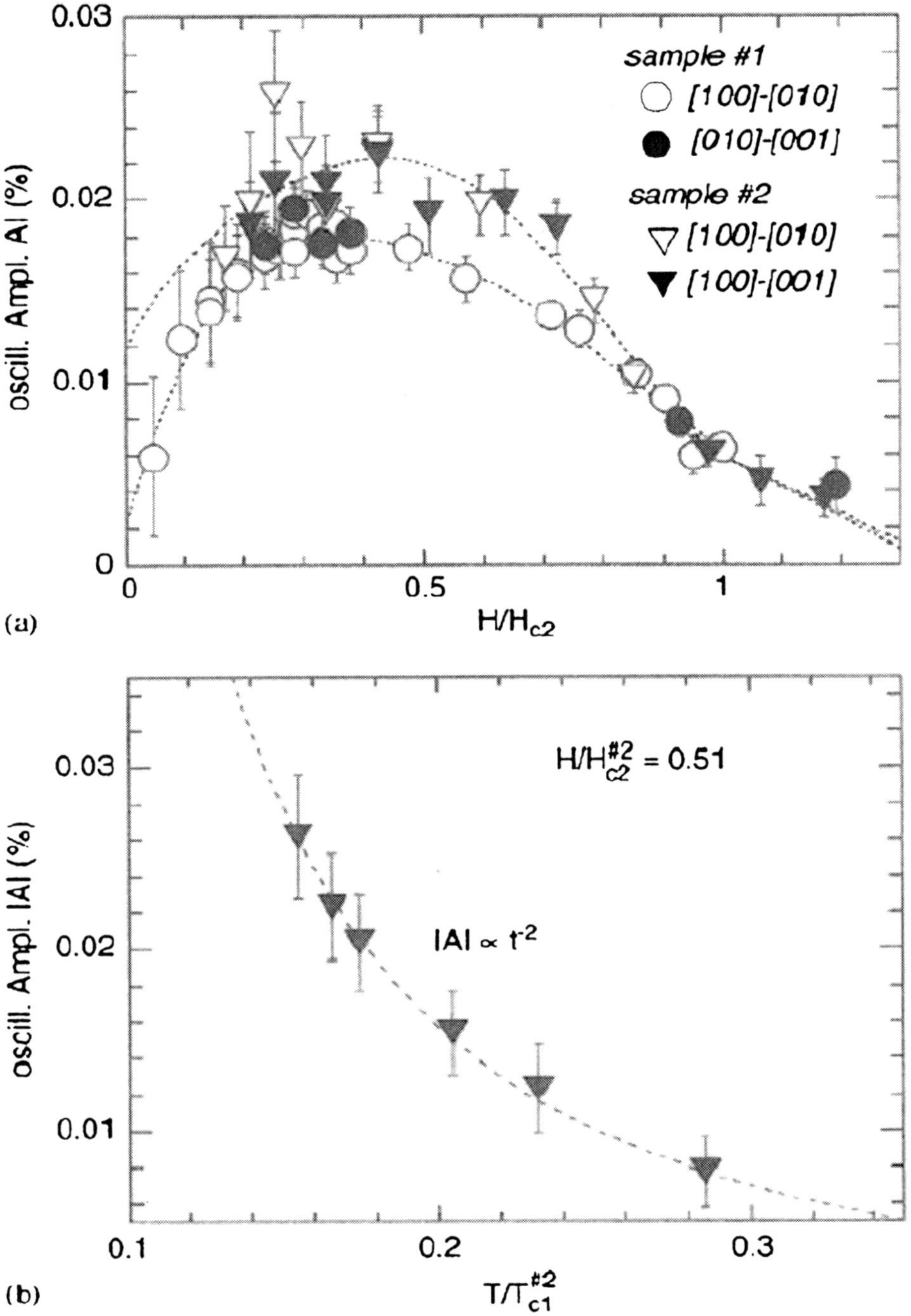

Figure 9.9. Field and temperature behavior of the amplitude |*A*| of the fourfold oscillation: (a) |*A*| versus H/H_{c2}. Dotted lines are guide to the eyes. Gray area marks the normal state; (b) |*A*| versus T/T_{c1} of S#2 measured in $\mu_0 H = 1.2$ T rotating in the *ac* plane. The dashed line is a fit to the data $C(t) \propto t^{-2}$, with $t = T/T^{*2}_{c1}$. Reproduced with the permission of Elsevier from Ref. [22].

of gap nodes or minima along these directions. The oscillation amplitude $|A|$ becomes maximum at $H/H_{c2} \approx 0.03$. Below, $|A| \to 0$. The temperature variation of $|A|$ follows a quartic dependence.

The other filled skutterdite superconductor is YFe_4P_{12} [23] with $T_c = 7$ K. However, since it is not a heavy fermion compound, we shall not discuss its properties in this chapter.

9.3. $PuCoGa_5$

The discovery of superconductivity in the transuranium compound $PuCoGa_5$ with $T_c \approx 18.5$ K has attracted considerable attention [24]. This is by far the highest critical temperature for any heavy fermion superconductor. $PuCoGa_5$ crystallizes in the $HoCoGa_5$ structure, the same type as the $CeMIn_5$ materials [25–27]. The existence of a local moment magnetic susceptibility in the normal state [24] and photoemission of partially localized 5f electrons [28] indicate that $CeMnIn_5$ and $PuCoGa_5$ have similar electronic structure.

The H–T phase diagram [29] of $PuCoGa_5$, inferred from the heat capacity data as a function of temperature in a magnetic field applied along the three orthogonal directions, is shown in Figure 9.10.

The data indicate that T_c is suppressed at an approximate rate of 10 T/K irrespective of field orientation. This initial slope indicates that the upper critical field H_{c2} is of the order of 75 T. In addition, there is no evidence of anisotropy.

Curro et al. [30] have performed nuclear magnetic resonance (NMR) experiments in $PuCoGa_5$ in the normal and superconducting states. Measurements of the Knight shift

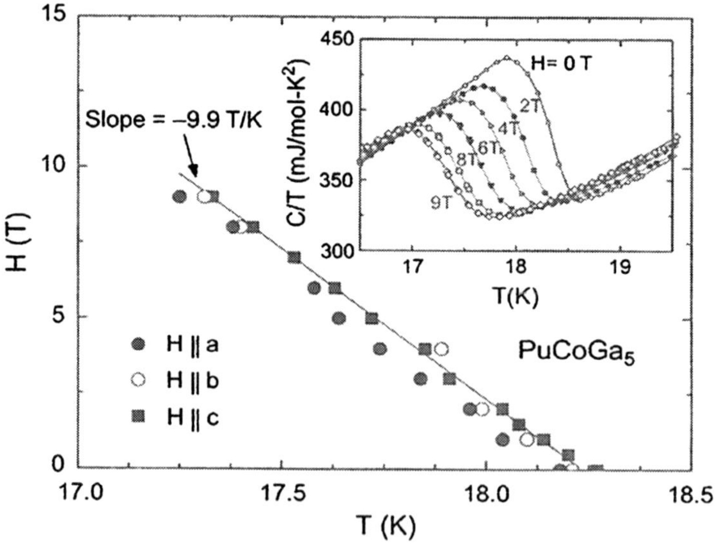

Figure 9.10. Magnetic field–temperature phase diagram of $PuCoGa_5$ inferred from single-crystal heat capacity measurements with magnetic field applied along three orthogonal directions. The inset shows representative heat capacity data from which $T_c(H)$ was inferred. Reproduced with the permission of Elsevier from Ref. [29].

and nuclear-spin-lattice-relaxation rate in the superconducting state show a spin-singlet with lines of nodes in the gap function. Their calculations indicate that their data are consistent with a strong-coupling d-wave gap function with $\Delta/k_B T_c = 4$. In the normal state, the temperature dependence of the spin-lattice-relaxation rate suggests the presence of antiferromagnetic correlations, and scales with T_c in the same manner as in the other heavy fermion superconductors and high-T_c superconductor $YBa_2Cu_3O_7$.

The resonance frequency of a nucleus in the superconducting state is given by $f = \gamma B_0(1+\Delta B)(1+K_0+\Delta\chi_s(T))$, where ΔB is given by the demagnetization field, K_0 is a temperature independent chemical shift of the particular nucleus, A the hyperfine coupling between the nucleus and the conduction electrons, and χ_s the spin susceptibility which is proportional to the Knight shift. K_0 and A are measured in the normal state and are known constants [31] while $\chi_s(T)$ and ΔB are unknown. The temperature dependence of $\chi_s(T)$ was obtained by Curro et al. [30] by measuring the frequencies of the ^{59}Co and the ^{71}Ga. Their results for temperature dependence of $\chi_s(T)$ is shown in Figure 9.11. They also determined that $\Delta B \approx 40$ G.

Curro et al. [30] have measured the spin-lattice-relaxation rate, T_1^{-1}, in the normal and superconducting states on a second powder sample of $PuCoGa_5$ in zero field at the ^{69}Ga NQR frequency. Their results are shown in Figure 9.12. $T_1^{-1} \sim T^3$ down to $\sim 0.3T_c$, then crosses over to a linear behavior below due to impurity scattering. The best fit to the data (the solid line) is given by a d-wave order parameter with $\Delta/k_B T_c = 4$ [31].

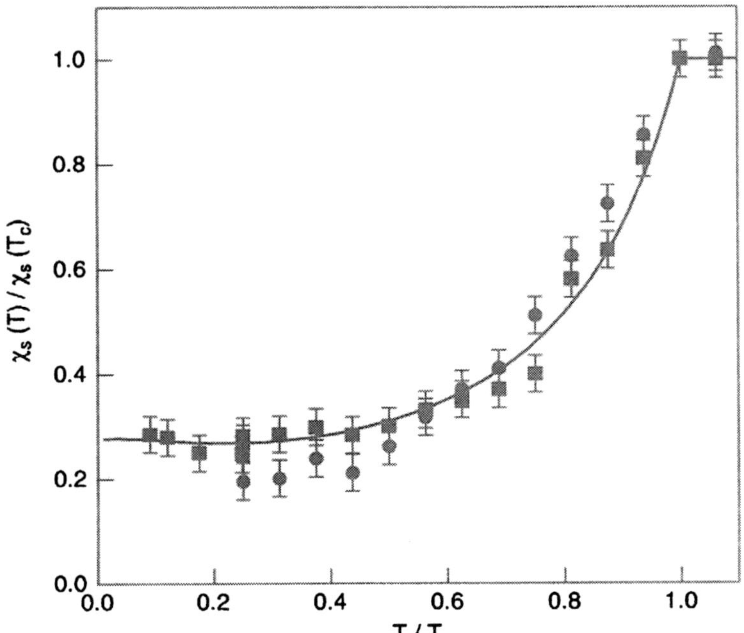

Figure 9.11. The normalized Knight shift of the ^{59}Co (●) and the ^{71}Ga (■) of $PuCoGa_5$ below T_c. The solid line is a fit to the dirty d-wave [31]. Reproduced with the permission of Elsevier from Ref. [30].

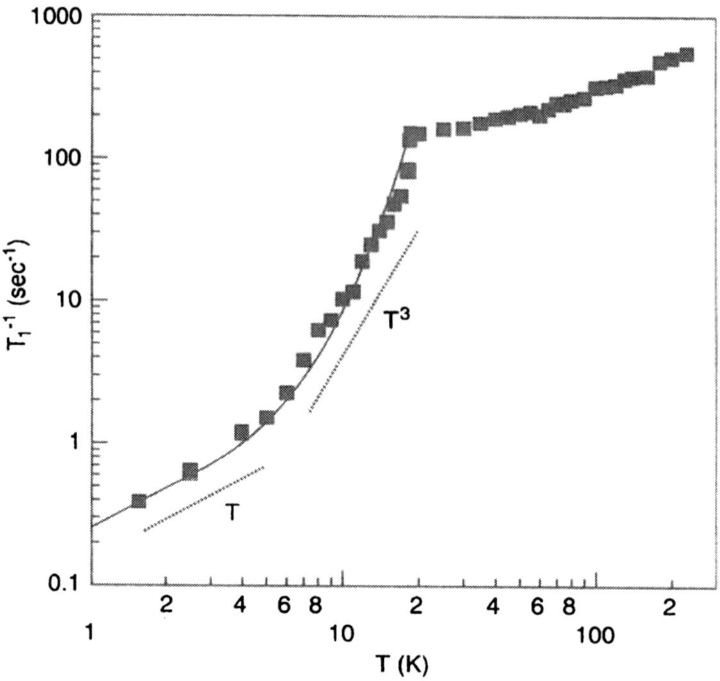

Figure 9.12. The spin-lattice relaxation of Ga(1) in $PuCoGa_5$ versus temperature. The solid line is given by a d-wave order parameter with $\Delta/k_BT_c = 4$ [31]. Reproduced with the permission of Elsevier from Ref. [30].

Figure 9.13 shows T_1T normalized to the value at T_c versus T/T_c for three d-wave superconductors, the high-T_c cuprate $YBa_2Cu_3O_7$, $CeCoIn_5$, and $PuCoGa_5$. Even though their transition temperatures vary by two orders of magnitude, the normal state relaxation scales T_c up to $5T_c$. This suggests that the same antiferromagnetic fluctuations present in $YBa_2Cu_3O_7$ and $CeCoIn_5$ and intimately connected with d-wave superconductivity must also be present in $PuCoGa_5$.

Recently, Joyce et al. [32] have examined the electronic structure of select Pu materials, by means of photoemission (PES) and mixed-level model (MLM) calculations [33, 34], of the heavy fermion superconductor $PuCoGa_5$ as well as $PuIn_3$ and δ-Pu metal. The MLM is a multi-electron extension of the generalized gradient approximation. The direct comparison of the PES data with the MLM calculations for the three Pu materials is shown in Figure 9.14.

The results indicate that there is excellent agreement of the PES data with MLM data for $PuCoGa_5$ and fairly good agreement for the δ-Pu and $PuIn_3$. The photoemission results indicate that the 5f electrons exhibit both localized and itinerant character. In fact, the dual nature of the Pu 5f electrons demarks the boundary between localized and itinerant 5f character. The MLM calculations were done by assuming that four of the five Pu 5f electrons are localized. The agreement between the PES data and the MLM calculations for $PuCoGa_5$ is remarkably good with the model and data in agreement with the width, intensity, and energy position of both the peak at the Fermi level and the

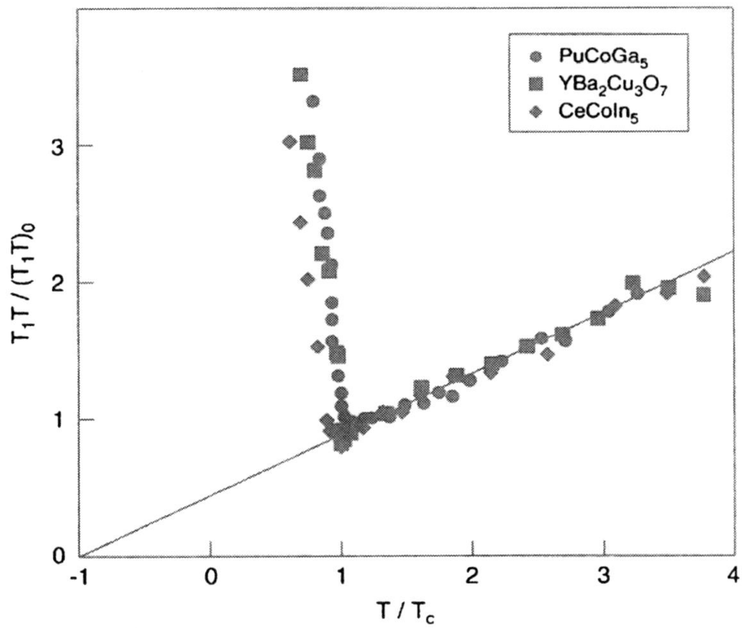

Figure 9.13. T_1T for several unconventional superconductors, normalized to the value at T_c versus T/T_c. The solid line is a linear fit. Reproduced with the permission of Elsevier from Ref. [30].

main 5f peak centered 1.2 eV below the Fermi level. However, we note that there are a large number of other models which have been proposed for Pu-based compounds [32].

Since superconductivity coexists with magnetic order in cerium isostructural systems, $PuCoGa_5$ could be close to a quantum critical point. In order to study this possibility, Colineau et al. [35] have investigated the influence of Pu substitution by U and Np on the electronic properties of superconductor $PuCoGa_5$ by SQUID magnetometry. The critical parameters are dramatically reduced compared to the pure sample. In $Pu_{0.9}U_{0.1}CoGa_5$ and $Pu_{0.9}Np_{0.1}CoGa_5$, the critical temperature drops from $T_c \approx 18.6$ K to 8.4 and 7.2 K and the upper critical field, from $H_{c2}(0) \approx 74$ T down to 8.7 and 3.4 T, respectively. Their results are shown in Figure 9.15.

In Figure 9.15, a sharp decrease to negative values of the zero-field-cooled magnetic susceptibility indicates the occurrence of superconductivity in $Pu_{0.9}U_{0.1}CoGa_5$ and $Pu_{0.9}Np_{0.1}CoGa_5$ below $T_c \approx 8.4$ and 7.2 K, respectively. While $Pu_{0.9}U_{0.1}CoGa_5$ is close to perfect diamagnetism, $Pu_{0.9}U_{0.1}CoGa_5$ reaches only ~5% of the ideal $-1/4\pi$ value which suggests that the latter may contain other non-superconducting phases.

9.4. $PuRhGa_5$

The discovery of superconductivity in $PuRhGa_5$ with $T_c \approx 9$ K was reported by Wastin et al. [36] in 2003. $PuRhGa_5$ crystallizes in the tetragonal $HoCoGa_5$ structure with the lattice parameters $a = 4.2354$ Å and $c = 6.7939$ Å. This structure has a two-dimensional feature,

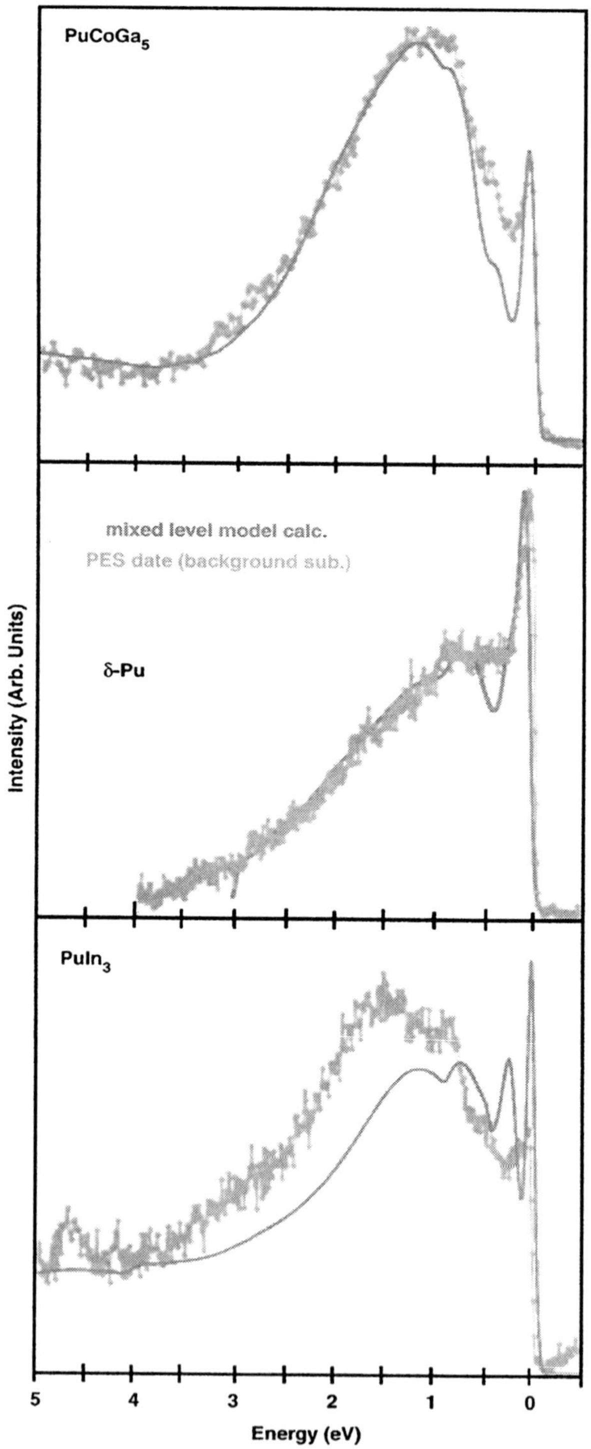

Figure 9.14. Photoemission (green diamonds) and MLM calculations (blue lines) for δ-Pu, $PuIn_3$, and $PuCoGa_5$. Reproduced with the permission of Elsevier from Ref. [32].

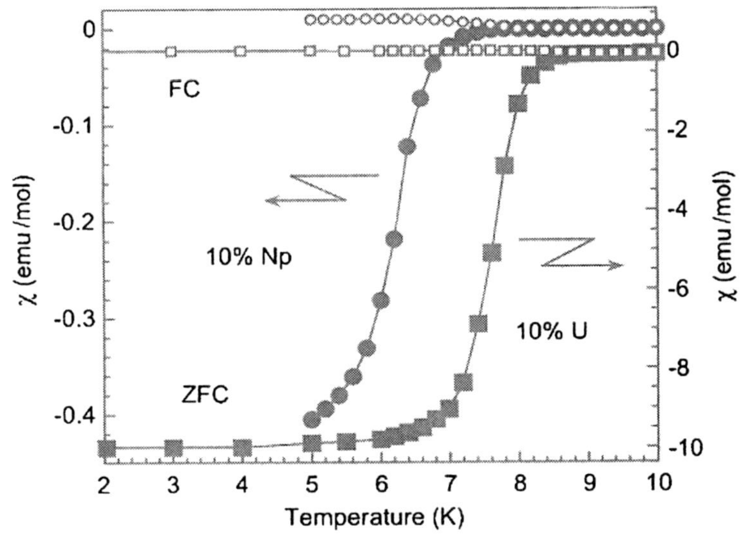

Figure 9.15. Zero-field-cooled (ZFC, full symbols) and field-cooled (FC, open symbols) magnetic susceptibility of $Pu_{0.9}U_{0.1}CoGa_5$ (squares) and $Pu_{0.9}Np_{0.1}CoGa_5$ (circles), measured at $H = 0.001$ T and $H = 0.01$ T, respectively. No diamagnetic transitio n is observed on FC curves due to the enhanced flux pinning of the material. Reproduced with the permission of Elsevier from Ref. [35].

where alternating $PuGa_3$ and $RhGa_2$ layers are stacked along the *c* axis. There are two crystallographically inequivalent Ga sites in this structure, which are denoted Ga(1) (the 1c site) and Ga(2) (the 4i site), respectively. The Ga(1) site is surrounded by four Pu atoms in the *c* plane while the Ga(2) site is surrounded by two Pu and two Rh atoms in the *a* plane.

The high-pressure measurements on $PuRhGa_5$ is shown in Figure 9.16 in which the electrical resistance has been plotted against temperature for pressures up to 18.7 GPa. It displays a metallic shape in the normal state but an NFL behavior ($\rho(T) \sim T^{1.3}$) develops up to 50–60 K. The inset of Figure 9.16 shows the plot of T_c of both $PuRhGa_5$ and $PuCoGa_5$ against pressure.

For $PuRhGa_5$, T_c increases rapidly with increasing pressure and reaches 16 K at 10 GPa. After that it decreases with pressure; $PuRhGa_5$ remains superconducting up to 20 GPa. There is a large broadening of T_c (~6 K) at 18.7 GPa. In contrast, T_c continues to increase with pressure up to 18.7 GPa for $PuCoGa_5$. This suggests that the pairing mechanism is differently affected by pressure for the two materials. The layered crystal structure associated with the quasi-2D Fermi surface calculated for these materials [38] suggests that anisotropic properties might be the cause for this difference.

The (*T*–*P*) phase diagram of $PuRhGa_5$ and $PuCoGa_5$ can be compared to that of $CeCoIn_5$. In these isostructural compounds, the superconducting transition temperature increases with increasing pressure and reaches a maximum before decreasing at higher pressure. The NFL behavior is also maintained over a large range of pressure. However, $PuIrGa_5$ is different in the sense that its resistance versus temperature plot displays Fermi liquidproperties and there is no indication of a superconducting transition even at very low temperatures.

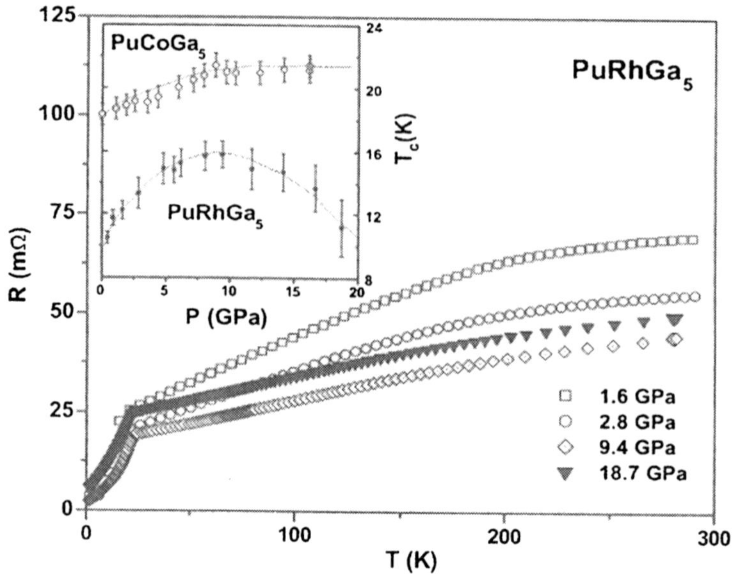

Figure 9.16. Evolution of the electrical resistance of $PuRhGa_5$ crystal up to 18.7 GPa. The inset shows T_c behavior of $PuCoGa_5$ as a function of the applied pressure. Reproduced with the permission of Elsevier from Ref. [37].

Recently, Saki et al. [39] have carried out NMR/NQR measurements on the Ga(2) site for $PuRhGa_5$. The $1/T_1T$ versus T plots for NMR/NQR on the ^{69}Ga(2) site is shown in Figure 9.17.

As Figure 9.17 shows, below T_c, $1/T_1T$ decreases sharply and paraboloically as T decreases. This result suggests that $PuRhGa_5$ has an unconventional superconductivity with an anisotropic gap. Saki et al. [39] have estimated that the gap value, $2\Delta_0(T \rightarrow 0) \approx 5k_BT_c$, and the ratio of the residual DOS to that of normal state, $N_{res}/N_0 \approx 0.25$, if one assumes a polar state $\Delta_0(T)\cos(\theta)$. This estimated gap is similar to that of $CeIrIn_5$ [41] while it is smaller than $8k_BT_c$ for $CeCoIn_5$ [42] and $PuCoGa_5$ [31]. This type of anisotropic superconductivity would take place on a cylindrical Fermi surface predicted by band calculation [43].

In the normal state, $1/T_1T$ for the NQR is almost temperature independent below 30 K. This Korringa behavior ($1/T_1T$ = constant) is a typical feature of Fermi liquid. Sakai et al. [39] measured $1/T_1T$ for the ^{69}Ga(2) NMR center line under an applied field (H_0) of ~110 kOe parallel to the c axis. T_c was suppressed from 8.5 K (zero field) to ~4 K in this applied field. As shown in Figure 9.17, the Korringa behavior of $1/T_1T$ of the NMR is also present in the normal state below ~30 K. This result indicates that superconductivity sets in $PuRhGa_5$ after a coherent Fermi-liquid state is established. The coherence in the Fermi-liquid state is slowly reduced above 30 K since $1/T_1T$ decreases gradually as T increases. The difference in NQR and NMR values of $1/T_1T$ is probably caused by different quantization axes, which are the a axis and the c axis, respectively. The $1/T_1T$ behavior in the normal state of $PuRhGa_5$ is thus different than that of $PuCoGa_5$, in which case, $1/T_1T$ increases monotonically from room temperature right down to T_c [31].

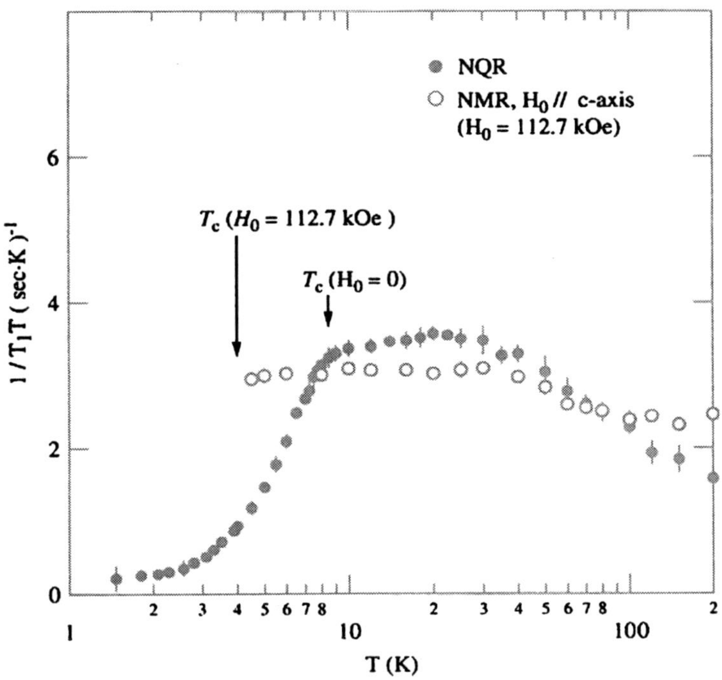

Figure 9.17. $1/T_1T$ versus T plot for the $^{69}Ga(2)$ NQR/NMR lines in $PuRhGa_5$ lines. Reproduced with the permission of Elsevier from Ref. [40].

In order to study the sensitivity of T_c to structural distortions which would clarify the role of 5f-electronic anisotropy in the related materials $PuMGa_5$ and $CeMIn_5$ (where M is an isoelectronic transition metal), Sarrao et al. [29] have plotted T_c as a function of the tetragonal distortion $(2zc\text{-}a)/a$. Here, a and c are the tetragonal lattice parameters of the $HoCoGa_5$ structure and z is a structural parameter locating the Ga(4i) position along the c axis. Their T_c versus tetragonal distortion $(2zc\text{-}a)/a$ for a set of $CeMIn_5$ and $PuMGa_5$ materials is shown in Figure 9.18.

From Figure 9.18, one observes that there is a linear relationship for both the families of superconductors. However, the slope of T_c versus $(2zc\text{-}a)/a$ is ~10 times larger for $PuMGa_5$ than for $CeCoIn_5$ which indicates that there is greater degree of freedom and hence stronger coupling of the 5f electrons when compared with the 4f case. Another interesting result is the observation that the maximum T_c in $CeMIn_5$ is obtained for $(2zc\text{-}a)/a > 0$, whereas all known values of $(2zc\text{-}a)/a$ for $PuMGa_5$ are negative. This suggests that further T_c optimization may be possible in this system.

9.5. Similarities between Cu and Pu containing "high T_c" superconductors

Recently Wachter [44] has compared the Cu "high T_c superconductors" with equivalent measurements on "high T_c" $PuCoGa_5$ and $PuRhGa_5$. He has observed the following common features. First, in all materials, spin pseudogaps have been observed which

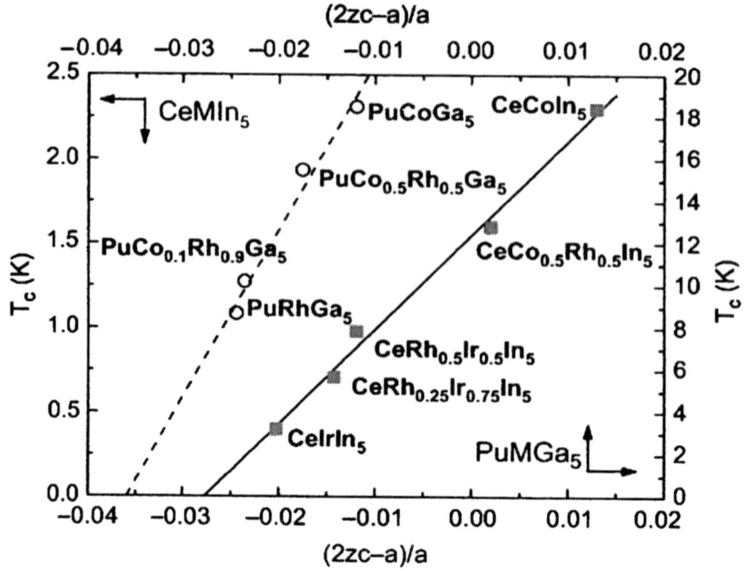

Figure 9.18. T_c versus tetragonal distortion $(2zc\text{-}a)/a$ for a set of $CeMIn_5$ and $PuMGa_5$ materials. Reproduced with the permission of Elsevier from Ref. [29].

necessitates at least antiferromagnetic short-range order, i.e., in clusters. Second, all Cu and Pu superconductors are of mixed valence as photoemission data have shown. The majority ions (Cu or Pu) are magnetic and the minority ions are non-magnetic and act as spin holes. Only short-range correlations remain since these spin holes have a concentration of 10% and hence dilute the antiferromagnetic order. According to Wachter [44], two dimensionality is not essential and n- or p-type conductivity is not important.

References

[1] W. Jeitschko and D. Braun, Acta Crystallogr. B **33**, 3401 (1977).

[2] B.C. Sales, in *Handbook on the Physics and Chemistry of Rare Earths, Vol. 33,* eds. K.A. Gschneidner, Jr., J.-C.G. Bunzli, and V.K. Pecharsky, p.1 (Elsevier, Amsterdam, 2003).

[3] Y. Aoki, H. Sugawara, H. Hisatomo, and H. Sato, J. Phys. Soc. Jpn. **74**, 209 (2005).

[4] B. Maple, J. Phys. Soc. Jpn. **74**, 222 (2005).

[5] H. Sato, H. Sugawara, D. Kikuchi, S. Sanada, K. Tanaka, H. Aoki, K. Kuwahara, Y. Aoki, and M. Kohgi, Physica B **378–380**, 46 (2006).

[6] K. Takegahara, H. Harima, and Y. Yanase, J. Phys. Soc. Jpn. **70**, 1190 (2001).

[7] T. Goto, Y. Nemoto, K. Onuki, K. Sakai, T. Yyamaguchi, M. Akatsu, T. Yanagisawa, H. Sugawara, and H. Sato, J. Phys. Soc. Jpn. **74**, 263 (2005).

[8] E.D. Bauer, N.A. Frederick, P.-C. Ho, V.S. Zapf, and M. B. Maple, Phys. Rev, B **65**, 100506(R) (2002).

[9] H. Kotegawa, M. Yogi, Y. Imamaura, Y. Kawasaki, G.-Q. Zheng, Y. Kitaoka, S. Ohsaki, H. Sugawara, M. Aoki, and H. Sato, Phys. Rev. Lett. **90**, 027001 (2003).

[10] R. Vollmer, A. Faiβt, C. Pfleiderer, H.v. Lohneysen, E.D. Bauer, P.-C. Ho, V. Zapf, and M.B. Maple, Phys. Rev. Lett. **90**, 057001 (2003).
[11] M.-A. Measson, J.P. Brison, G. Seyfarth, D. Braithwaite, G. Lapertot, B. Salce, J. Flouquet, E. Lhotel, C. Paulsen, H. Sugawara, H. Sato, P.C. Canfield, K. Izawa, and Y. Matsuda, Physica **359–361**, 827 (2005).
[12] K. Izawa, Y. Nakajima, J. Goryo, Y. Matsuda, S. Osaki, H. Sugawara, H. Sato, P. Thalmeier, and K. Maki, Phys. Rev. Lett. **90**, 117001 (2003).
[13] T. Tayama, T. Sakakibara, H. Sugawara, Y. Aoki, and H. Sato, J. Phys. Soc. Jpn. **72**, 1516 (2003).
[14] Y. Matsuda, K. Izawa, T. Watanabe, Y. Kasahra, Y. Nakajima, T. Sakakibara, C.J. van der Beek, M. Nohara, T. Hanaguri, H. Takagi, J. Goryo, K. Maki, P. Thalmier, S. Osaki, H. Sugawara, H. Sato, H. Shishido, R. Settai, and Y. Onuki, J. Phys. Chem. Solids **66**, 1365 (2005).
[15] S. Drobnik, K. Grube, C. Pfleiderer, H.v. Lohneysen, E.D. Bauer, and M.B. Maple, Physica **359–361**, 901 (2005).
[16] K. Kuwahara, K. Iwasa, M. Kohgi, K. Kaneko, N. Metoki, S. Raymond, M.-A. Measson, J. Flouquet, H. Sugawara, Y. Aoki, and H. Sato, Phys. Rev. Lett. **95**, 107003 (2005).
[17] M. Nishiyama, T. Kato, H. Sugawara, D. Kikuchi, H. Sato, H. Harima, and G.-Q. Zheng, J. Phys. Soc. Jpn. **74**, 1938 (2005).
[18] M. Nishiyama, T. Kato, H. Sugawara, D. Kikuchi, H. Sato, H. Harima, and G.-Q. Zheng, Physica B **378–380**, 192 (2006).
[19] R. Fehrenbacher and M.R. Norman, Phys. Rev. B **50**, 3495 (1994).
[20] N.A. Frederick, T.D. Do, P.C. Ho, N.P. Bulch, V.S. Zapf, and M.B. Maple, Phys. Rev. B **69**, 024523 (2004).
[21] S. Ozcan, W.M. Yuhasz, N.A. Frederick, and M.B. Maple, Physica B **378–380**, 182 (2006).
[22] J. Custers, Y. Namai, T. Tayama, T. Sakakibara, H. Sugawara, Y. Aoki, and H. Sato, Physica B **378–380**, 179 (2006).
[23] I. Shirotani, Y. Shimaya, K. Kihou, C. Seikene, N. Takeda, M. Ishikawa, and T. Yogi, J. Phys. Condens. Matter **15**, S2201 (2003).
[24] J.L. Sarrao, L.A. Morales, J.D. Thompson, B.L. Scote, G.R. Stewart, F. Wastin, J. Rebizant, P. Boulet, E. Colineau, and G.H. Lander, Nature **420**, 297 (2002).
[25] H. Hegger, C. Petrovic, E.G. Moshopoulou, M.F. Hundley, J.L. Sarrao, Z. Fisk, and J.D. Thompson, Phys. Rev. Lett. **84**, 4986 (2000).
[26] C. Petrovic, P.G. Pagliuso, M.F. Hundley, R. Movschovich, J.L. Sarrao, Z. Fisk, and A. Monthoux, J. Phys. Condens. Matter **13**, L337 (2001).
[27] C. Petrovic, R. Movschovich, M. Jaime, P.G. Pagliuso, M.F. Hundley, J.L. Sarrao, Z. Fisk, and J.D. Thompson, Europhys. Lett. **53**, 354 (2001).
[28] J.J. Joyce, J.W. Willis, T. Durakiewicz, M. T. Butterfield, E. Guziewicz, J.L. Sarrao, L.A. Morales, A.J. Arko, and O. Erikssion , Phys. Rev. Lett. **91**, 157001 (2003).
[29] J.L. Sarrao, E.D. Bauer, L.A. Morales, and J.D. Thompson, Physica B **359–361**, 1144 (2005).
[30] N.J. Curro, T. Caldwell, E.D. Bauer, L.A. Morales, M.J. Graf, Y. Bang, A.V. Balatsky, J.D. Thompson, and J.L. Sarrao, Physica B **378–380**, 915 (2006).

[31] N.J. Curro, T. Caldwell, E.D. Bauer, L.A. Morales, M.J. Graf, Y. Bang, A.V. Balatsky, J.D. Thompson, and J.L. Sarrao, Nature **434**, 622 (2005).
[32] J.J. Joyce, J.M. Wills, T. Durakiewicz, M.T. Butterfield, E. Guziewicz, D.P. Moore, J.L. Sarrao, L.A. Morales, A.J. Arko, O. Eriksson, A. Delin, and K.S. Graham, Physica B **378–380**, 920 (2006).
[33] A.J. Arko, J.J. Joyce, L. Morales, J. Wills, L. Lashley, P. Westin, and J. Rebizant, Phys. Rev. B **62**, 1773 (2000).
[34] O. Erikssion et al., J. Alloy Compds **287**, 1 (1999).
[35] E. Colineau, F. Wastin, P. Boulet, P. Javorski, and J. Rebizant, Physica B **359–361**, 1075 (2005).
[36] F. Wastin, P. Boule, J. Rebizant, E. Colinneau, and G.H. Lander, J. Phys. Condens. Matter **15**, S2279 (2003).
[37] J.-C. Griveau, P. Boulet, E. Collineau, F. Wastin, and J. Rebizant, Physica B **359–361**, 1093 (2005).
[38] T. Hotta and K. Ueda, Phys. Rev. B **67**, 104518 (2003).
[39] H. Sakai, Y. Tokunaga, T. Fujimoto, S. Kambe, R.E. Walstedt, H. Yasuoka, D. Aoki, Y. Homma, E. Yamamoto, A. Nakamura, Y. Shikoawa, K. Nakajima, Y. Arai, T.D. Matsuda, Y. Haga, and Y. Onuki, J. Phys. Soc. Jpn. **74**, 1710 (2005).
[40] H. Sakai, Y. Tokunaga, T. Fujimoto, S. Kambe, R.E. Walstedt, H. Yasuoka, D. Aoki, Y. Homma, E. Yamamoto, A. Nakamura, Y. Shiokawa, K. Nakajima, Y. Arai, T.D. Matsuda, Y. Haga, and Y. Onuki, Physica B **378–380**, 1005 (2006).
[41] G.-Q. Zheng, K. Tanabe, T. Mito, S. Kawasaki, Y. Kitaoka, D. Aoki, Y. Haga, and Y. Onuki , Phys. Rev. Lett. **86**, 4664 (2001).
[42] Y. Kohori, Y. Yamato, Y. Iwamoto, T. Kohara, E.D. Bauer, M.B. Maple, and J.L. Sarrao, Phys. Rev. B **64**, 134526 (2001).
[43] T. Maehira, T. Hotta, K. Ueda, and A. Hasegawa, Phys. Rev. Lett. **90**, 207007 (2003).
[44] P. Wachter, Physica C **453** (2007).

Subject Index

AC susceptibility, 173
AFM quantum critical point, 68
AFQ, 285
Anderson lattice, 13, 57, 115–117, 326
Anderson model, 9–11, 13–14, 17, 24, 26, 30, 41, 118–119, 121, 277, 280
antiferromagnetic fluctuations, 246, 269
antiferromagnetism, 40, 65, 89, 104, 159, 206, 266–269, 274

BCS energy gap, 272
BCS superconductor, 141, 170
BCS theory of anisotropic superconductivity, 256
Berry-phase, 38
Bloch states, 54, 273
Bogoliubov transformation, 257, 282
bosonic bath, 37, 40, 77–78

canonical ensemble, 15
cavity method, 26–27, 37, 45
CDMFT, 48
$CeAl_2$, 5
$CeAl_3$, 2–3, 6, 53
$Ce_3Bi_4Pt_3$, 5, 32, 293–296
$CeCoIn_5$, 1, 6, 142, 163–172, 175, 246, 249–251
$CeCu_2Ge_2$, 142, 146, 150–151, 154–155, 279
$CeCu_2Si_2$, 2–6, 53, 66, 115, 141–150, 155, 175, 261, 266, 276, 279
$CeCu_6$, 6, 35, 70, 98–101
$CeCu_{6-x}Au_x$, 98
CEF levels, 235
$CeFe_4P_{12}$, 311–312
$CeIn_3$, 142, 156–159, 167–168
$CeIr_3Si_2$, 110–111
$CeIrIn_5$, 168–172, 175, 250
$Ce_{1-x}La_xRu_2Si_2$, 89–90
$CeNi_2Ge_2$, 70, 142, 155–158
CeNiSn, 5, 32, 292, 296–297, 300–305, 311
$CeOs_4Sb_{12}$, 313–316
$CePd_{2+x}Ge_{2-x}$, 153–154
$CePd_2Si_2$, 71, 106, 108, 142, 150–156, 158–159
$CePt_3B$, 159
$CePt_3Si$, 159–162, 225
$CePt_3Si_{1-x}Ge_x$, 160–162
CeRhAs, 296–300
$CeRhIn_5$, 167–169, 175
CeRhSb, 296–298, 300–302
$CeRh_2Si_2$, 142, 154–155, 159
$CeRu_2Ge_2$, 91
$CeRu_4P_{12}$, 311–313
$CeRu_2Si_2$, 6, 87–93, 95, 99, 101, 103–104, 120, 123–124, 136
$CeRu_4Sn_6$, 306–308
chemical potential, 12, 15, 30, 39, 45, 51, 75, 120–121, 123, 132, 256, 275, 277, 295
Clebsch-Gordon coefficients, 54
Cluster calculations, 14
Cluster DMFT, 23
cluster perturbation theory (CPT), 47
coherence length, 151, 174, 189, 194
coherence temperature, 1, 4, 9, 37, 60, 81, 141, 307
cold Fermi surface, 76
collective modes, 265–266, 285
Cooper pairs, 141, 144–145, 151, 163, 215, 225, 255, 260, 265
critical temperature, 2, 163, 185, 224, 236, 244, 247, 261, 271, 275, 328
crystal symmetry, 225, 260
Curie temperature, 219, 222–223, 225, 316

DCA, 47–48
de Haas-van Alphen oscillations, 2
density of states, 25–26, 30, 32, 39, 41, 52, 87, 108, 117, 123, 153, 214, 240, 260, 263–265, 268, 297, 304, 316, 321
dHvA experiment, 97
DMFT, 23, 28, 34–35, 37–39, 42–45, 48, 76–78, 123
DMRG method, 280
δ-Pu, 246
d-wave superconductivity, 246
dynamical cluster approximation, 47

EDMFT, 23, 35, 37–39, 42–44, 76–78
effective action, 24, 27–28, 30, 33, 38, 75, 77, 275
effective Hamiltonian, 10, 256
effective mass, 1–2, 6, 9, 13, 43, 53, 57, 62, 81, 88, 97, 102, 106, 123, 134, 136, 141, 144, 150–151, 154, 235, 255, 276, 279, 281

electrical resistivity, 2, 35–37, 51, 64, 110, 145, 151, 160–161, 169, 175, 177, 188–190, 196, 300, 306, 309–310, 312, 314–315, 320–321
electron-hole pairs, 129
EPR shift, 135
ESP, 283
exact diagonalization, 29

Fermi-Dirac distribution, 51
Fermi gas, 51, 257, 266
Fermi level, 12, 14–15, 51, 60–62, 116–117, 246–247, 255, 291, 304, 310, 313–314, 318, 323
Fermi liquid, 6, 24, 51, 57, 62, 66, 68, 72, 75, 80, 82–83, 106, 144, 207, 249–250
Fermi-liquid theory, 13, 51, 53, 57, 59
Fermi sphere, 51
fermionic bath, 37, 40, 77–78
ferrimagnetic, 110
FFLO superconducting state, 163, 165
Filled skutterdites, 235
FL state, 67, 76, 151, 157, 208
FL* state, 63, 65–67, 72–73, 75–76, 106, 142, 145, 151–152, 156–157, 163–166, 169, 183, 188, 208, 213, 271
fractionalized Fermi liquids, 72

gap equation, 260–262, 265, 271, 283
Ginzburg-Landau theory, xiii, 260-261, 263
Grassmann variables, 58
Green's function, 23, 25–30, 32, 42, 45, 51, 74, 121, 132–133, 268, 271, 275, 282

Hall effect, 91, 93, 207
heavy electrons, 279
heavy Fermi liquids, 57
heavy fermions, 1–2, 5–6, 11, 13, 28, 44, 53, 57, 104, 115, 118–119, 144, 255–256, 269
heavy-fermion superconductors, 142, 156, 175, 178, 183, 189, 215, 246, 255–256, 261
heavy Fermion surface, 61
HFL, 106
hidden order, 102, 106, 141, 201, 205, 207
Higgs phase, 63
high-T_c superconductors, 2
$HoCoGa_5$ structure, 244, 247, 251
hot Fermi surface, 75
hot lines, 70–71
Hund's rule, 10, 66

inelastic neutron scattering, 99, 240, 285, 323
INS, 183, 214–215, 285
Ising-exchange, 39
itinerant electrons, 131, 135

KKRZ equations, 53
Knight shift, 56, 136, 216, 218, 244–245, 320
Kondo alloys, 11
Kondo-boson, 57–58
Kondo energy, 35, 78, 298, 300
Kondo exchange, 35, 62
Kondo insulators, 30–32, 291–293, 296–297, 299–300, 307, 310, 315, 320, 326, 329–330
Kondo lattice, 9, 32–33, 38–39, 41, 53, 57, 68, 72, 77–80, 88, 123, 150, 291–292, 301, 307
Kondo screening, 57, 60–61, 75, 79–80
Kondo singlets, 53, 81, 144

Lagrangian, 13, 33, 57
Landau quasiparticles, 97
lanthanide, 1, 67, 311
Lifshitz transition, 81
LISA, 23–26, 29–30, 34
LMAF, 196–199, 203, 205–206
local effective action, 24, 27
Luttinger theorem, 57, 62, 76
Luttinger-Ward functional, 23, 28

magnetic excitons, 215, 281
magnetoresistance, 91, 94, 209, 307
magnons, 115, 145, 157, 214–215
Matsubara frequency, 43–44
mean free path, 174
Mean-field approximation, 13, 125, 328
mean-field theory, 23–24, 35, 37, 48, 58, 72, 74–75, 120, 123, 273, 329
Meissner effect, 173, 194, 227–228
metamagnetic behavior, 120, 137, 219
metamagnetic crossover, 103, 120
metamagnetic transition, 88, 115, 117–120, 123, 207, 317, 319
metamagnetic quantum criticality, 105
metamagnetism, 6, 87, 93, 95, 97, 104, 110, 115, 119, 124, 136
mixed valence, 252, 280, 317, 325
MMT, 88–89, 91–92, 94–96, 98–99, 101, 106, 108, 110
MnSi, 95
Mottness, 48
MR, 91–94, 106, 108, 142, 160, 166, 184, 198, 201, 203, 206–207, 216, 220, 244, 250–251, 280, 299, 307, 311–312, 314–316, 320, 322
multichannel Kondo model, 66

Neel cusp, 43–44
Neel temperature, 5, 41, 65, 90, 99, 101, 150, 152, 154, 156, 160, 169, 175, 198, 200, 202, 319

neutral spinons, 73
neutron diffraction, 110, 198
NMR, 142, 160, 166, 184, 198, 201, 203, 206–207, 216, 220, 244, 250–251, 299, 311–312, 316, 320, 322
non-crossing approximation, 29
non-Fermi liquid behavior, xi, 35, 63
NQR, 142, 175, 207, 220, 236, 240, 245, 250–251, 313–314, 316

OP, 31, 34, 190, 194, 198, 207, 221
order parameter, 39–40, 78, 156, 159, 163, 166, 190, 201, 205, 237, 245–246, 261, 263, 265–269, 271, 282, 284
OSP, 283

pairing mechanism, 146, 215, 249, 281
Pauli spin matrices, 10, 62, 258
Pauli spin susceptibility, 2, 56, 131
particle-hole symmetry, 128
perturbation theory, 23, 27, 29, 31–32, 47, 51, 56, 123
PFL, 106
phase diagram, 41–43, 45–46, 63, 67, 69, 75, 79–80, 82, 87, 89, 91, 93, 101, 103–109, 141, 144–148, 153–158, 161, 163, 165–167, 169, 171, 174–175, 183, 185–188, 191, 193–194, 196, 198–199, 203, 205–206, 209, 211–212, 219, 222–223, 225–228, 236–239, 242, 244, 249, 270, 315, 318–319
point-group symmetry, 263
polarized phase, 91, 221
$PrNi_2Al_3$ structure, 208, 215
$PrOs_4Sb_{12}$, 1, 142, 235–242, 284–287
pseudospin, 116, 260–261, 282
p-T phase diagram, 154–155, 157, 175, 206, 219, 223, 226–228, 319
$Pu_{0.9}Np_{0.1}CoGa_5$, 247, 249
$Pu_{0.9}U_{0.1}CoGa_5$, 247, 249
$PuCoGa_5$, 142, 244–251
$PuIn_3$, 246, 248
$PuMGa_5$, 251–252
$PuRhGa_5$, 247, 249–251
p-wave superconductivity, 266

QCP, 39, 45, 67–70, 76, 78–79, 81, 83, 89, 95, 97, 99, 104–106, 108, 145–146, 158, 171, 188, 207, 225, 227–228, 269–270, 276, 280
QMC, 29, 32, 39–40, 42, 45
QPT, 67, 78–79, 144–145
quadrupolar exciton exchange, 285
quadrupolar Kondo model, 66–67
quantum critical point, 6, 89, 105, 188, 247, 269
quantum-critical regime, 78
quantum Monte Carlo, 29
quantum phase transition, 39, 67, 78–79, 144, 223
quasiparticle, 5, 13, 35, 43–44, 51–53, 57–59, 62, 67–68, 70–72, 74–76, 79, 81, 97, 102, 115–117, 120, 122–123, 141, 144, 151, 167, 183–184, 186, 215, 235–236, 255–256, 263–264, 266, 274–277, 279, 282–283, 287, 307, 320, 322, 327–328

resistivity, 2–3, 31, 35–37, 51, 57, 59, 63–65, 70–71, 83, 92–93, 95–97, 104, 106, 109–110, 142, 144–161, 167–169, 171, 173–175, 177, 183–184, 188–190, 196, 205, 207–211, 216, 220–222, 224–225, 227, 236, 276–280, 291–293, 300–304, 306, 309–310, 312, 314–315, 317, 320–321, 323–325
resonant level, 53
RKKY interaction, 9, 34–35, 39–42, 77–79, 150, 153, 174
RT_4X_{12}, 235
Ru-O bilayer, 93

self-energy, 23, 26, 28, 30, 33, 43–44, 47, 58, 116–117, 119, 121–122, 132–133, 275–277, 279, 282, 329
singular Fermi liquid, 82–83
skutterdites, 141, 235
slave bosons, 329
slave-particle, 72
SMAF, 195–199, 201, 205
small electron Fermi surface, 74
SmB_6, 32, 324–326, 329
SmS, 325–327
specific heat, 1–2, 4, 9, 14–15, 18, 51–52, 57, 59, 61, 69–70, 82–83, 88–89, 91, 95, 104, 106, 108, 110, 115, 118, 123, 142, 144, 147–149, 155, 159–160, 163–164, 166–171, 183–185, 187–191, 193–194, 196, 198, 201, 203–204, 209, 212–213, 215–216, 221, 224, 235–239, 242, 276, 294–295, 302, 306–309, 313, 316, 318, 324, 326–327
spherical Bessel functions, 54
spin color, 61
spin excitons, 129
spin fluctuations, 45, 69–70, 78, 119–121, 123, 125, 145, 154–155, 157, 175, 178, 213, 219, 228, 273, 313–314, 316
spin liquid state, 150
spinon Fermi surface, 72–75
spin-orbit effects, 53–54, 56
spin-orbit interaction, 194, 260, 267
spin quasiparticle, 116–117
SQUID magnetometry, 247

$Sr_3Ru_2O_7$, 92–98, 103–104
strong coupling, 60, 214
superconducting transition, 142, 146–148, 150–152, 154–155, 157, 159–160, 163, 168, 171, 183–184, 187, 194, 196, 201, 203, 206, 208, 213, 215–216, 218, 238, 242, 249, 262, 278–279, 284
superconducting transition temperature, 151, 154, 157, 160, 163, 201, 206, 208, 213, 218, 249, 278–279, 284
superexchange, 117–118, 123
symmetry-breaking, 194

tetracritical point, 193–194
tetragonal distortion, 159, 251–252
tetrahedral harmonics, 286
t-J model, 117–119
$ThCr_2Si_2$ structure, 150, 266
theory of Kondo insulators, 330
thermal expansion, 91–92, 188, 190–191, 206, 209, 291, 326
time-reversal symmetry, 184, 186, 205, 227–228, 260, 262
TmSe, 317–319

UAl_2, 1
UBe_{13}, 4–6, 66–67, 183–188, 261
$U(Be_{1-x}Th_x)_{13}$, 183
UFe_4P_{12}, 316–318
UGe_2, 142, 219–222, 225, 227
UIr, 225, 227–228
underscreened Kondo model, 81–83
UNi_2Al_3, 141, 183, 215–218
UPd_2Al_3, 101–102, 141, 183, 208–215, 284
UPt_3, 1, 5–6, 99–101, 103–104, 115, 141, 183, 188–196, 199, 237, 256, 261, 267
URhGe, 142, 221–225, 227
URu_2Si_2, 101–105, 141, 183, 201–208, 266
$U(Ru_{1-x}Rh_x)_2Si_2$, 106–107
U_2Ru_2Sn, 308–311, 320–322
U_2Zn_{17}, 5

valence fluctuations, 1, 14, 146, 156, 175, 276, 279

wavefunction, 52
weak coupling, 60
Weiss function, 24, 27–28, 42, 45
Wilson ratio, 2, 57, 59

YbB_{12}, 322–323, 329
$YbRh_2(Si_{0.95}Ge_{0.05})_2$, 83
$YbRh_2Si_2$, 35–37, 45, 108–110

Printed in the United States
128625LV00003B/17/P